Lecture Notes in Computer Science 14530

The series Lecture Notes in Computer Science (LNCS), including its subseries Lecture Notes in Artificial Intelligence (LNAI) and Lecture Notes in Bioinformatics (LNBI), has established itself as a medium for the publication of new developments in computer science and information technology research, teaching, and education.

LNCS enjoys close cooperation with the computer science R & D community, the series counts many renowned academics among its volume editors and paper authors, and collaborates with prestigious societies. Its mission is to serve this international community by providing an invaluable service, mainly focused on the publication of conference and workshop proceedings and postproceedings. LNCS commenced publication in 1973.

Shivam Bhasin · Thomas Roche
Editors

Smart Card Research and Advanced Applications

22nd International Conference, CARDIS 2023
Amsterdam, The Netherlands, November 14–16, 2023
Revised Selected Papers

Editors
Shivam Bhasin
Nanyang Technological University
Singapore, Singapore

Thomas Roche
NinjaLab
Montpellier, France

ISSN 0302-9743 ISSN 1611-3349 (electronic)
Lecture Notes in Computer Science
ISBN 978-3-031-54408-8 ISBN 978-3-031-54409-5 (eBook)
https://doi.org/10.1007/978-3-031-54409-5

This Springer imprint is published by the registered company Springer Nature Switzerland AG
The registered company address is: Gewerbestrasse 11, 6330 Cham, Switzerland

Paper in this product is recyclable.

Preface

These are the proceedings of the 22nd International Conference on Smart Card Research and Advanced Applications (CARDIS 2023). This year's CARDIS was held in Amsterdam, the Netherlands, and took place from November 14 to November 16, 2023. The Complex Cyber Infrastructure group of the University of Amsterdam in Amsterdam organized the conference this year. A half-day pre-conference tutorial accompanied CARDIS 2023.

CARDIS has been the venue for security experts from industry and academia to discuss developments in the security of smart cards and related applications since 1994. Smart cards play an increasingly important role in our daily lives through their use in banking cards, SIM cards, electronic passports, and devices of the Internet of Things (IoT). Thus, it is naturally important to understand their security features and develop sound protocols and countermeasures while maintaining reasonable performance. In this respect, CARDIS aims to gather security experts from industry, academia, and standardization bodies to make strides in the field of embedded security.

The present volume contains 13 papers that were selected from 28 submissions following a double-blind peer review process. The 30 members of the Program Committee and 9 external reviewers evaluated the submissions, wrote 84 reviews (3 by submission), and engaged in extensive discussions on the score of each article. Two invited talks were part of the technical program. Axel Poschmann, from PQShield Ltd, discussed the role of hardware security from a geo-political aspect in his Keynote talk titled "*Hardware Security and the Rise of the Bloc-Cipher*". Jean-Sebastien Coron, from the University of Luxembourg, presented recent advancements in post-quantum algorithms and the associated challenge of safeguarding them from side-channel attacks in his Keynote talk titled "*Post-quantum algorithms and side-channel countermeasures*". CARDIS 2023 also provided a platform for PhD students to present their work in the PhD forum. With 5 accepted presentations, students presented their ongoing work through a lightning talk followed by a poster presentation for one-to-one interaction with the audience. CARDIS 2023 was preceded by 2 tutorials. A tutorial titled "*Everything You Always Wanted to Know About Payment Terminals Security, But Were Afraid to Ask*" was delivered by David Samyde and Jean-Jacques Quisquater. The second tutorial on "*Side-channel cryptanalysis of a masked AES with SCALib*" was given by Olivier Bronchain and Gaëtan Cassiers.

Organizing a conference can be challenging, but we are happy to report that this year's conference went surprisingly smoothly. We express our deepest gratitude to the General Chair, Francesco Regazzoni, to the Local Arrangements co-Chairs, Kostas Papagiannopoulos and Marco Brohet, to Grace Millerson, and to all the members of the organization team who enabled CARDIS 2023 to succeed. We thank the authors who submitted their work and the reviewers who volunteered to review and discuss the submitted articles. The authors greatly appreciate the amazing work of our invited speakers, who gave entertaining and insightful presentations. We thank Springer for publishing the accepted papers in the LNCS series and the sponsors NLNCSA (part of the AIVD), Riscure, and PQShield for their generous financial support and Ninjalab for supporting the paper submission system. We are grateful to the CARDIS steering committee for allowing us to serve as the program chairs of such a well-recognized conference. Finally, we thank all presenters, participants, and session chairs, physically and online, for their support in making this CARDIS edition a great success.

December 2023

Shivam Bhasin
Thomas Roche

Organization

General Chair

Francesco Regazzoni	University of Amsterdam, The Netherlands

Program Committee Chairs

Shivam Bhasin	Nanyang Technological University, Singapore
Thomas Roche	NinjaLab, France

Steering Committee

Sonia Belaïd	CryptoExperts, France
Begül Bilgin	Rambus Cryptography Research, The Netherlands
Ileana Buhan	Radboud University, The Netherlands
Thomas Eisenbarth	University of Lübeck, Germany
Jean-Bernard Fischer	Kudelski, Switzerland
Vincent Grosso	CNRS, France
Tim Güneysu	Ruhr-University Bochum, Germany
Marc Joye	Zama, USA
Konstantinos Markantonakis	Royal Holloway, UK
Nele Mentens	KU Leuven, Belgium
Amir Moradi	TU Darmstadt, Germany
Svetla Nikova	KU Leuven, Belgium
Thomas Pöppelmann	Infineon Technologies, USA
Jean-Jacques Quisquater	UC Louvain, Belgium
Francesco Regazzoni	University of Amsterdam, The Netherlands
Tobias Schneider	NXP Semiconductors, Austria
François-Xavier Standaert	UC Louvain, Belgium
Yannick Teglia	ThalesDIS, France

Program Committee

Jan-Pieter D'Anvers	KU Leuven, Belgium
Aydin Aysu	North Carolina State University, USA

Melissa Azouaoui	NXP Semiconductors, Germany
Debapriya Basu Roy	IIT Kanpur, India
Davide Bellizia	Telsy, Italy
Jakub Breier	Silicon Austria Labs, Graz, Austria
Olivier Bronchain	NXP Semiconductors, Germany
Ileana Buhan	Radboud University, The Netherlands
Eleonora Cagli	CEA-Leti, Université Grenoble Alpes, France
Łukasz Chmielewski	Masaryk University, Czech Republic
Elke De Mulder	Google Inc., USA
Naghmeh Karimi	University of Maryland Baltimore County, USA
Elif Bilge Kavun	University of Passau, Germany
Mustafa Khairallah	Seagate Research, Singapore
Juliane Krämer	University of Regensburg, Germany
Ben Marshall	PQShield Ltd, UK
Philippe Maurine	University of Montpellier, France
David Oswald	University of Birmingham, UK
Kostas Papagiannopoulos	University of Amsterdam, The Netherlands
Guilherme Perin	Leiden University, The Netherlands
Romain Poussier	ANSSI, France
Prasanna Ravi	Nanyang Technological University, Singapore
Pascal Sasdrich	Ruhr-University Bochum, Germany
Patrick Schaumont	Worcester Polytechnic Institute, USA
Tobias Schneider	NXP Semiconductors, Austria
Sujoy Sinha Roy	TU Graz, Austria
Marc Stöttinger	RheinMain University of Applied Science, Germany
Rei Ueno	Tohoku University, Japan
Srinivas Vivek	IIIT Bangalore, India
Nusa Zidaric	Leiden University, The Netherlands

Additional Reviewers

Thomas Aulbach
Akira Ito
Dirmanto Jap
Georg Land
Suraj Mandal
Guenael Renault
Peter Pessl
Okan Seker
Patrick Struck

Contents

Fault Attacks

Microarchitectural Insights into Unexplained Behaviors Under Clock Glitch Fault Injection ... 3
Ihab Alshaer, Brice Colombier, Christophe Deleuze, Vincent Beroulle, and Paolo Maistri

An In-Depth Security Evaluation of the Nintendo DSi Gaming Console ... 23
pcy Sluys, Lennert Wouters, Benedikt Gierlichs, and Ingrid Verbauwhede

A Differential Fault Attack Against Deterministic Falcon Signatures ... 43
Sven Bauer and Fabrizio De Santis

Fault Attacks Sensitivity of Public Parameters in the Dilithium Verification ... 62
Andersson Calle Viera, Alexandre Berzati, and Karine Heydemann

Side-Channel Analysis

Attacking at Non-harmonic Frequencies in Screaming-Channel Attacks ... 87
Jeremy Guillaume, Maxime Pelcat, Amor Nafkha, and Rubén Salvador

Bernoulli at the Root of Horizontal Side Channel Attacks ... 107
Gauthier Cler, Sebastien Ordas, and Philippe Maurine

Blind Side Channel Analysis Against AEAD with a Belief Propagation Approach ... 127
Modou Sarry, Hélène Le Bouder, Eïd Maalouf, and Gaël Thomas

Leveraging Coprocessors as Noise Engines in Off-the-Shelf Microcontrollers ... 148
Balazs Udvarhelyi and François-Xavier Standaert

Smartcards and Efficient Implementations

The Adoption Rate of JavaCard Features by Certified Products and Open-Source Projects ... 169
Lukas Zaoral, Antonin Dufka, and Petr Svenda

PQ.V.ALU.E: Post-quantum RISC-V Custom ALU Extensions on Dilithium and Kyber ... 190
Konstantina Miteloudi, Joppe W. Bos, Olivier Bronchain, Björn Fay, and Joost Renes

Side-Channel and Neural Networks

Keep It Unsupervised: Horizontal Attacks Meet Simple Classifiers ... 213
Sana Boussam and Ninon Calleja Albillos

Deep Stacking Ensemble Learning Applied to Profiling Side-Channel Attacks ... 235
Dorian Llavata, Eleonora Cagli, Rémi Eyraud, Vincent Grosso, and Lilian Bossuet

Like an Open Book? Read Neural Network Architecture with Simple Power Analysis on 32-Bit Microcontrollers ... 256
Raphaël Joud, Pierre-Alain Moëllic, Simon Pontié, and Jean-Baptiste Rigaud

Correction to: PQ.V.ALU.E: Post-quantum RISC-V Custom ALU Extensions on Dilithium and Kyber ... C1
Konstantina Miteloudi, Joppe W. Bos, Olivier Bronchain, Björn Fay, and Joost Renes

Author Index ... 277

Fault Attacks

Microarchitectural Insights into Unexplained Behaviors Under Clock Glitch Fault Injection

Ihab Alshaer[1,2(✉)], Brice Colombier[3], Christophe Deleuze[1], Vincent Beroulle[1], and Paolo Maistri[2]

[1] Univ. Grenoble Alpes, Grenoble INP, LCIS, 26000 Valence, France
ihab.alshaer@univ-grenoble-alpes.fr
[2] Univ. Grenoble Alpes, CNRS, Grenoble INP, TIMA, 38000 Grenoble, France
[3] Université Jean Monnet Saint-Etienne, CNRS, Institut d Optique Graduate School, Laboratoire Hubert Curien UMR 5516, 42023 Saint-Etienne, France

Abstract. With the widespread use of embedded system devices, hardware designers and software developers started paying more attention to security issues in order to protect these devices from potential threats. Physical attacks represent an important threat to these devices, and fault injection is one of the major physical attacks. However, misunderstanding the effects of the fault injection would lead to proposing either over-protections or under-protections for these devices, thus affecting the performance/cost ratio and/or the security of the device. In this article, we provide a better representation of occurring fault, as a result of clock glitch, through novel models, in order to better understand the effects of fault injection. Also, we examine their dependencies with respect to the target device and the target program. Finally, we make use of the presented fault models to break the control-flow integrity of a program by altering the value of the program counter, in order to provide an actual application example.

Keywords: Fault injection attacks · Clock glitch · Fault model

1 Introduction

Given how frequently embedded systems are used in various spheres of life, securing them from malicious activities is fundamental. Sensitive data in embedded systems can be efficiently protected using cryptographic algorithms, which are frequently implemented in software on embedded microprocessors. However, such solutions can be vulnerable to attacks that seek to gain access to this private data. In particular, they might be vulnerable to physical attacks.

Fault injection is a major and powerful active physical attack. Since the well-known Boneh *et al.* attack [7], where the authors were able to break an implementation of CRT-RSA by inducing faults into the computations, it has been an attractive research topic.

S. Bhasin and T. Roche (Eds.): CARDIS 2023, LNCS 14530, pp. 3–22, 2024.
https://doi.org/10.1007/978-3-031-54409-5_1

It is possible to perform the fault injection in a variety of ways: by exposing a digital device to radiations [6], laser beams [11], or an electromagnetic pulse [13], by causing perturbations in the power supply [20] or in the clock signal [2], by altering the environment's temperature [18], *etc.*

1.1 Fault Injection Effects

Several works [8,10,13,15,19,22] claimed that the effect of a fault injection or the success of a fault injection attack is somehow random. In some cases [8,10,19], the corruption is expressed as random bit flips or random byte faults. On the other hand, other works [13,15,22], described it as random data corruptions that are applied at the instruction set architecture (ISA) level, either on the instruction data or on the contents of the registers.

Based on such variety, analyzing the vulnerabilities that fault injection can exploit is extremely challenging, and thus, developing countermeasures is significantly more complex. This will definitely result in either over-protections, which will affect the performance and the cost of the device, or conversely, in under-protections too, which will affect the security of the device.

Recent studies [2,3,11,12,21] tried to explain the effects of the fault injection by looking at the lower levels of abstraction of digital systems. In particular, they focus on the register transfer level (RTL), microarchitectural level and/or binary encoding level of the instructions. In some cases [12,21], however, the authors only conducted simulations at ISA and RTL levels: they did not confirm the realism of their analysis with physical fault injections. In contrast, [3,11] performed physical fault injections, but they only focused on two kinds of faulty behaviors: complete-instructions skip and complete-instructions replay faults. In [2], authors offered a thorough analysis and justification of the experimental findings that show how the alignment of the instructions in the flash memory can affect the obtained faulty behaviors. Based on that, they proposed two fault models at the binary encoding level: *Skip* a specific number of bits and *Skip and Repeat* a specific number of bits, whose value is strictly related to the flash memory access size. However, in their work, they explicitly said that these two fault models explain many of the obtained faulty behaviors, but *not all* of them.

1.2 Contributions

In this article, we propose a new inferred fault model, the *partial update fault model*, that is applied to the binary encoding of the instructions. This new fault model aims at explaining different faulty behaviors that are obtained when performing clock glitch fault injection campaigns on a 32-bit microcontroller. Therefore, it improves the vulnerability analysis process against fault injection, and hence, allows developers to design cost-effective countermeasures. We also show how a subcase of the new fault model is instruction-independent with high probability, and its manifestation is highly device-dependent. We show how we can execute new instructions even with a different length of encoding as a result of a clock glitch, by exploiting the variable-length capabilities of the target microcontroller. Finally, we make use of the presented fault model to modify the program counter to a chosen address, whose value is stored in a general-purpose register.

1.3 Outline

This article is organized as follows: Sect. 2 briefly describes the methodology we followed to explain the obtained results. Section 3 presents the inferred binary encoding fault models. Section 4 describes the experimental setup, then experimental results are reported and discussed in Sect. 5. Section 6 presents different practical scenarios to modify the program counter, based on the presented fault models. The article is concluded along with future perspectives in Sect. 7.

2 Fault Model Inference

The method we followed in this work to describe and characterize the fault injection results is comparable to the methods used in [3,9,12,13]. The core of the analysis consists in comparing the outcomes of executions, both the simulations and the actual fault injections, at various levels of digital system abstraction. In this work, we focus our analysis on two abstraction levels: ISA level and binary encoding of the instructions. We also enrich these descriptions by providing insights at the microarchitectural level.

On one side, physical fault injections are performed, with appropriate injection parameters, on a target device that is executing a simple target program, which is formed of a sequence of assembly instructions (step ① in Fig. 1). Then, from the physical fault injection results, we infer fault models at the binary encoding level of the instructions (step ② in Fig. 1). Applying the inferred binary fault models to the simulated execution of the same target program, that was used in step ①, is the next step (step ③ in Fig. 1). The outcomes of the physical fault injection and the software execution are then compared in order to provide better characterization of the impact of the fault injection and validate the inferred fault models (step ④ in Fig. 1). The comparison is carried out on the output values of the processor's general-purpose registers. Each of these registers has a known value at the beginning, and any change can be detected after performing step ④ in Fig. 1.

3 Partial Update Fault Model

This section presents the inferred binary encoding fault models that are applied to the target programs. These fault models seek to explain the faulty behaviors that have been observed after physical fault injection campaigns have been carried out on the target device that is running these target programs.

It has been observed through these physical fault injection experiments that not all of the observed faulty behaviors can be explained by the binary encoding fault models described in [2]. There are in fact other faulty behaviors, which can be explained with the new fault models described in this section.

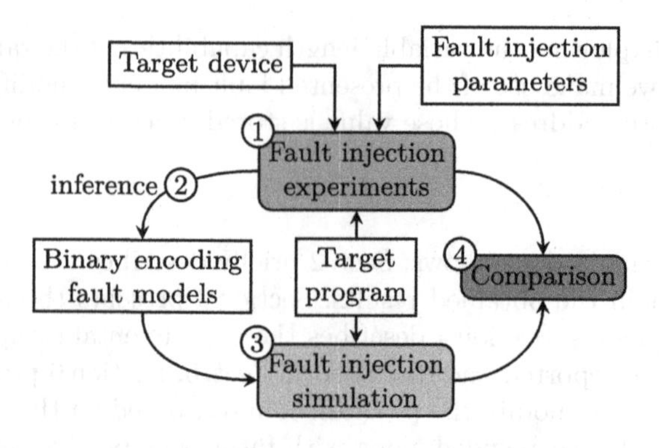

Fig. 1. Fault model inference methodology.

3.1 Partial Update from the Precharge Value

This fault model corresponds to a fault that happens while the fetched data or instruction is propagated between internal registers from the flash memory to the core, as shown in Fig. 2.

The hypothesis behind this fault model is based on the fact that not all bits of the data are propagated at the same speed from an internal register to another through a bus or combinational logic. Consequently, not all flip-flops within the destination register will get the update at the same time at a rising edge of a new clock cycle.

In nominal conditions, the clock period is defined such that all signals can be correctly sampled (i.e., the critical path has a positive slack). In case of a clock glitch, however, this behavior is disrupted by the fact the clock edge occurs quite sooner than expected. Thus, with the suitable injection parameters, it may happen that some flip-flops will receive the correct update, while some will receive the precharge value of the bus. Assuming that the precharge value of a bus or a wire between two registers is zero, then the correct update of a flip-flop means receiving the correct logic one or zero, while not receiving the correct update means capturing the precharge value of the bus, *i.e.* zero.

This model is observed as a reset on some bits while the instructions are transferred through the fetch data path in Fig. 2, as shown experimentally in Subsect. 5.1.

It should be mentioned that in [13], the authors claimed that some of the observed faults, as a result of electromagnetic fault injection, might be related

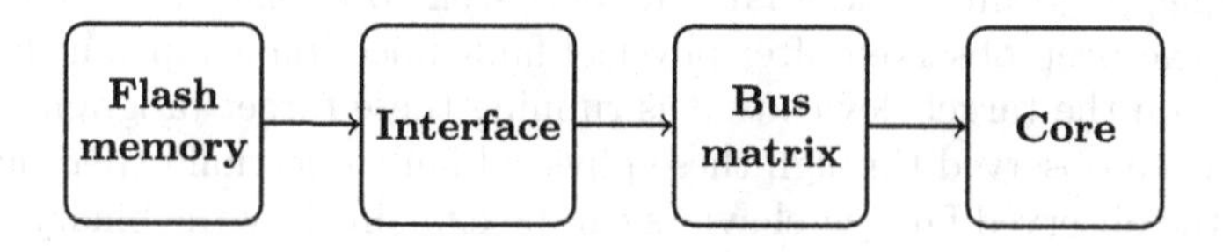

Fig. 2. Fetch data path in a microcontroller.

to the precharge value of the target microcontroller's bus. However, they didn't have a clear model that could explain their observed faults.

3.2 Partial Update from the Previous Value

This fault model is somehow similar to the previous one. It occurs on the same path shown in Fig. 2. However, instead of receiving the precharge value from the bus, some flip-flops within a destination register will keep their old values, either because the values have not been changed or because the corresponding wire still keeps the old values, and hence, the updated values are similar to the old ones. Conversely, and at the same time, other flip-flops in the destination register will receive the correct updated value.

This model is formally described as a bitwise OR between the old value and the new value of an internal register. This merge might be a full merge or a partial merge, as shown experimentally in Subsect. 5.2. Thus, in each flip-flop, the resulting value can be either the previous value or the correct value *i.e.* the value that the flip-flop should receive under normal execution, without any fault injection.

When looking at the instructions execution in this case, we observe the following behavior. The instruction(s) fetched at clock cycle i is executed normally. However, the instruction(s) fetched at clock cycle $i+1$ is not the one being executed. Instead, the observed instruction(s) is a full or partial merge between the fetched data at clock cycle i and the fetched data at clock cycle $i+1$. More details about this behavior are provided in Subsect. 5.2.

3.3 Discussion

Exploiting the transition value of a wire or a bus from a precharge (or a previous) value to a new value is a well-established modelling approach in power analysis attacks [1,16,17], which employ the so-called Hamming weight (or distance) leakage model. Likewise, our approach shows a similar pattern: depending on the type of register transition occurring (from previous or precharge value), the corresponding partial update fault model applies.

In Sect. 5, we show that both cases can occur for the same device. This is not in contrast with our modeling, as depending on the actual element that is affected by the fault injection (in our case, the clock glitch) and the fine-tuning of the injection parameters, we may see different outcomes. Further details ahead.

4 Experimental Setup

The target device and the fault injection method we employed are presented in this section. Section 5 includes the target programs, the corresponding experimental outcomes, and the discussion that follows.

4.1 Clock Glitch Fault Injection

An effective method of fault injection is to introduce perturbations on the clock signal. Compared with other fault injection methods like laser or electromagnetic pulses, clock glitch is known to be effective and the least expensive method. Also, it can offer a respectable level of controllability thanks to its temporal accuracy and, consequently, the location of the injection within the target program. In clock glitch fault injection, the glitch interferes with the normal operation of the clock signal, possibly causing a timing violation that results in a variety of erroneous behaviors. Additionally, since the glitch is introduced into the global clock, it is uncertain which microarchitectural component might be affected by the fault injection.

The following settings, illustrated in Fig. 3, are tuned when performing clock glitch fault injection:

- Delay: the time between the rising edge of a trigger signal and the rising edge of the target clock cycle.
- Shift: the time between the rising edge of the glitch and the rising edge of the target clock cycle.
- Width: the duration of the glitch.

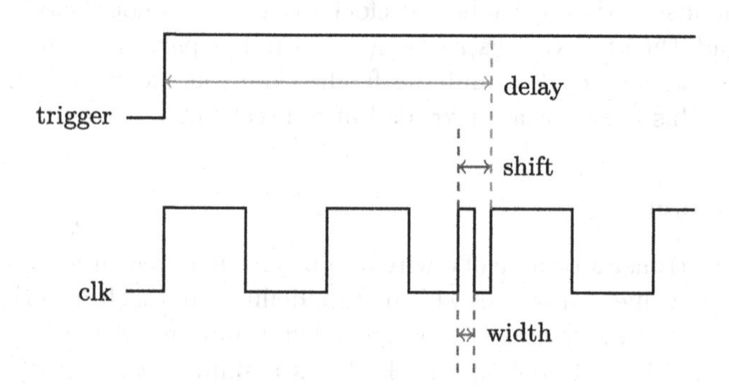

Fig. 3. Clock glitch parameters

In this work, the clock glitch fault injection campaigns have been carried out using the ChipWhisperer environment [14].

4.2 Target Device

The target device is a 32-bit microcontroller that embeds an Arm Cortex-M4 processor. The Arm Cortex-M4 has a 3-stage pipeline: fetch, decode and execute. It has 13 general-purpose 32-bit registers, `R0` to `R12`. Arm Cortex-M4 is based on ARMv7-M architecture [5] and supports the Thumb-2 instruction set [4].

Thumb-2 is a variable-length instruction set that offers two encoding lengths: 16 and 32 bits. The instruction has a 32-bit encoding if the most significant five bits of a 32-bit word have one of the following values [5]: `0b11101`, `0b11110` or `0b11111`.

The flash memory access size in this microcontroller is 64 bits. Therefore, up to two 32-bit or four 16-bit instructions can be fetched simultaneously. Additionally, as the supported instruction set is a variable-length instruction set, misaligned instructions can be fetched in several configurations as described in [2]. For example, the first half of a 32-bit instruction may be fetched at a given clock cycle, while the second half is fetched at the next clock cycle.

In the experiments, the processor is put in a known state before each fault injection. This is done by initialization instructions, that are located before the target instructions. After each execution, the values of the general purpose registers are transferred to a control computer via a serial communication in order to analyze the results. The target programs are presented in the next section.

5 Experimental Results

The result of any fault injection experiment is assigned to one of these classes:

- Crash: we obtain a crash, reset, or failure when attempting to read the target final state via the serial communication,
- Silent: the final state of the target is the so-called golden state, *i.e.* as if no fault was injected,
- Fault: the final state of the target is different from the golden state.

The results of the different clock glitch fault injection campaigns are discussed separately with respect to each fault model in the following subsections.

5.1 Partial Update from the Precharge Value

This section demonstrates how the *partial update from the precharge value* fault model explains many of the faulty behaviors observed during the fault injection campaigns. Also, it demonstrates the relation between this fault model and both the target instruction and the target device. To put it another way, it determines whether some bits in the fetched data are more sensitive to this fault model than other bits and, if so, whether the target instruction or the target device is to blame. The following subsections provide detailed results when targeting different instructions, and also when targeting a new device, identical to the already used one.

High-Hamming Weight Instruction. Since the *partial update from the precharge value* fault model causes some bits of the target instruction to be reset, it makes sense to choose an instruction with a large Hamming Weight in order to maximize the occurrence of the considered fault model. Under this assumption, we chose the instruction `SUBS R6, 0xff`, whose encoding in Thumb-2 is `0x3eff`. Our rationale is twofold: the instruction has a comparatively large Hamming

Weight (13) given its size. Secondly, since the *partial update from the precharge value* fault model causes some bits of the instruction to be reset, applying it on `0x3eff` results in an instruction that can be discriminated with high probability.

The objective behind these experiments is to show that several faulty behaviors can be explained using the *partial update from the precharge value* fault model. In addition, we want to see if some bits are more vulnerable than others to be reset within a target instruction. Finally, were it the case, we need to know if this is because of the target instruction or of the target device. Since the fetch size in the target device is 64 bits, a 16-bit instruction may reside in any of four different positions within these 64 bits. Therefore, four injection campaigns have been performed, where the position of `0x3eff` is different from one campaign to another. The main reason of changing the position of the target instruction is to find out if the fault model depends on the target instruction, or it depends on its position within the fetched 64 bits, and hence, depends on the physical implementation of the target device. The remaining three positions are filled with three instructions with the encoding `0x0000`, in order to minimize possible side effects from other instructions and make the analysis easier. This encoding corresponds to the `MOVS R0, R0` instruction, which is equivalent to a `NOP`.

Table 1 gives the target part code of each fault injection campaign. It also shows the glitch parameters that are used. These parameters are chosen in order to maximize the number of faults that can be classified under the *partial update from the precharge value* fault model. Position refers to the location of `0x3eff` within the fetched 64 bits. The values of shift and width are provided as a percentage of a single clock cycle: the glitch is introduced before the rising edge of the target clock cycle if shift is negative. ChipWhisperer provides an additional parameter, called fine-width, which is used to offer fine-tuning of the width parameter. It has been noticed that fine-width provides better reproducibility of the results when it is used. Repetitions is the number of executions for each combination of parameters. For each fault injection campaign, the total number of experiments corresponds therefore to more than 20 000 executions, as summarized in the last row of the table. The same value of delay is used in all the campaigns, and it depends on the number of initialization instructions that precede the target part.

The results of the four injection campaigns on `0x3eff` with respect to the three classes (*i.e.*, Crash, Silent and Fault) are presented in Table 2. All the resulting faulty behaviors can be classified under two fault models: *Skip* (all the general purpose registers keep their initial values), or *partial update from the precharge value* . Table 2 also provides the number of observed behaviors linked to each fault model among the faulty executions.

Figure 4 shows the encoding of the **executed** instructions for each injection campaign, along with the number of times each of them is observed. All of these faulty behaviors are classified under the *partial update from the precharge value* fault model. This is because all of them can be seen as a reset on some bits of the original instruction `0x3eff`. It should be noticed that resetting all the

Table 1. Experimental parameters

Position	1st	2nd	3rd	4th
Target	`0x3eff`	`0x0000`	`0x0000`	`0x0000`
part	`0x0000`	`0x3eff`	`0x0000`	`0x0000`
code	`0x0000`	`0x0000`	`0x3eff`	`0x0000`
	`0x0000`	`0x0000`	`0x0000`	`0x3eff`
Shift	−13			
Width	{6, 10}	{6, 10}	{3, 4}	{3, 4}
Fine width	[−255, 255]			
Repetitions	20			
Total	20440			

Table 2. Fault obtained when targetting the `0x3eff` instruction at four different positions

Class	Position			
	1st	2nd	3rd	4th
Crash	0	0	23	1
Silent	33	1574	2273	158
Fault	20 407	18 866	18 144	20 281
Skip	11 523	8295	11 107	7901
Partial update from the precharge value	8884	10 571	7037	12 380

bits of `0x3eff` will result in executing `0x0000`, which is an instruction with no effect as mentioned earlier, and thus classified under the skip fault model.

Additionally, Table 2 and Fig. 4 show that the number of faulty behaviors and the observed executed instructions depend on the position of the instruction in the fetched 64 bits. Thus, the results depend on the position rather than the instruction. Furthermore, the results of each position show that some instructions are more probable to be executed than others as a result of the fault injection.

To better understand the effect of the fault at each position, and hence, on each bit in the position, we define a metric called *bit sensitivity*. It measures the probability for a bit to be reset as a result of the clock glitch fault injection over the faulty behaviors that are classified under the *partial update from the precharge value* fault model at a specific position. The *bit sensitivity* $S_p(f, b)$ of bit b to a given fault model f at position p is defined in Eq. (1).

$$S_p(f,b) = 1 - \frac{P(b = 1 \mid p)}{P(\text{fault model} = f \mid p)} \quad (1)$$

Figure 5 presents the bit sensitivity values for the results obtained during the fault injection campaigns on `0x3eff` at all positions. Obviously, bits that are

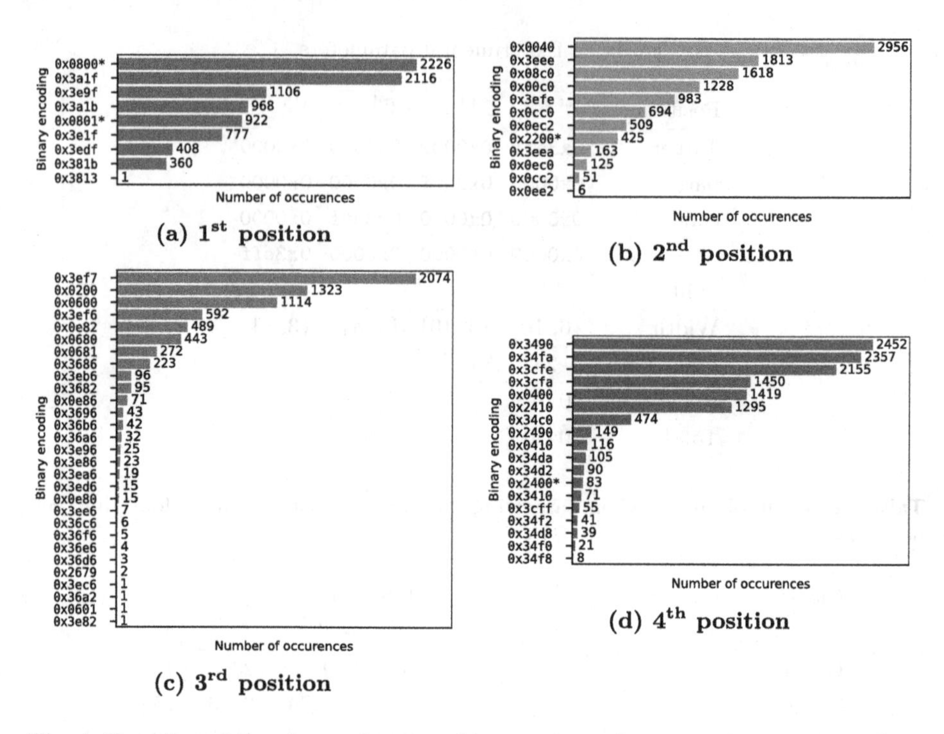

(a) 1st position

(b) 2nd position

(c) 3rd position

(d) 4th position

Fig. 4. Encoding of the observed executed instructions when targeting `0x3eff` at four different positions within the target programs.

zero in `0x3eff` (bits 8, 14, and 15) have no corresponding bit sensitivity value. We can see that the bit sensitivity is different from one position to another and from one bit to another at the same position. Thus, under the *partial update from the precharge value* fault model, some instructions are more probable than others.

Subsection 5.1 presents the results of targeting a different instruction, to confirm that the *partial update from the precharge value* fault model depends on the physical implementation of the device, and not on the target instruction.

It is important to note that whenever there is a doubt about the execution of an instruction, results are confirmed using alternative initial register values. Nonetheless, in rare circumstances, more than one instruction may produce the same outcome, for instance, when the value of a register is zero. For example, this might happen because of moving zero to the register, or by shifting its value by 32 bits. In Fig. 4, when the encoding is starred, it means that there is an alternative instruction that could lead to the same outcome, and we selected one based on other observed encoding at the same position. It is important to stress that this is happening only in a few cases (4 times), and does not affect the measurements or the general conclusion.

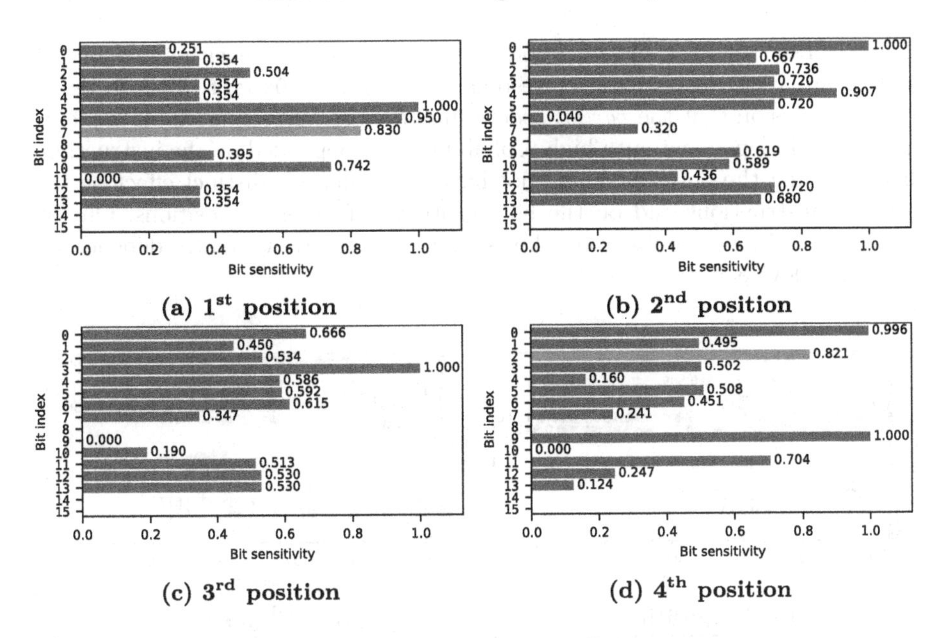

(a) 1st position **(b) 2nd position**

(c) 3rd position **(d) 4th position**

Fig. 5. Bit sensitivity values obtained when targeting `0x3eff`.

Confirming Sensitive Bits We carried out extra experiments with the value `0x3b7d`, which is the encoding of the `SUBS R3, 0x7d` instruction. Again, we chose this instruction since it has a relatively high Hamming weight, and allows recognizing the encoding of the executed instructions as a result of the *partial update from the precharge value* fault model with high probability. However, we specifically took care to have ones in the most sensitive positions from Fig. 5 to see if these measurements are reproducible when targeting a different instruction.

The experimental parameters for the fault injection campaigns on `0x3b7d` are identical to that of `0x3eff`, given in Table 1. The only difference is that the target program has `0x3b7d` instead of `0x3eff`. The classification results are presented in Table 3, while the bit sensitivity values are plotted in Fig. 6.

Table 3. Fault obtained when targeting the `0x37bd` instruction at four different positions

Class	Position			
	1st	2nd	3rd	4th
Crash	0	0	0	0
Silent	39	2304	2589	197
Fault	20 401	18 136	17 851	20 243
Skip	11 694	8386	10 528	7606
Partial update from the precharge value	8707	9750	7323	12 637

It is clear that the classification results and the bit sensitivity values of `0x3b7d` are very close to the corresponding results of `0x3eff`. This leads us to the conclusion that the *partial update from the precharge value* fault model is instruction-independent with high probability. If it depended on the instruction, then changing the position should not have an observable distinct effect on the executed instructions and on the bit sensitivity of different positions. On the other hand, the next subsection shows that bit sensitivity greatly depends on the target device.

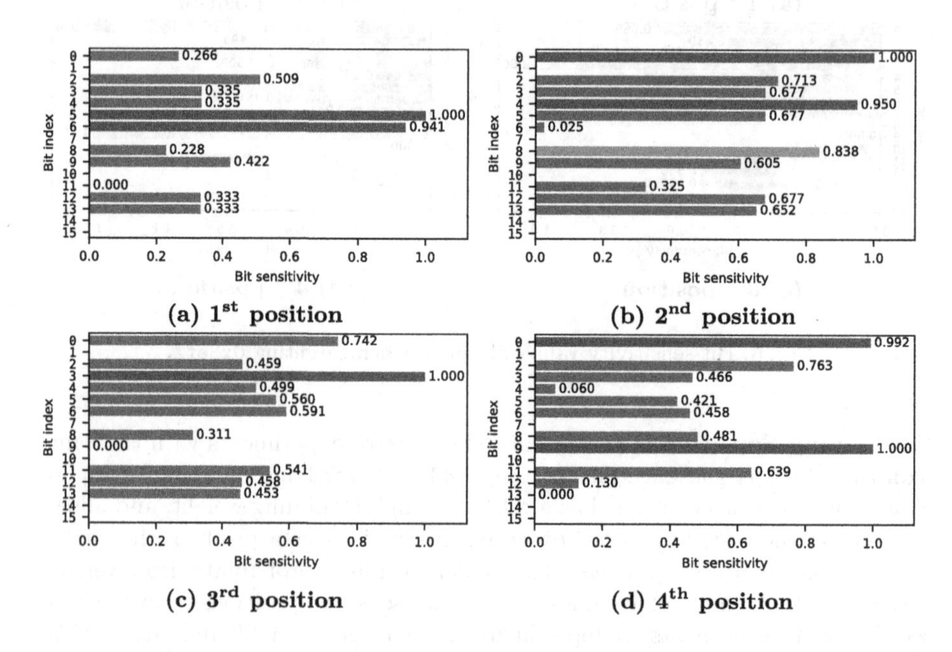

Fig. 6. Bit sensitivity values obtained when targeting `0x3b7d`.

`0x3eff` Experiments on a New Microcontroller. In this section, we present the results of targeting the `0x3eff` instruction again while using a brand new device, which we did not use to perform any experiment previously. In any other means, it is identical to the one we used in the previous experiments. This is done to better understand the dependency of the *partial update from the precharge value* fault model on the target device. The experimental parameters of this campaign are identical to those in Table 1.

For our purposes, it is enough to present the results on the 2nd and 4th positions to see that they are very different between the two devices. The results are presented in Table 4 and Fig. 7.

A very interesting observation is that the number of faults and the bit sensitivity were much higher when performing the fault injection campaigns on the old device. This is clear for the bit sensitivity of the 4th position in Fig. 7b, as we can see the distribution of the bit sensitivities is similar to that in Figs. 5d

Table 4. Fault obtained when targeting the `0x3eff` instruction using the *new* device

Class	Position	
	2nd	4th
Crash	0	2
Silent	18 058	16 995
Fault	2382	3443
Skip	345	0
Partial update from the precharge value	2037	3443

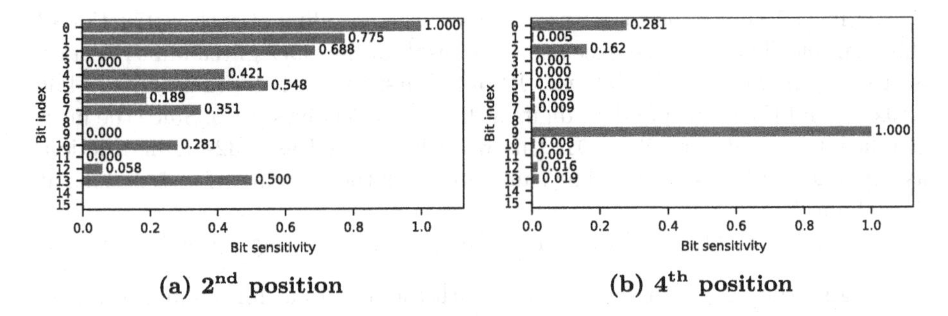

(a) 2nd position **(b) 4th position**

Fig. 7. Bit sensitivity values obtained when targeting `0x3eff` on the *new* device at the 2nd and the 4th positions.

and 6d, however, on the old device, the sensitivity is much higher. This could be explained as an aging effect, since the old device has been used for fault injection experiments for a few months. We speculate that the bit sensitivity could increase over time (a common consequence of performance degradation due to aging), but further research is needed to confirm this observation.

Conclusion on Bit Sensitivity. To summarize, the bit sensitivity figures illustrate that, as a result of the *partial update from the precharge value* fault model, the probability distribution of the corrupted instruction is not random, and it depends on several features that are mostly device-dependent. The probability of executing a given instruction differs from the probability of executing another. This discrepancy is determined by both the instruction's position inside the target program and the target device itself. This is of prime importance if the instruction results in a security vulnerability, as will be highlighted in Sect. 6.

5.2 Partial Update from the Previous Value

In this section, we focus on the occurrence of faulty behaviors that can be classified under the *partial update from the previous value* fault model. In this case, there is no precharge value and the transition occurs from the value that was previously stored in the register. In the time while the register is updating its value, a transient situation may occur when some bits already have the new value, whereas others are still to be updated. This behavior can be seen a merge between the previous and the new instruction. The following subsections give examples of observed faulty behaviors that are classified as full or partial merge.

Full Merge. The merge is considered full if and only if the observed executed instruction(s) can be expressed as a bitwise OR between the data fetched at clock cycle i and the data fetched at clock cycle $i + 1$. In the following, an example is provided, which illustrates how *two new* 32-bit instructions are executed as a result of a full merge between eight different 16-bit instructions.

Listing 1.1 shows the target program and the encoding of each instruction of this example. The observed execution as a result of the clock glitch fault injection is given in Listing 1.2. The bitwise OR of the first two hexadecimal digits at line 1 (`0xa9`) and the corresponding digits at line 5 (`0x42`) gives `0xeb`. Since the most significant five bits are `0b11101`, this word is decoded as a 32-bit instruction, as explained in Subsect. 4.2. The same holds for the merging of instructions at lines 3 and 7.

Listing 1.2 is obtained by a full merge applied on Listing 1.1 as follows:

- Merging the 32 bits at lines 1 and 2 with the 32 bits at lines 5 and 6 respectively: `0xa9000000 | 0x42000305 = 0xeb000305`.
- Merging the 32 bits at lines 3 and 4 with the 32 bits at lines 7 and 8 respectively: `0xa9000000 | 0x42020405 = 0xeb020405`.

```
1 ADD  R1, SP, 0x0    // 0xa900
2 MOVS R0, R0         // 0x0000
3 ADD  R1, SP, 0x0    // 0xa900
4 MOVS R0, R0         // 0x0000
5 TST  R0, R0         // 0x4200
6 LSLS R5, R0, 0xc    // 0x0305
7 TST  R2, R0         // 0x4202
8 LSLS R5, R0, 0x10   // 0x0405
```

Listing 1.1. Target program to execute two new 32-bit instructions as a result of full merge.

```
1 ADD  R1, SP, 0x0   // 0xa900
2 MOVS R0, R0        // 0x0000
3 ADD  R1, SP, 0x0   // 0xa900
4 MOVS R0, R0        // 0x0000
5 ADD  R3, R0, R5    // 0xeb000305
6 ADD  R4, R2, R5    // 0xeb020405
```

Listing 1.2. Observed execution as a result of full merge on Listing 1.1.

Partial Merge. In this case, only part of the fetched data at clock cycle i and the data fetched at clock cycle $i + 1$ is merged. The target code that is used for this example is shown in Listing 1.3.

```
1 ADD  R1, R1, 0x4    // 0xf1010104
2 ANDS R2, R0         // 0x4002
3 MOVS R0, R0         // 0x0000
4 ADD  R2, R2, 0xa    // 0xf102020a
5 MOVS R4, R0         // 0x0004
6 MOVS R0, R0         // 0x0000
```

Listing 1.3. Target program for partial merge experiment.

One of the observed executions that can be classified under Partial merge is as the following: We observed that not all the 32 bits at lines 1 (`0xf1010104`) are systematically merged with the corresponding 32 bits at line 4 (`0xf102020a`): only the destination and source registers are merged. In addition to this behavior, another partial merge occurred in the following instructions, only over the least significant digit, between the 16 bits at line 2 (`0x4002`) and the 16 bits at line 5 (`0x0004`). As a consequence, only the destination register at line 5 is affected. The observed execution of this example is shown in Listing 1.4. It should be mentioned that we cannot discriminate on the opcode values (0xf1), as it is the same in both `ADD` instructions. It is worth mentioning that a Full merge was also observed for the target program in Listing 1.3

```
1 ADD  R1, R1, 0x4    // 0xf1010104
2 ANDS R2, R0         // 0x4002
3 MOVS R0, R0         // 0x0000
4 ADD  R3, R3, 0xa    // 0xf103030a
5 MOVS R6, R0         // 0x0006
6 MOVS R0, R0         // 0x0000
```

Listing 1.4. Observed execution as a result of partial merge after targeting Listing 1.3.

6 Program Counter Modification

In this section, we exploit the proposed fault models to change the value of the program counter to an address stored in one of the general purpose registers. Being able to modify the program counter allows to break the control flow integrity of a program. This is leveraged in various attacks, such as privilege escalation or secure-boot violation [20].

In the following subsections, we measure the probability of modifying the program counter under different scenarios for the target program in Listing 1.5. The results of the different scenarios, along with the fault models that led to the success of the attack, and the glitch parameters that allowed observing the results are summarized in Table 5. The success rate is computed over 10 000 executions for each clock glitch fault injection scenario. The glitch parameters are tuned to maximize the success rate.

```
1  R8 = address of line 11
2  // series of 0x0000
3  ADD R6, R1, 0x4c7    // 0xf20146c7
4  ADD R3, R3, 0xa
5  ADD R4, R4, 0xb
6  ADD R5, R6, R3
7  ADD R3, R3, 0xf
8  // series of 0x0000
9  ADD R5, R5, 0x5
10 // series of 0x0000
11 ADD R1, R1, 0x3
12 ADD R9, R0, R6
```

Listing 1.5. Target program for PC modification experiments.

6.1 Misaligned Code

In [2], the authors were able to modify the program counter to an address stored in `R8` as a result of the skip fault model in a misaligned code. This is done by executing the least significant 16 bits of a misaligned 32-bit instruction. The first half of the 32-bit instruction is fetched at clock cycle i and its second half is fetched at clock cycle $i + 1$. Therefore, skipping the fetched data at clock cycle i results in decoding the remaining half that is fetched at clock cycle $i + 1$, and executing when it is a valid encoding for a 16-bit instruction. The same thing can happen for the `ADD R6, R1, 0x4c7` instruction shown in Listing 1.5. Its least significant 16 bits (`0x46c7`) are the encoding of `MOV PC, R8`, which stores the value of `R8` into the program counter. Thus, executing `MOV PC, R8` leads to a jump from line 3 to line 11, since `R8` stores the address of line 11.

We reproduced their attack on Listing 1.5. Many useful and dummy instructions are used in Listing 1.5 to make sure of detecting the execution of `MOV PC, R8`. The success rate in this scenario was 100 %. We noticed that 9996 of the executions can be classified under the skip fault model. However, *four* executions can be classified under the *partial update from the precharge value* fault model. This is because resetting some bits of the most significant 16 bits of `ADD R6, R1, 0x4c7`, will lead to execute two 16-bit instructions, as the most significant five bits do not identify a valid encoding for a 32-bit instruction (as detailed in Sect. 4.2). For these four executions, we confirmed this is happening by observing the values of the registers that the `MOVS R1, R0` instruction, of encoding `0x0001`, had been executed. Thus, the instructions `MOVS R1, R0` and `MOV PC, R8` are executed in sequence.

6.2 Aligned Code

The aforementioned attack relies on the misalignment of the code in memory, as explained in [2]. We add a single `MOVS R0, R0 (0x0000)` to the target program, just before `ADD R6, R1, 0x4c7` instruction, to realign it. The code is now aligned, all bits of `0xf20146c7` are fetched in a single clock cycle. In this case, the fault model that we can rely on to create new instructions (and thus modify the program counter) is the *partial update from the precharge value* fault model. The aim is to reset bits over the most significant 16 bits while not touching the least significant 16 bits, in order to keep the encoding of `MOV PC, R8`, *i.e.*, `0x46c7`. The success rate of the clock glitch fault injection campaign in this case was 0.71 %. However, no side effect is observed along with executing `MOV PC, R8`, but this is normal as resetting some bits of the most significant 16 bits of `0xf20146c7` could lead to execute many 16-bit instructions with no observable effect like `MOVS R0, R0 (0x0000)`, or `TST R0, R0 (0x4200)` for example.

This result is an improvement over the state of the art, since one could imagine that making the code aligned will protect from the misaligned faulty behaviors that are described in [2]. Thus, aligning the code cannot be considered a sufficient countermeasure against clock glitch attacks, that might focus on misaligned codes. However, aligning the sensitive instructions can effectively decrease the success rate, as demonstrated experimentally.

6.3 Countermeasure: Register Substitution

In this scenario the code is misaligned, but we changed the destination register in `ADD R6, R1, 0x4c7` from `R6` to `R2`. Other occurrences of `R6` are replaced with `R2` in the rest of the program. Now, the least significant 16-bit word for `ADD R2, R1, 0x4c7` is `0x42c7`. The success rate in this scenario was *zero*: no fault led to modify the program counter to the value in `R8`, even when we used the same experimental parameters that previously led to a success rate of 100 %. The `R2` register was chosen because `2` in the encoding can not be turned into a `6` by resetting bits. Thus, we avoid obtaining the encoding of `MOV PC, R8`.

This scenario shows that a clear understanding of the fault effect led to the design of a very simple and cost-effective countermeasure. This proposal clearly has no overhead and is easily implemented by the compiler, except in rare cases where registers might be under a lot of pressure.

6.4 Trojan

In this case, dummy code with no effect on the target program is added just before `ADD R2, R1, 0x4c7`, where the code is protected against executing `MOV PC, R8`. This dummy code is shown in Listing 1.6. It implements a Trojan that can be activated by clock glitch fault injection in order to controllably execute the `MOV PC, R8` instruction. It is clear that the *partial update from the previous value* fault model in the full merge setting will lead to execute `MOV PC,`

Table 5. Experimental results obtained, and fault injection parameters used when attempting to modify the program counter.

	Fault injection scenario			
	Misaligned	Aligned	Protected	Trojan
Success rate	100 %	0.71 %	0.0 %	95.11 %
Fault models	Skip [2] (99.96 %) *partial update from the precharge value* (Sect. 5.1) (0.04 %)	*partial update from the precharge value* (Sect. 5.1)	–	*partial update from the previous value* (Sect. 5.2)
Shift	−12	−13	–	−9
Width	3	10	–	4

R8, since we have that 0x4281 | 0x0446 = 0x46c7 (MOV PC, R8). The experimental success rate of this scenario was 95.11 %.

This scenario is possible if we assume that the attacker is the software developer himself. Alternatively, the compiler used to compile the code may be untrusted and thus represent the attacker. As a countermeasure, a code review, based on the presented fault models, should be able to detect such Trojans.

```
1 CMP  R1, R0          // 0x4281
2 MOVS R0, R0          // 0x0000
3 MOVS R0, R0          // 0x0000
4 MOVS R0, R0          // 0x0000
5 LSLS R6, R0, 0x11    // 0x0446
6 MOVS R0, R0          // 0x0000
7 MOVS R0, R0          // 0x0000
8 MOVS R0, R0          // 0x0000
```

Listing 1.6. Dummy code implementing a Trojan.

7 Conclusion and Future Works

A new binary encoding fault model has been presented and defined: the partial update fault model, which comes in two variations: the *partial update from the precharge value* and the *partial update from the previous value* fault models. These fault models allow explaining a wide range of the faulty behaviors that are obtained when performing clock glitch fault injection campaigns. Therefore, they can be used to perform vulnerability analysis of software codes against these fault attacks, and help in better designing efficient and low-cost countermeasures. We have also given an exploitation example: modifying the program counter can be achieved and explained through these fault models. Following that, we proposed a simple yet effective countermeasure against such vulnerability. We also examined the dependency of *partial update from the precharge value* fault model with respect to the target device and program.

In terms of future works, proper formalization of protections against the presented fault models will be very necessary. At software level, an automated framework made of vulnerability assessment followed by automatic code protection would greatly improve the security of the targeted application. At a lower level, several approaches might be envisioned at different abstraction levels, from ISA down to transistor-level. Also, targeting other architectures will be important to see if the proposed models can be generalized to various architectures.

Finally, although clock glitch has been used in this article to perform the fault injection, the presented results may be generalized to other fault injection techniques that rely on timing violations. This includes, for example, voltage glitch and electromagnetic fault injection.

Acknowledgments. This work has been supported by the LabEx PERSYVAL-Lab (ANR-11-LABX-0025-01) and the French National Research Agency in the framework of the "Investissements d'avenir" program (ANR-15-IDEX-02).

References

1. Alioto, M., Poli, M., Rocchi, S.: Differential power analysis attacks to precharged buses: a general analysis for symmetric-key cryptographic algorithms. IEEE Trans. Dependable Secure Comput. **7**(3), 226–239 (2010)
2. Alshaer, I., Colombier, B., Deleuze, C., Beroulle, V., Maistri, P.: Variable-length instruction set: feature or bug? In: 25th Euromicro Conference on Digital System Design, pp. 464–471. IEEE, Maspalomas (2022)
3. Alshaer, I., Colombier, B., Deleuze, C., Maistri, P., Beroulle, V.: Cross-layer inference methodology for microarchitecture-aware fault models. Microelectron. Reliab. **139**, 114841 (2022)
4. ARM Limited: ARM architecture reference manual Thumb-2 supplement. https://developer.arm.com/documentation/ddi0308/d. Accessed 24 February 2023
5. ARM Limited: Armv7-m architecture reference manual. https://developer.arm.com/documentation/ddi0403/latest. Accessed 24 February 2023
6. Baumann, R.: Radiation-induced soft errors in advanced semiconductor technologies. IEEE Trans. Device Mater. Reliab. **5**(3), 305–316 (2005)
7. Boneh, D., DeMillo, R.A., Lipton, R.J.: On the importance of eliminating errors in cryptographic computations. J. Cryptology **14**, 101–119 (2001)
8. Buhren, R., Jacob, H.N., Krachenfels, T., Seifert, J.: One glitch to rule them all: Fault injection attacks against amd's secure encrypted virtualization. In: Kim, Y., Kim, J., Vigna, G., Shi, E. (eds.) ACM SIGSAC Conference on Computer and Communications Security, pp. 2875–2889. ACM, Virtual Event, Republic of Korea (2021)
9. Dureuil, L., Potet, M.-L., de Choudens, P., Dumas, C., Clédière, J.: From code review to fault injection attacks: filling the gap using fault model inference. In: Homma, N., Medwed, M. (eds.) CARDIS 2015. LNCS, vol. 9514, pp. 107–124. Springer, Cham (2016). https://doi.org/10.1007/978-3-319-31271-2_7
10. Khelil, F., Hamdi, M., Guilley, S., Danger, J., Selmane, N.: Fault analysis attack on an FPGA AES implementation. In: Aggarwal, A., Badra, M., Massacci, F. (eds.) International Conference on New Technologies, Mobility and Security, pp. 1–5. IEEE, Tangier (2008)

11. Khuat, V., Danger, J., Dutertre, J.: Laser fault injection in a 32-bit microcontroller: from the flash interface to the execution pipeline. In: Workshop on Fault Detection and Tolerance in Cryptography, pp. 74–85. IEEE, Milan (2021)
12. Laurent, J., Deleuze, C., Pebay-Peyroula, F., Beroulle, V.: Bridging the gap between RTL and software fault injection. ACM J. Emerg. Technol. Comput. Syst. **17**(3), 38:1–38:24 (2021)
13. Moro, N., Dehbaoui, A., Heydemann, K., Robisson, B., Encrenaz, E.: Electromagnetic fault injection: towards a fault model on a 32-bit microcontroller. In: Fischer, W., Schmidt, J. (eds.) 2013 Workshop on Fault Diagnosis and Tolerance in Cryptography, Los Alamitos, CA, USA, August 20, 2013, pp. 77–88. IEEE Computer Society (2013)
14. O'Flynn, C., Chen, Z.D.: ChipWhisperer: an open-source platform for hardware embedded security research. In: Prouff, E. (ed.) COSADE 2014. LNCS, vol. 8622, pp. 243–260. Springer, Cham (2014). https://doi.org/10.1007/978-3-319-10175-0_17
15. Proy, J., Heydemann, K., Berzati, A., Majéric, F., Cohen, A.: A first ISA-level characterization of EM pulse effects on superscalar microarchitectures: a secure software perspective. In: International Conference on Availability, Reliability and Security, pp. 7:1–7:10. ACM, Canterbury (2019)
16. Randolph, M., Diehl, W.: Power side-channel attack analysis: a review of 20 years of study for the layman. Cryptography **4**(2), 15 (2020)
17. Shelton, M.A., Samwel, N., Batina, L., Regazzoni, F., Wagner, M., Yarom, Y.: ROSITA: towards automatic elimination of power-analysis leakage in ciphers. In: Annual Network and Distributed System Security Symposium. The Internet Society, Virtual event (2021)
18. Skorobogatov, S.P.: Local heating attacks on flash memory devices. In: Tehranipoor, M., Plusquellic, J. (eds.) IEEE International Workshop on Hardware-Oriented Security and Trust, pp. 1–6. IEEE Computer Society, San Francisco (2009)
19. Spensky, C., et al.: Glitching demystified: analyzing control-flow-based glitching attacks and defenses. In: IEEE/IFIP International Conference on Dependable Systems and Networks, pp. 400–412. IEEE, Taipei (2021)
20. Timmers, N., Spruyt, A., Witteman, M.: Controlling PC on ARM using fault injection. In: Workshop on Fault Diagnosis and Tolerance in Cryptography, pp. 25–35. IEEE Computer Society, Santa Barbara (2016)
21. Tollec, S., Asavoae, M., Couroussé, D., Heydemann, K., Jan, M.: Exploration of fault effects on formal RISC-V microarchitecture models. In: Workshop on Fault Detection and Tolerance in Cryptography, pp. 73–83. IEEE, Virtual Event/Italy (2022)
22. Trouchkine, T., Bouffard, G., Clédière, J.: EM fault model characterization on SoCs: from different architectures to the same fault model. In: Workshop on Fault Detection and Tolerance in Cryptography, pp. 31–38. IEEE, Milan (2021)

An In-Depth Security Evaluation of the Nintendo DSi Gaming Console

pcy Sluys(✉), Lennert Wouters, Benedikt Gierlichs, and Ingrid Verbauwhede

COSIC, KU Leuven, Kasteelpark Arenberg 10, Heverlee, Belgium
{pcy.sluys,lennert.wouters,benedikt.gierlichs, ingrid.verbauwhede}@esat.kuleuven.be

Abstract. The Nintendo DSi is a handheld gaming console released by Nintendo in 2008. In Nintendo's line-up the DSi served as a successor to the DS and was later succeeded by the 3DS. The security systems of both the DS and 3DS have been fully analyzed and defeated. However, for over 14 years the security systems of the Nintendo DSi remained standing and had not been fully analysed. To that end this work builds on existing research and demonstrates the use of a second-order fault injection attack to extract the ROM bootloaders stored in the custom system-on-chip used by the DSi. We analyse the effect of the induced fault and compare it to theoretical fault models. Additionally, we present a security analysis of the extracted ROM bootloaders and develop a modchip using cheap off-the-shelf components. The modchip allows to jailbreak the console, but more importantly allows to resurrect consoles previously assumed irreparable.

Keywords: Nintendo DSi · boot ROM · fault injection · secure boot · modchip · embedded security

1 Introduction

The Nintendo DSi is a handheld gaming console released by Nintendo in 2008. It was designed as a small upgrade to its predecessor, the Nintendo DS, adding extra 'multimedia' features such as two cameras and a web browser. These consoles still have active *homebrew* communities of people making their own self-published games and programs. Additionally, these communities are interested in reverse-engineering the publicly undocumented hardware. This is done to develop more precise emulators, write code that is capable of rendering more spectacular graphical effects, or simply as a reason in and of itself.

The security systems of the DS [6] and the 3DS [8,30,31] have been fully analyzed and defeated. This has not yet been the case for the DSi.

The goal of this research is to analyze the previously unexamined parts of the security system of the Nintendo DSi, more specifically, its *boot ROMs*. This would allow for better hardware preservation and brick recovery: current exploits [17]

S. Bhasin and T. Roche (Eds.): CARDIS 2023, LNCS 14530, pp. 23–42, 2024.
https://doi.org/10.1007/978-3-031-54409-5_2

rely on the second-stage bootloader residing in eMMC (Embedded MultiMedia Card) memory existing and having a correct digital signature. This eMMC, a Samsung moviNAND chip [16], has a low erase-write-cycle lifetime, and might be affected by bugs in the wear-levelling management firmware (this is the case for other eMMCs made by the same manufacturer in that era [24]). Simply replacing the eMMC chip with another would not fix the situation, as the console uses the eMMC's CID (Card IDentifier, a uniquely identifying number of every eMMC memory) to derive cryptographic keys used to encrypt the FAT32 filesystem [16]. Furthermore, this work can be used as inspiration on how to tackle other, similar targets.

This paper is structured as follows: Sect. 2 provides background information and covers related work. In Sects. 3 and 4, we show how the boot ROMs can be extracted. Section 5 reflects on the fault injection campaign and proposes an explanation of what type of faults actually occur when performing the readout attack. A security analysis is then performed in Sect. 6. The results of this analysis are used in Sect. 7 to build a modchip capable of jailbreaking the console. Section 8 then provides a conclusion.

1.1 Contributions

This work presents the following contributions: We extract the boot ROMs of the DSi using a second-order EMFI attack. We then look into the security aspects of these ROMs, completing the security analysis of the Nintendo DSi. Finally, we develop a modchip able to jailbreak the system in its very first bootstage. This modchip attack can be used to revive bricked consoles with a broken eMMC.

1.2 Responsible Disclosure

We did not disclose our research results to Nintendo ahead of submission for several reasons.

First, Nintendo appears to only accept vulnerability reports through their HackerOne bug bounty program. At the time of writing Nintendo only accepts submissions for the Switch console. Note that the 3DS (the DSi's successor) is explicitly listed as out of scope[1]. Secondly, by submitting a report through the HackerOne program we would agree to not publish our findings, even if the report is considered out of scope. Finally, Nintendo discontinued the DSi[2], and no new game titles have been released since 2016[3]. Due to the above factors, vulnerability disclosure to the vendor is currently not considered.

[1] See https://hackerone.com/nintendo/updates, https://archive.ph/Yh7YV.

[2] The exact date is unclear. Nintendo never announced an official date when the DSi would go out of support, instead changing the console's status silently. The 3DS was discontinued in 2020.

[3] *Crazy Train*, a downloadable DSiWare title.

2 Background and Related Work

This section gives an overview of the hardware of the Nintendo DSi, fault injection, and previous attacks on the DSi. These elements are necessary to understand the attacks used in this work, and the current state-of-the-art regarding DSi exploits.

2.1 The Nintendo DSi Gaming Console

The DSi is an interesting hybrid between the DS and 3DS: it keeps the former's CPU and GPU, while the peripherals, chipset, boot process and overall security architecture resemble those of the 3DS much more closely.

The DSi, much like the DS, has two CPU cores, an ARM7TDMI and an ARM946E-S, typically shortened to respectively ARM7 and ARM9. The ARM7 is used for I/O tasks and has exclusive access to many I/O peripherals, while the ARM9 is much faster and more powerful, and has exclusive access to the GPU. It has SoC-internal SRAM specific to each separate CPU core with configurable mapping options, and external DRAM shared between the two cores. These CPUs can also communicate using a FIFO interface. Unlike the DS, it boots from eMMC NAND, which contains the second-stage bootloaders (in raw eMMC blocks) as well as the system menu and various apps (on an encrypted FAT32 filesystem). It has extra peripherals such as cameras and an SD card slot. More information about the DS and DSi can be found in [14].

The DS only used symmetric-key Blowfish encryption without authentication, and a system menu residing in external flash without cryptographic protection mechanisms. This naturally lead to the proliferation of 'flashcarts' [6], on which homebrew (and pirated) games can be loaded and played. Compared to this, the DSi and 3DS both use a full secure boot chain using digital signatures and AES encryption, from the first bootloader [5,8] down to individual games and applications [26].

2.2 Fault Injection

Fault injection is an attack method targeting the physical implementation of a device, by actively tampering with its operation. By bringing one or more environmental parameters (such as supply voltage, clock signal, incident electromagnetic field, etc.) outside the operating range of the device for a short amount of time, the target can be made to malfunction without crashing or shutting down. After these parameters return to their normal range, the effects of this malfunctioning can still propagate logically, possibly subverting the security properties of the device [1,34,41]. Naturally, people have proposed countermeasures to stop such attacks [2], resulting in an arms race between attackers and defenders.

Fault injection can be used to circumvent security checks [12], tamper with cryptographic algorithms to obtain secrets [9,35], and even directly take control over the execution flow of a processor [28,37]. In the context of gaming consoles, this often translates to obtaining decryption keys [10,18] and arbitrary code execution capabilities in a high-privilege environment (e.g. a bootloader, hypervisor, or security coprocessor) [8,11].

2.3 Earlier Work on the DSi

Interestingly enough, it was the 3DS of which the security system was fully broken first. From software exploits in the operating system [26] to bootloader 'unlocks' [20]. Finally, the boot ROMs were extracted as well [8], leading to the discovery of fatal vulnerabilities [30,31].

The ROM extraction method presented by derrek et al. is very relevant to this work [8]. Their method relies on two quirks: SRAM is not cleared across resets, and some exception vectors in the ROM are hardcoded to jump into SRAM. Using fault injection, it is possible to cause an undefined instruction exception early during boot ROM execution. By first poisoning SRAM with a payload, resetting the SoC and then quickly injecting faults, an attacker can thus obtain code execution while the boot ROMs are executing [8]. Once ROM images had been obtained, it became clear that the boot ROMs contain a vulnerability in the PKCS#1 ASN.1 parsing code [31].

Meanwhile, the DSi had survived earlier attempts at breaking into its security system, and still stood strong at the time the 3DS was released. For example, Micah Elizabeth Scott built a setup to trace all DRAM accesses [32], but this only lead to exploits in specific games, not in the full system. As DRAM is initialized only by the second-stage bootloader before loading the System Menu, DRAM probing could not be used against any bootloader of the DSi. In addition, the `SCFG` control registers [15] are used to mitigate such attacks as well: they can be used to prohibit the CPU from accessing I/O registers related to eMMC, the SD card, WiFi, etc., until a reboot happens. This way, a cartridge-based game cannot access the eMMC filesystem, for example.

However, as the 3DS included a DSi backwards compatibility mode (including emulating the DSi bootchain starting from the second-stage bootloader), the defeat of the 3DS opened a new avenue for analyzing its predecessor. As it was now possible to decrypt and reverse-engineer the DSi's second-stage bootloader, a vulnerability was discovered here [17], allowing for persistent arbitrary code execution capabilities, at cold boot.

One downside of this jailbreak is that it targets the second-stage bootloader, rather than the ROM bootloader. This means the console's eMMC memory still needs to contain a valid cryptographic signature along with an intact second-stage bootloader. As already mentioned in Sect. 1, this eMMC memory is prone to failure, rendering the console unable to boot.

3 ARM7 ROM Extraction

The 3DS ROM extraction method described in the previous section can be used to extract one of the two boot ROMs of the Nintendo DSi. This section describes how we extracted this boot ROM, and what information is contained within.

3.1 Method

The method used here is similar to the one used for the 3DS. SRAM contents persist across resets, and the ROM is hardcoded to jump into SRAM when an undefined instruction exception occurs. By first filling SRAM with a payload, then resetting the SoC and injecting a fault, an attacker can obtain arbitrary code execution capabilities on the ARM7.

3.2 Practical Considerations

It was already theorized to be possible to extract the ARM7 boot ROMs using fault injection, though earlier attempts using VFI led to no results [7,23]. Instead we opted to use EMFI here, as it seemed more practical with the wire-bond BGA package of the SoC, and the complex layer stackup and connection layout of the power supply rails on the PCB. The used fault injection setup consists of a NEWAE ChipSHOUTER as EMFI injector, and a NEWAE ChipShover as positioning stage.

The jailbreak exploit described in Sect. 2.3 is a prerequisite for this attack. It is used to run custom code that fills SRAM with payload code, from which the attack can be performed. More specifically, a region of memory called *WiFi RAM* is used to store the ROM extraction payload. This RAM serves as a queue for WiFi packets, and is untouched by any bootloader. It is thus guaranteed to survive during execution of the boot ROM. The rest of SRAM is filled with NOP sleds (valid as both ARM and Thumb code) that jump to the payload in WiFi RAM.

However, this setup comes with a few downsides. EMFI requires moving the WiFi daughterboard and shielding that are normally placed above the SoC and DRAM. However, the SPI bus of the daughterboard does need to remain connected, as it contains a SPI flash memory used by the boot ROM (cf. Sect. 6.2). It is required to exist in order for the system to reach the state where the second-stage bootloader jailbreak attack is executed. This is worked around by moving the daughterboard to the side, and soldering thin wires to reconnect the SPI bus. A photo of the modified target can be found in Fig. 1a.

The Raspberry Pico was chosen as the controller for the setup. It is fast, can be controlled on a low level, supports USB as a communication method, and its PIO state machines allow for fast and precise control of glitch pulses. These pulses are sent to the ChipSHOUTER using signals from the target as trigger inputs. It is able to assert the reset line of the target, and watches the `GPIO330` and `CAM_LED` lines as trigger and success signals. The Pico also acts as a new device on the target's I2C bus as a backchannel for `printf`-style debugging of payloads. A desktop computer sends FI parameters to the Pico (using a USB-UART connection) and controls the ChipShover positioning.

To find the optimal combination of fault injection parameters, a divide-and-conquer approach is used. A parameter sweep determines the probe positioning, coil voltage and pulse width to raise an undefined instruction exception in the ARM7. This sweep is done using a test payload injected using the attack from

[17] that tries to replicate the situation when the boot ROM is running. For these tests, the ARM7 runs a test payload that fills SRAM with payload code and sets up the undefined instruction exception handler to jump to it. The payload signals success using the CAM_LED line whenever an undefined instruction exception occurs, while sending out a CPU register dump over I2C. The optimal timing to inject a fault is then discovered by simply sweeping through the entire range (between reset release and the first activity on the eMMC bus), using a payload that performs the ARM7 ROM extraction attack.

(a) The modified DSi with a relocated WiFi daughterboard (A). This configuration makes it possible to target the SoC using EMFI. The large square IC next to this connector footprint is the SoC (B). Next to the SoC are the DRAM (C) and eMMC (D).

(b) A photo of the EMFI setup to extract the boot ROM. The target DSi sits on the stepper table, with a ChipSHOUTER hanging above. In front sits a breadboard with a Raspberry Pico and supporting components (e.g. a level shifter), under which a logic analyzer is placed to inspect the whole system.

Fig. 1. Photos of the modified target and the EMFI setup

3.3 Results and Analysis

The attack worked, and the ARM7 boot ROM has been extracted successfully. It was possible to obtain a dump approximately once every 90 s, with one attempt made per second.

Static analysis of this boot ROM reveals that it mostly contains driver code for various non-volatile memories. The second-stage bootloader is read from one of these memories, depending on a configuration byte in SPI flash. Additionally, it became clear that the cryptographic verification of the second stage bootloader is performed by the ARM9 processor. Communication between the two cores happens using the FIFO interface. The ARM9 ROM image is thus needed to perform a security analysis of the system.

Nevertheless, the ARM7 ROM contains some information that will be useful later on. Right before jumping to the code of the second-stage bootloader (after the latter has been decrypted and verified by the boot ROMs), both ROMs are completely detached from the system memory buses. More specifically, the ARM7 ROM writes to an MMIO register to disable both ROMs [15], while the ARM9 waits for this transaction to be completed.

4 ARM9 ROM Extraction

In this section, a method of extracting the ARM9 boot ROM is described. This method, using a second-order fault injection attack, is then used in practice.

4.1 Method

The fault injection exploit used to extract the ARM7 ROM can be used as a starting point. Once the console boots, a first fault is injected to take control of the ARM7. The payload then continues booting normally, but 'forgets' to detach the ROMs from the system buses. This will leave the ARM9 stuck in an infinite loop (cf. Sect. 3.3). A second fault can be injected to break the ARM9 core out of this loop and continue the boot process normally, while both ROMs are still readable. As soon as execution ends up in an applet or game (running custom code inserted using a pre-existing exploit such as [17]), the ROM can be read out and e.g. be saved to the SD card.

4.2 Practical Considerations

The method described above is a second-order fault injection attack, i.e. it requires *two* successful faults. While such attacks tend to be seen as difficult to pull off (e.g. the authors of [38] call it 'unrealistic'), and countermeasures rarely exist, they have already been performed successfully before [4,10,13].

The same EMFI setup used to extract the ARM7 ROM image is used here as well. Similarly, the ARM7 parameter sweeps for the ARM7 takeover part of the ARM9 ROM extraction attack can be reused as-is. Only the probe positioning, coil voltage and pulse width for breaking the ARM9 out of an infinite loop still needs to be determined. A test payload is used here as well: the ARM9 is placed in an infinite loop, while timer interrupts signal a 'heartbeat' message to the Pico to detect crashes. It is, just like for the ARM7 parameter sweep, injected using the Unlaunch exploit. Once these steps have been completed, the parameters can be combined for the final boot ROM extraction attack. A photo of the setup is provided in Fig. 1b.

4.3 Results

The attack worked, and the ARM9 boot ROM has been extracted successfully. The timeline of a successful exploit is shown in Fig. 2. The success rate was high enough to obtain a dump once every $\approx$90 min.

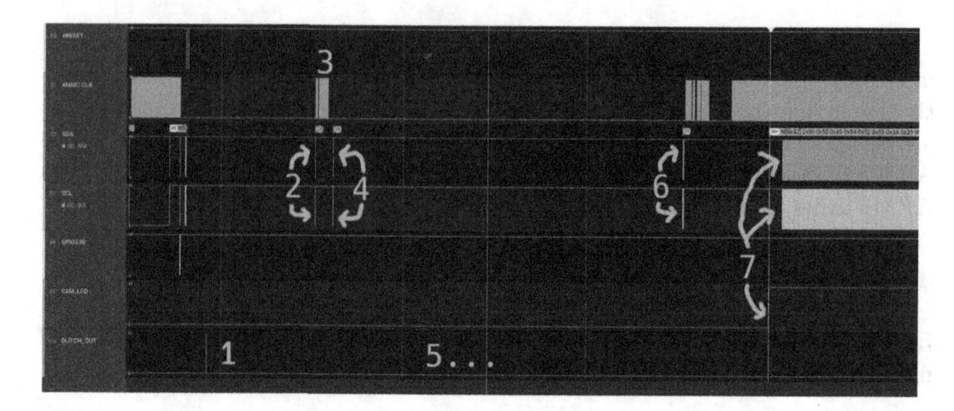

Fig. 2. Logic analyzer capture of the ARM9 boot ROM extraction process. Some time after reset release, a first fault is injected (1). The ARM7 payload then starts mimicking the regular boot ROM execution (2–4). After this, the SoC hangs as the ARM9 waits for the ROMs to be locked away (5), which the ARM7 payload does not do. After injecting a fault into the ARM9 successfully (not pictured on the `GLITCH_OUT` line), the console continues booting (6) and the result is transferred over I2C (7).

5 Fault Model Analysis

This section looks into the faults injected in the previous sections, and proposes an explanation of the real faults occurring (rather than the ones aimed for in Sect. 3.1), based on observations made during the parameter search.

5.1 Method

Observations come from two sources. One source is the influences of variations in fault parameters during the parameter search. The second is the state of the ARM7 CPU right after takeover: the payload used dumps the CPU and SCFG MMIO registers to the I2C backchannel.

5.2 Observations

One expects a fault on the ARM7 to corrupt an instruction, turning it into an undefined opcode. When decoding such an instruction, the CPU will then jump to the undefined instruction exception (UND) handler, where the payload resides. At one point, the boot ROM clears and reinitializes the UIE vector. The fault must thus be injected before this point in time.

First of all, some irregularities occur in the register dumps. The dump would normally show a link register (`lr`, `r14`) pointing to an instruction that gets executed *before* the clearing of the UIE vector. Similarly, the mode field of `cpsr` would indicate the CPU to be in undefined instruction mode (`0x1b`) when running the payload. However, `lr` sometimes points to code running *after* the UIE vector clear occurs. Similarly, the `cpsr` mode field is equal to `0x1f` (system mode) most of the time, rather than `0x1b`.

Secondly, the fault timing seems to have a rather large window in which successes are observed. More specifically, the window starts with a high peak in the success rate, after which a long trail can be seen. This is shown in Fig. 3.

A third interesting pattern emerges in the influence of the pulse injector coil voltage (and thus H-field strength) on the success rate. As one would expect (c.f. [25]), a threshold exists below which no faults will occur. However, when increasing the voltage even further, the success rate seems to *decrease*.

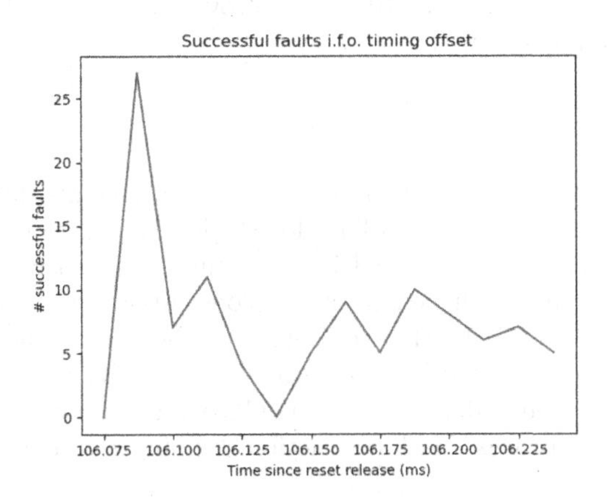

Fig. 3. Graph depicting the number of successful faults when attacking the ARM7 during boot ROM execution, in function of the time offset of the injected pulse. It starts out with zero successes, peaks slightly before 106.1 ms, after which it becomes much smaller again. Every possible moment was attempted 1400 times, the peak thus corresponds to a success rate of about 2%.

5.3 Explanation

From the `cpsr` information, it is clear that the real fault mechanism is *not* causing an UIE. Instead, a direct program counter corruption seems more likely. As the rest of SRAM is filled with NOP sleds that jump to the payload, it is not unlikely that a `pc` corruption would end up there.

The 'long tail' of the success rate i.f.o. the fault timing confirms this. The boot ROM clears SRAM upon starting up. The later the fault is injected (and thus, the more SRAM has been cleared), the lower the chance of a `pc` corruption ending up in the NOP sled.

Furthermore, the authors of [21] provide a possible explanation pointing towards `pc` corruption as well: increasing the coil voltage ends up corrupting more bits in an instruction word. This is desirable when trying to effect a large change (needed for e.g. an UIE), but less so for smaller ones (e.g. changing the destination register into the program counter, while keeping the rest intact). If the latter is what is needed instead of the former to cause successful faults and take over the ARM7 CPU core, increasing the voltage is counterproductive, which is what we observed.

6 ARM9 ROM Analysis

This section demonstrates how the security analysis of a boot ROM can be conducted, and what vulnerabilities lie in the ARM9 boot ROM. As this is the first work to reverse-engineer this ROM, we provide a functional description as well, as a reference for others.

6.1 Method

To properly analyze the functioning of the boot ROMs, a combination of multiple tools is needed. Ghidra[4] serves as a base for static analysis to discover possible vulnerabilities. These are then tested in small 'unit tests' using Unicorn[5] and Python scripting. However, this environment does not suffice to emulate the full boot procedure with both ARM cores active at the same time. To overcome this, we extended the Nintendo DS emulator melonDS[6] to support booting from the boot ROMs and enable debugging using GDB[7]. This way, an exploit can be tested in the full boot process, with instruction stepping and memory inspection.

6.2 Functional Description

The boot ROMs load, decrypt and verify the second-stage bootloader as follows: first, the ARM7 reads configuration bytes from an external SPI flash. Depending on this configuration, it will boot from either eMMC or the SPI flash itself. If a special button combination is pressed, the game cartridge will be booted from instead, just like the 3DS [31]. Then, a 512-byte boot header is read from the boot medium. This header contains information on the offset, size, load address, SRAM mapping configuration, and optional compression flags of the second-stage payload binaries, as well as an RSA-1024 signature. The ARM7 sends this boot header to the ARM9 over the FIFO interface, and the ARM9 then verifies the RSA signature, sending the result back to the ARM7.

The RSA signature format is rather peculiar: instead of using PKCS#1, a custom format is used. The RSA signature appendix contains the hashes of the boot header and of the decrypted second-stage binaries, a partial AES key, and a hash of all the previous items concatenated. The RSA public key resides in the

[4] https://ghidra-sre.org/.
[5] https://www.unicorn-engine.org/.
[6] https://melonds.kuribo64.net/.
[7] Available at https://github.com/melonDS-emu/melonDS/pull/1583.

ARM9 ROM. PKCS#1 v1-style padding is used, but without ASN.1 encoding. A diagram of the boot header and signature formats is shown in Fig. 4.

The payload binaries are encrypted using AES-128-CTR. The key is derived from two 128-bit partial keys (`keyX` and `keyY`). `keyX` is hardcoded in the boot ROM, `keyY` comes from the RSA signature appendix as described above. The IV is 96 bits in size, and consists of the corresponding binary's size repeated three times (the first time as-is, the second time its binary complement, the third time its two's complement).

Using this construction, the RSA public key (stored in the protected half of the ARM9 boot ROM) is needed to obtain the AES key for decrypting the second bootstage. Nintendo leaked the RSA public key, `keyX` and `keyY` by having the 3DS be less secure than its predecessor, as described in Sect. 2.3. Furthermore, with any system update, Nintendo can also change this `keyY` in the RSA signature appendix, rendering previous leaks useless as long as the public key remains secret.

The ARM7 then sets up its AES accelerator peripheral and DMA engine to read and decrypt the first payload (i.e. the ARM7 binary of the second-stage bootloader). The plaintext is then sent to the ARM9, which then calculates its SHA-1 hash and compares it against the corresponding one in the RSA signature appendix. The procedure is repeated for the second payload (the ARM9 binary). Once the two hashes have been checked, the ARM9 communicates back the result to the ARM7. If everything verified correctly, both ROMs then jump to their respective second-stage binaries.

Both second-stage payloads can optionally be compressed using an LZ variant, this can be configured separately per payload binary. Normally, the payload is sent by writing it to a specific SRAM bank that can be mapped privately between both CPU cores. However, when this compression option is used, another option becomes available. It allows the data to be sent over the FIFO, and the ARM9 then decompresses it upon reception (instead of the ARM7 after reading from its AES accelerator). This new option applies to both binaries at once. The compression options are ignored when booting from a game cartridge, such payloads are always uncompressed. This implementation detail will become relevant in Sect. 7.

6.3 Vulnerabilities

The cryptography of the ROMs seems rather interesting, with its custom signature format and use of primitives that have now become outdated [3,22]. However, no straightforward way of forging such a signature appears possible, without either factoring the modulus or creating a second preimage of a SHA-1 hash. Only *collision* attacks against SHA-1 have been demonstrated in practice [36]. The method of deriving the AES IV could be problematic: if both binaries are equal in size, the CTR keystream would be reused, making decryption much easier. Though, this does not occur in practice. Secondly, the verification code never confirms whether the padding and data block together span the entire 128-byte RSA message. This again does not spell doom of the scheme, as the

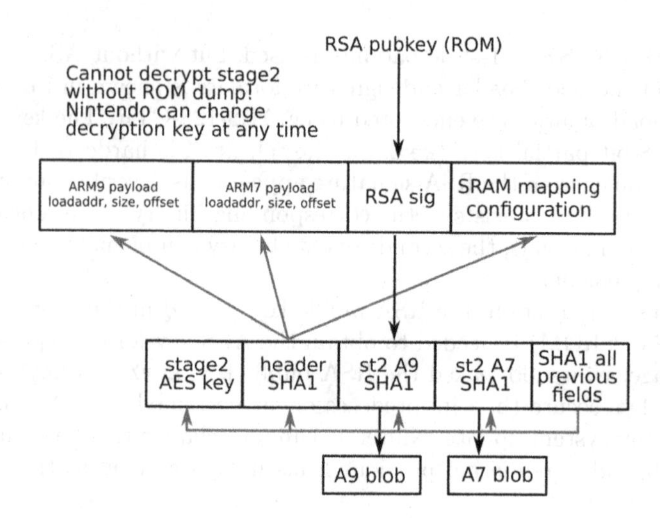

Fig. 4. Header format of the second bootstage and verification chain of the boot ROM: the RSA signature's message contains several hashes, not only for the code of the next bootstage, but also of the header information, and for the other information in the RSA message. Next to these hashes it also contains the partial AES used to decrypt the next bootstage.

data block itself is 116 bytes in size, and the padding must be at least 8 bytes. This leaves 32 bits of data inside the RSA message that can be ignored. This on its own is too little to give an adversary any significant practical advantage to forge a signature.

The ROM software does not seem to include any 'obvious' vulnerabilities, as (unlike the 3DS ROM) it performs no parsing. The boot header format is a C struct with fixed offsets and sizes, and no ASN.1 encoding is used in the RSA signatures. There seem to be a few oversights, though on their own they do not compromise the security of the system. The first one is that the return value of the function checking the RSA signature padding is ignored, as shown in Fig. 5. Improper padding checks were exploited in the 3DS [31] by brute-forcing an RSA signature with a corresponding appendix that triggers the bug. However, the 'hash of hashes' used here (which is not present in the 3DS) makes this computationally infeasible. The second oversight is that the payload binary sizes and load addresses are never sanity-checked. This was also exploited for the 3DS in a later stage of the jailbreak exploit [31]. The validity of these load addresses rely entirely on the authenticity of the RSA signature.

One particular spot in the ARM9 ROM code seems to be particularly vulnerable to fault injection. Right after the RSA signature verification, the ROM checks the 'hash of hashes' and the hash of the boot header. The corresponding assembly code is constructed such that, upon hash mismatch, a single 'load-bearing' `mov` instruction changes the return value in both cases, as shown in Fig. 6. If this `mov` instruction were to be faulted, *both* hash checks can be skipped. This however still leaves the second-stage payload binary hashes in place.

```
NANDboot_RSA_verify(&rsa_verif_outbuf,boothdr_NAND);
// ^ return value not checked! rsa_verif_out in .bss
if (!NANDboot_check_boothdr_sha1_hashes(&rsa_verif_out,boothdr_NAND)) {
  ipc_notifyID(0xf);
  return -2;
}
// continue...
```

(a) Part of the `NANDboot_do_verify_header` ARM9 boot ROM routine, which verifies the RSA signature and the hashes inside the RSA message.

```
bool swi_RSA_Decrypt_Unpad(RSA_heap *heap,byte *dest,byte *src,byte *key)
{ // snip: local variable declarations
  memset(dest_with_pad,0,0x80);
  memset(unpad_output,0,0x80);
  dest_ptrnfo.dst = (byte *)dest_with_pad;
  dest_ptrnfo.src = src;
  dest_ptrnfo.key = key;
  if (swi_RSA_Decrypt(heap,&dest_ptrnfo,&lenout) &&
      RSA_parse_padding((byte *)unpad_output,&len_unpad_out, (byte *)
          ↪ dest_with_pad,lenout, 0x80)) {
    memcpy(dest,unpad_output,len_unpad_out);
    // if bad padding: not copied, dest is zero-initialized
    return true;
  }
  return false;
}
```

(b) `swi_RSA_Decrypt_Unpad` routine, called by `NANDboot_RSA_verify`

Fig. 5. Code snippets of the ARM9 boot ROM showing how the ROM code forgets to check whether the RSA signature had well-formed padding. If this is not the case, it ends up using zero-initialized memory as the signature appendix.

```
bool NANDboot_check_boothdr_sha1_hashes(RSA_appendix *sig,NAND_boothdr *)
{ // snip...
  result = true;
  if (!swi_SHA1_Compare(sig->SHA1_boothdr,boothdr_digest)
      || !swi_SHA1_Compare(sig->SHA1_hash_of_hash,all_flds_digest)) {
    result = false; // skip this -> bypass both checks
  }
  return result;
}
```

Fig. 6. Code snippets of the ARM9 boot ROM showing how the ROM's checks for both SHA1 hashes in the RSA message end up in the same code path. This makes it such that only one instruction needs to be skipped in order to bypass both hash checks. `NANDboot_check_boothdr_sha1_hashes` is called by the code in Fig. 5a.

7 Practical Exploitation

With the elements from the previous sections, we can now devise an exploitation method. Once this exploit has been tested, we can design a cheap modchip that performs the attack. We then select suitable crowbar MOSFETs and perform another parameter search. After this is done, we insert the modchip into the DSi and evaluate the performance of this modchip.

7.1 Method

This attack uses a single injected fault to take control over both ARM cores. When the attack succeeds, both boot ROMs are still mapped into the address space. It works as follows:

1. Use fault injection to bypass both the boot header hash check and the 'hash of hashes' check at once.
2. Set the load address of the second-stage ARM7 binary (which gets loaded first) to an ARM9 stack address (in its DTCM, Data Tightly Coupled Memory). Let this binary be LZ-compressed, with the FIFO option enabled. This will make the ARM7 send the decrypted, compressed binary to the ARM9 over the FIFO interface. The ARM9 then decompresses it and writes it to the destination address in its own address space (i.e. the DTCM). This allows an attacker to control return addresses (and thus the program counter) of the ARM9. This all happens before the ARM9 has a chance to calculate the hash of the decrypted and decompressed second-stage binary.
3. DTCM is marked as no-execute by the MPU. Thus, use ROP [33] to copy the payload to ITCM (Instruction Tightly Coupled Memory, which is read-write-execute), and jump to it. The attacker now fully controls the ARM9.
4. The 'ARM7' payload (running on the ARM9) uses the FIFO interface to instruct the ARM7 to now load the next second-stage binary (meant for the ARM9).
5. Set the load address of the second-stage ARM9 binary to an ARM7 stack address, and do not use compression for this binary.
6. The ARM7 will load the decrypted ARM9 binary into its address space. It intends to send it to the ARM9 by remapping the SRAM bank backing it. Instead, it will take over control of the ARM7. The ARM7 stack is executable, thus no further steps need to be taken. The attacker is now in control of both cores, with the boot ROMs still mapped into memory.

This attack has been tested to work in melonDS and with the EMFI setup. The following sections deal with how a modchip can be built using this attack.

7.2 Design

Before a PCB can be designed and firmware can be written, several questions still need to be answered.

The first is the choice of the boot medium. A custom game cartridge—akin to a flashcart—would be ideal in terms of practicality, but this is not possible: compression cannot be used with a game cartridge, while the exploit relies on its use. The eMMC memory is also not a possibility: modifying its contents in-situ is difficult (it requires BGA rework skills). The final option is thus to use the SPI flash as boot medium. Luckily, this IC resides on a daughterboard instead of being directly soldered onto the motherboard. Replacing the daughterboard with one with malicious contents is thus relatively easy.

This replacement daughterboard can thus serve as the modchip. For practicality reasons, voltage glitching will have to be the fault injection mechanism, EMFI or LFI modchips have never been made, as far as the authors know. This can be done by using a crowbar MOSFET [19] on the 1.2V SoC core power rail. As the Raspberry Pico was used in the ROM extraction setup, reusing the RP2040 here is also a straightforward choice. Several other modchips use this microcontroller as well [39,40].

7.3 Evaluation

The modchip is capable of injecting faults successfully to take over both CPU cores of the target. Though, the success rate is low enough that success occurs roughly once every ten minutes. A photo of the modchip can be seen in Fig. 7.

The success rate could be increased by experimenting more with the setup, such as using more different MOSFETs, using MOSFET drivers, activating multiple MOSFETs at once, and so on. Nintendo Switch modchips [39] use two IRFHS8342 MOSFETs, for example. These are also positioned very close to the supply rails, while here, it is placed on the modchip PCB.

Fig. 7. A photo of the modchip installed on the target DSi. The modchip is placed on top of the motherboard, in the front of the photo. The beige connector on the modchip could be used to mount the WiFi board on top of the modchip. Though, doing so would make it impossible to close the plastic shell of the console. The loose SD card connector and breadboards are remnants from the boot ROM extraction process. In this photo, the crowbar MOSFET has been placed on a separate small breadboard as a workaround for signal integrity issues.

8 Conclusion

The DSi used a much more 'ad-hoc' and 'home-built' security system compared to the 3DS, with its custom signature format and lack of operating system being able to enforce security boundaries. Nevertheless, it ended up being more difficult to actually break, from finding an entrypoint to extracting and exploiting the boot ROMs. This stands in contrast to the fact that, especially with its use of outdated primitives, it looks much weaker on paper.

Though, this does not mean using the DSi's security system in a different context is a good idea: it is very specialized towards gaming, its main concern is piracy. The security system depends on custom hardware features not present in typical SoCs to achieve its security objectives. For example, the use of two different CPUs with different memories working in tandem on the same task, and the `SCFG` registers able to disallow games to access certain hardware. Getting one of these elements slightly wrong would have had bad consequences. For new designs, using trusted execution environments e.g. ARM TrustZone seems like a better idea. These systems are much better understood and provide less room for mistakes. Similarly, the boot ROM was hard to attack simply because it has little attack surface. More modern devices typically implement a USB-based 'recovery mode' in their boot ROM, with all the consequences stemming from implementing this complex protocol (see e.g. [11]).

Secondly, this work shows that second-order fault injection attacks on complex SoCs are feasible without too much trouble. Such attacks should thus also be part of the attacker model (when applicable), instead of dismissing them as 'unfeasible' or 'impractical'.

Thirdly, countermeasures against such attacks are difficult to implement. As a rather complex fault model is used to extract the ROM image, it cannot easily be mitigated in software. Furthermore, as this work has shown second-order attacks to be feasible, such countermeasures against simple models (such as instruction skips) can also still be defeated. Similarly, ASLR, the typical countermeasure against ROP, would be very difficult to implement in the context of a boot ROM. Implementing code relocations would add more attack surface than it would remove. Instead, countermeasures at a lower level need to be used: fault detectors can detect FI attacks, builtin redundancy and error correction can stop them, and microarchitectural security techniques such as CFI and pointer authentication can stop ROP attacks as well. Finally, though, much of this work was only possible thanks to [17], which only happened because the 3DS leaked information about the security system of the DSi. Vendors leaking hardware secrets is very difficult to defend against.

Overall, this work shows how to resurrect consoles with a broken eMMC chip by breaking the security system. This was needed because normally, due to the secure boot implementation, the console would not be able to boot with such a broken component.

While in this setting, the user can still be seen as an attacker, it is not necessarily a *bad thing* that the user can perform such attacks. This question is relevant in the context of IoT security, where users may want to use such attacks to disable features that harm their privacy or safety [29]. This is already demonstrated by their reluctance to even let such devices be connected to the Internet [27].

An archive of the source code of the ROM extraction setup, modchip firmware and exploit payload, can be found at
https://gitlab.ulyssis.org/pcy/dsi-hacking-stuff.

Acknowledgments. We would like to thank Arthur Beckers for his practical help with setting up and conducting the fault injection run against the ARM7 boot ROM.

This work was supported by CyberSecurity Research Flanders with reference number VR20192203 and by the Research Council KU Leuven C1 on Security and Privacy for Cyber-Physical Systems and the Internet of Things with contract number C16/15/058. In addition this work was supported by the European Commission through the Horizon 2020 research and innovation program under grant agreement Belfort ERC Advanced Grant 101020005 695305, through H2020 Twinning SAFEST 952252, through the Horizon Europe research and innovation program under grant agreement HORIZON-CL3-2021-CS-01-02 101070008 ORSHIN.

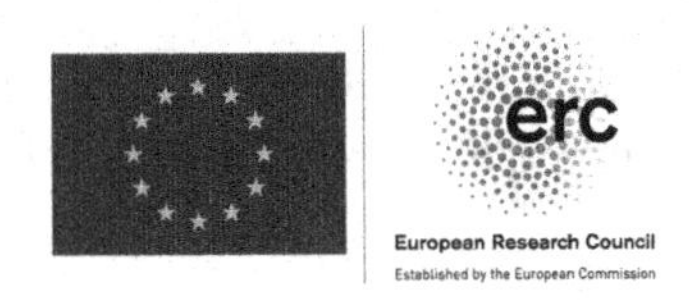

References

1. Bar-El, H., Choukri, H., Naccache, D., Tunstall, M., Whelan, C.: The sorcerer's apprentice guide to fault attacks. Cryptology ePrint Archive, Paper 2004/100 (2004). https://eprint.iacr.org/2004/100
2. Barenghi, A., Breveglieri, L., Koren, I., Naccache, D.: Fault injection attacks on cryptographic devices: theory, practice, and countermeasures. Proc. IEEE **100**(11), 3056–3076 (2012). https://doi.org/10.1109/JPROC.2012.2188769
3. Barker, E., Dang, Q.: NIST Special Publication 800-57 Part 3 Revision 1: Recommendation for Key Management: Application-Specific Key Management Guidance. National Institute of Standards and Technology (2015). https://doi.org/10.6028/NIST.SP.800-57pt3r1. http://nvlpubs.nist.gov/nistpubs/SpecialPublications/NIST.SP.800-57Pt3r1.pdf
4. Blömer, J., da Silva, R.G., Günther, P., Krämer, J., Seifert, J.-P.: A practical second-order fault attack against a real-world pairing implementation. In: 2014 Workshop on Fault Diagnosis and Tolerance in Cryptography, pp. 123–136 (2014). https://doi.org/10.1109/FDTC.2014.22
5. Copetti, R.: Nintendo 3DS architecture - a practical analysis (2023). https://archive.li/cGFtC. https://www.copetti.org/writings/consoles/nintendo-3ds/#anti-piracy-and-homebrew. Accessed 28 Sept 2023
6. Copetti, R.: Nintendo DS architecture - a practical analysis (2020). https://archive.ph/28Jmb. https://www.copetti.org/writings/consoles/nintendo-ds/%5C#security-mechanisms. Accessed 27 Mar 2022
7. dark samus, Worklog - Getting the DSi bootroms - BitBuilt, Online forum (2017). https://archive.ph/AvDsQ. https://bitbuilt.net/forums/index.php?threads/.948/. Accessed 27 Mar 2022
8. derrek, nedwill, naehrwert, Nintendo hacking 2016: game over. In: 33rd Chaos Communications Congress: 'Works for Me' (2016). https://media.ccc.de/v/33c3-8344-nintendo_hacking_2016. Accessed 27 Mar 2022
9. Dusart, P., Letourneux, G., Vivolo, O.: Differential fault analysis on A.E.S. Cryptology ePrint Archive, Paper 2003/010 (2003). https://eprint.iacr.org/2003/010
10. fail0verflow, PS4 Aux Hax 2: Syscon (2018). https://archive.ph/mt3YK. https://fail0verflow.com/blog/2018/ps4-syscon/. Accessed 09 Oct 2022
11. Galauner, A., Bazanski, S.: Glitching the switch. In: OpenChaos 2018 (2018). https://media.ccc.de/v/c4.openchaos.2018.06.glitching-the-switch. Accessed 05 Oct 2022
12. Gerlinsky, C.: Breaking code read protection on the NXP LPC-family microcontrollers. In: REcon 2017 Brussels Hacking Conference (2017). https://doi.org/10.5446/32392
13. den Herrewegen, J.V., Oswald, D., Garcia, F.D., Temeiza, Q.: Fill your boots: enhanced embedded bootloader exploits via fault injection and binary analysis. IACR Trans. Cryptogr. Hardw. Embed. Syst. **2021**(1), 56–81 (2020). https://doi.org/10.46586/tches.v2021.i1.56-81. https://tches.iacr.org/index.php/TCHES/article/view/8727
14. Korth, M.: GBATEK (2021). https://archive.ph/Ws1cO. https://problemkaputt.de/gbatek.htm. Accessed 04 Oct 2022
15. Korth, M.: GBATEK DSi Control Registers (SCFG) (2021). https://archive.ph/rPwKB. https://problemkaputt.de/gbatek-dsi-control-registersscfg.htm. Accessed 04 Oct 2022

16. Korth, M.: GBATEK DSi SD/MMC Internal NAND layout, See 'boot info blocks' (2021). https://archive.li/I7S9E. https://problemkaputt.de/gbatek-dsi-sd-mmc-internal-nand-layout.htm. Accessed 27 Mar 2022
17. Korth, M.: Unlaunch (2018). https://archive.ph/g5Qv0. https://problemkaputt.de/unlaunch.htm. Accessed 27 Mar 2022
18. Lu, Y.: Attacking hardware AES with DFA (2019). https://doi.org/10.48550/ARXIV.1902.08693. https://arxiv.org/abs/1902.08693. Supplementary text available at https://yifan.lu/2019/02/22/attacking-hardware-aes-withdfa/. https://archive.ph/oQlE7. Accessed 06 Oct 2022
19. Lu, Y.: Injecting software vulnerabilities with voltage glitching (2019). https://doi.org/10.48550/ARXIV.1903.08102. https://arxiv.org/abs/1903.08102
20. McClintic, M., Maloney, D., Scires, M., Marcano, G., Norman, M., Wright, A.: Keyshuffling attack for persistent early code execution in the Nintendo 3DS secure bootchain (2018). https://doi.org/10.48550/ARXIV.1802.00092. https://arxiv.org/abs/1802.00092
21. Moro, N., Dehbaoui, A., Heydemann, K., Robisson, B., Encrenaz, E.: Electromagnetic fault injection: towards a fault model on a 32-bit microcontroller (2013). https://doi.org/10.1109/FDTC.2013.9. http://arxiv.org/abs/1402.6421
22. National Institute of Standards and Technology, NIST Retires SHA-1 Cryptographic Algorithm (2022). https://archive.ph/zUJQk. https://www.nist.gov/news-events/news/2022/12/nist-retires-sha-1-cryptographicalgorithm. Accessed 27 Apr 2023
23. nocash, ApacheThunder, dark samus, Get BOOTROM/Key Scrambler? - 4dsdev, Online forum (2016). https://4dsdev.kuribo64.net/thread.php?id=130. https://archive.ph/qdu9x. Accessed 27 Mar 2022
24. oranav, eMMC hacking, or: how I fixed long-dead Galaxy S3 phones. In: 34th Chaos Communications Congress: 'tuwat' (2017). https://media.ccc.de/v/34c3-8784-emmc_hacking_or_how_i_fixed_long-dead_galaxy_s3_phones. Accessed 24 Feb 2023
25. Ordas, S., Guillaume-Sage, L., Maurine, P.: Electromagnetic fault injection: the curse of flip-flops. J. Cryptogr. Eng. **7**(3), 183–197 (2017). https://doi.org/10.1007/s13389-016-0128-3. https://hal-lirmm.ccsd.cnrs.fr/lirmm-01430913
26. plutoo, derrek, smea, console hacking: breaking the 3DS. In: 32nd Chaos Communications Congress: 'Gated Communities' (2015). https://media.ccc.de/v/32c3-7240-console_hacking. Accessed 04 Oct 2022
27. Purdy, K.: Appliance makers sad that 50% of customers won't connect smart appliances. Ars Technica (2023). https://arstechnica.com/gadgets/2023/01/halfof-smart-appliances-remain-disconnected-from-internet-makers-lament/. Accessed 17 May 2023
28. Raelize, Espressif ESP32: Controlling PC during Secure Boot (2020). https://archive.li/6vEgT. https://raelize.com/blog/espressif-systemsesp32-controlling-pc-during-sb/. Accessed 09 Oct 2022
29. Ren, J., Dubois, D.J., Choffnes, D., Mandalari, A.M., Kolcun, R., Haddadi, H.: Information exposure from consumer IoT devices: a multidimensional, network-informed measurement approach. In: Proceedings of the Internet Measurement Conference. IMC 2019, pp. 267–279. Association for Computing Machinery, Amsterdam (2019). https://doi.org/10.1145/3355369.3355577
30. Scire, M., Mears, M., Maloney, D., Norman, M., Tux, S., Monroe, P.: Attacking the nintendo 3DS boot ROMs (2018). https://arxiv.org/abs/1802.00359. https://doi.org/10.48550/ARXIV.1802.00359

31. SciresM, Myria, Normmatt, TuxSH, Hedgeberg, Sighax and Boot9strap (2017). https://web.archive.org/web/20211105063611/. https://sciresm.github.io/33-and-a-half-c3/. Accessed 27 Mar 2022
32. Scott, M.E.: DSi RAM tracing (2009). https://archive.ph/lhMYa. https://scanlime.org/2009/09/dsi-ram-tracing/. Accessed 27 June 2023
33. Shacham, H., Buchanan, E., Roemer, R., Savage, S.: Return-oriented programming: exploits without code injection. Black Hat USA 2008 Briefings (2008)
34. Shepherd, C., et al.: Physical fault injection and side-channel attacks on mobile devices: a comprehensive analysis. Comput. Secur. **111**, 102471 (2021). https://doi.org/10.1016/j.cose.2021.102471
35. Sidorenko, A., van den Berg, J., Foekema, R., Grashuis, M., de Vos, J.: Bellcore attack in practice. Cryptology ePrint Archive, Paper 2012/553 (2012). https://eprint.iacr.org/2012/553
36. Stevens, M., Bursztein, E., Karpman, P., Albertini, A., Markov, Y.: The first collision for full SHA-1. In: Katz, J., Shacham, H. (eds.) CRYPTO 2017. LNCS, vol. 10401, pp. 570–596. Springer, Cham (2017). https://doi.org/10.1007/978-3-319-63688-7_19
37. Timmers, N., Mune, C.: Escalating privileges in Linux using voltage fault injection. In: 2017 Workshop on Fault Diagnosis and Tolerance in Cryptography (FDTC) (2017)
38. Whelan, C., Scott, M.: The importance of the final exponentiation in pairings when considering fault attacks. In: Takagi, T., Okamoto, E., Okamoto, T., Okamoto, T. (eds.) Pairing 2007. LNCS, vol. 4575, pp. 225–246. Springer, Heidelberg (2007). https://doi.org/10.1007/978-3-540-73489-5_12
39. Wololo, Picofly: The $3 Nintendo Switch hacking modchip is real, and it's now available (2023). https://archive.li/Puo2C. https://wololo.net/2023/03/21/picofly-the-3-nintendo-switch-hacking-modchip-is-realand-its-now-available/. Accessed 17 Apr 2023
40. Wouters, L.: Glitched on earth by humans: a black-box security evaluation of the SpaceX starlink user terminal. In: DEF CON 2022 (2022)
41. Yuce, B., Schaumont, P., Witteman, M.: Fault attacks on secure embedded software: threats, design and evaluation. CoRR abs/2003.10513 (2020). arXiv: 2003.10513. https://arxiv.org/abs/2003.10513

A Differential Fault Attack Against Deterministic Falcon Signatures

Sven Bauer(✉) and Fabrizio De Santis

Siemens AG, Technology, Munich, Germany
{svenbauer,fabrizio.desantis}@siemens.com

Abstract. We describe a fault attack against the deterministic variant of the FALCON signature scheme. It is the first fault attack that exploits specific properties of deterministic FALCON. The attack works under a very relaxed and realistic single fault random model. The main idea is to inject a fault into the pseudo-random generator of the pre-image trapdoor sampler, generate different signatures for the same input, find reasonably short lattice vectors this way, and finally use lattice reduction techniques to obtain the private key. We investigate the relationship between fault location, the number of faults, computational effort for a possibly remaining exhaustive search step and success probability.

Keywords: Fault attack · Post-quantum cryptography · Digital signature schemes · Lattice-based cryptography · Falcon

1 Introduction

In July 2022, the U.S. Department of Commerce's National Institute for Standard and Technology (NIST) announced the first four winners of the Post-Quantum Cryptography (PQC) standardization project after a six-year long competition and three rounds of evaluation [23]. The goal of the NIST PQC standardization project is to standardize public-key post-quantum algorithms for public-key encryption (PKE), key encapsulation mechanisms (KEM), and digital signatures, as currently specified in the NIST SP800-56A/B and FIPS 186 documents. The first drafts are expected to be released in 2023. NIST has selected only CRYSTALS-Kyber [30] for PKE/KEM, while three algorithms CRYSTALS-Dilithium [21], FALCON [27] and SPHINCS+ [17] were selected for digital signatures. The security of CRYSTALS-Kyber, CRYSTALS-Dilithium, and FALCON is based on hardness problems on structured lattices, while the security of SPHINCS+ is based on the security of classical hash functions, e.g., SHA-2, SHA-3. In general, cryptographic algorithms mainly based on the hardness of structured lattice problems offer comparatively small keys and ciphertexts as well as high computational performance.

A provable approach to design post-quantum digital signature schemes was first described by Gentry, Peikert, and Vaikuntanathan [14] and it is based on the hash-and-sign paradigm. This approach is referred to as the GPV framework.

S. Bhasin and T. Roche (Eds.): CARDIS 2023, LNCS 14530, pp. 43–61, 2024.
https://doi.org/10.1007/978-3-031-54409-5_3

Falcon instantiates the GPV framework [14] over NTRU lattices to achieve an efficient post-quantum digital signature scheme with very short signature size. In particular, NIST recommends Falcon for applications that require smaller signatures than CRYSTALS-Dilithium. The current version of Falcon [27] submitted to the NIST standardization project is non-deterministic, i.e., signing the same message twice results in different signatures with a very high probability. As pointed out in [14], any instance of the GPV framework must never output two different signatures for the same message hash. Falcon gets around this issue by randomizing the hash of the message that is to be signed. So signing the same message more than once will almost certainly lead to different message hash values and hence different signatures. However, a deterministic version of Falcon is desirable for some use-cases, cf. [18], and may also be considered for future standardization. Deterministic Falcon de-randomizes both the message hashing and the trapdoor sampler to support a fully deterministic signing mode as specified in [18]. However, it is very well known that deterministic digital signature schemes are prone to Differential Fault Analysis (DFA) (see, for example, [7,24]). Generally speaking, a DFA attacker can alter the signature generation process by physical means, e.g., by injecting voltage glitches, laser, or electromagnetic radiation. In this way, an adversary is able to collect pairs of correct and faulty outputs, i.e., digital signatures, and recover the secret private key used for signature generation.

Our DFA on deterministic Falcon signatures targets the pseudo-random generator in the pre-image trapdoor sampler, essentially reintroducing randomness, thus forcing the trapdoor sampler to produce different signatures for the same input, hence violating a central security assumption of Falcon.

1.1 Previous Work

Fault Attacks Against Deterministic ECDSA/EdDSA Signatures. A first DFA against deterministic ECDSA and EdDSA has been proposed in [4]. The paper presents two attacks exploiting faults injected during the calculation of the scalar multiplication and nonce during the digital signature generation routine. The adversary can recover the secret key by solving linear equations obtained from correct and faulty pairs of signatures. The attacks have subsequently been improved in [3,25,28,29], where different fault injection targets are exploited and an evaluation of real hardware platforms is performed. A further fault attack against deterministic ECDSA/EdDSA using lattice-based techniques to recover the secret key has recently been proposed in [8].

Fault Attacks Against Lattice-Based Signatures. The work of [6,12] investigated fault attacks on non-deterministic lattice-based signature schemes, such as BLISS [10], GLP [16], PASSSign [19], and ring-TESLA [1]. In particular, the work of [12] presented a fault attack exploiting early loop abort faults against discrete Gaussian sampling in the secret trapdoor lattice of the GPV-based hash-and-sign signature scheme of [11]. The possibility of lattice-based deterministic signatures was already hinted in [5,20]. However, the applicability of differential fault attacks against deterministic versions of CRYSTALS-Dilithium and qTESLA were investigated in a nonce-reuse scenario using a single random faults model [7].

Fault Attacks Against FALCON*Signatures.* In [22], the authors propose two types of fault attack against FALCON. First, they adapt an attack from [12] to FALCON. In this attack, an injected fault causes the trapdoor sampler to stop prematurely. The resulting faulty signatures is then just a vector in a low-dimensional lattice generated by the secret key component F. Hence, lattice reduction can be applied on a small number of faulty signatures to find F. From F, the other secret key components can be found using the public key and the NTRU equation that defines the relationship between the secret key components. The second attack assumes the attacker can set large parts of the output of the trapdoor sampler to zero, resulting in the same type of faulty signatures as the first attack. The processing of the faulty signatures to obtain the private key is hence the same.

1.2 Contributions

To the best of our knowledge, we describe the first fault attack on deterministic FALCON that exploits specific algorithmic properties. The attack works under a very generic and realistic fault model using random faults injected during the execution of the pre-image trapdoor sampler. We investigate the relationship between fault location, the number of faults, computational effort for a possibly remaining exhaustive search step and success probability. We provide simulation results both on a PC and an ARM Cortex-M4 processor to validate our claims. The exact success probability of the attack depends on the type and precise location of the fault and is typically above 80%, as we will see in Sect. 5.1.

Compared to the attacks in [12] and [22], the attack described in this paper has a more relaxed fault model and will hence presumably be considerably easier to perform in practice. The loop-abort faults in [12] require precise timing of the injected fault. The attacks in [22] also require either precise timing to achieve an instruction skip or the zeroing of parts of memory. Zeroing memory seems much more difficult to achieve with a fault attack than a simple instruction skip. In contrast, the attack presented in the following sections, only requires a fault that changes the output of a pseudo-random number generator at a certain time instant during its execution. This can be achieved by either skipping one or several instructions or by injecting faults to randomize register values. Because the fault can be injected almost anywhere during the execution of the pseudo-random number generator, the timing requirements for the fault injection are very relaxed, making our attack highly effective in practice and we believe it easier to realize than all previous fault attacks.

1.3 Outline

In Sect. 2, we define the notation and provide some background information on lattice-based cryptography and the FALCON digital signature scheme. In Sect. 3, we give a brief algorithmic description of FALCON. The proposed fault attacks are described in Sect. 4. A practical evaluation of the attack is provided in Sect. 5. Countermeasures to protect against the proposed fault attack are described in Sect. 6. Conclusion and outlook are provided in Sect. 7.

2 Preliminaries

In this section, we briefly introduce the necessary background on lattice-based cryptography and the FALCON digital signature scheme.

Notation. We denote row vectors by bold lowercase letters and matrices by bold uppercase letters. The inner product of two vectors $\mathbf{x}, \mathbf{y}$ is denoted by $\langle \mathbf{x}, \mathbf{y} \rangle$ and the Euclidean norm of a vector $\mathbf{x}$ is denoted by $\|\mathbf{x}\| = \sqrt{\langle \mathbf{x}, \mathbf{x} \rangle}$. Let $\phi(x) = x^n + 1$ for $n = 2^\kappa$ a power of two and let $\mathcal{R} = \mathbb{Z}[x]/(\phi)$. The ring $\mathcal{R}$ is a truncated polynomial ring that inherits its ring structure from $\mathbb{Z}[x]$. The elements $f \in \mathcal{R}$ are represented as polynomials, e.g., $f(x) = \sum_{i=0}^{n-1} f_i x^i$, or vectors, e.g., $\mathbf{f} = (f_0, \dots, f_{n-1})$. Note that $xf(x)$ corresponds to a negacyclic rotation of the corresponding coefficient vector, i.e. viewed as a vector $xf(x)$ is $(-f_{n-1}, f_0, f_1, \dots, f_{n-2})$.

Any polynomial $g \in \mathcal{R}$ induces a linear map $M_g : \mathcal{R} \rightarrow \mathcal{R}$ given by polynomial multiplication, $M_g(f) = fg$. Hence, we can also view polynomials as linear maps or matrices.

For any polynomial f, we denote its Fourier transform by $\hat{f}$ and by $\hat{\mathbf{f}}$ the corresponding coefficient vector. Let $q \in \mathbb{N}^\star$. We denote the lattice generated by the basis $\mathbf{B} = (\mathbf{b}_0, ..., \mathbf{b}_{m-1}) \in \mathbb{Z}^{m \times n}$ by $\Lambda(\mathbf{B}) = \left\{ \sum_{i=0}^{m-1} x_i \mathbf{b_i}, x_i \in \mathbb{Z} \right\}$ and the corresponding orthogonal lattice given by $\Lambda^\perp(\mathbf{B}) = \{\mathbf{x} \in \mathbb{Z}^n : \mathbf{x}\mathbf{B}^* = 0\}$.

The q-ary lattice is denoted by $\Lambda_q(\mathbf{B}) = \{\mathbf{y} \in \mathbb{Z}^n : \mathbf{y} = \mathbf{x}\mathbf{B} \bmod q, \mathbf{x} \in \mathbb{Z}^n\}$ and the corresponding orthogonal q-ary lattice is denoted by $\Lambda^\perp(\mathbf{B})_q = \{\mathbf{x} \in \mathbb{Z}^n : \mathbf{x}\mathbf{B}^* = 0 \bmod q\}$. Let $\sigma \in \mathbb{R}$ with $\sigma > 0$. For any $\mathbf{c} \in \mathbb{R}^n$, the Gaussian function on $\mathbb{R}^n$ with center at $\mathbf{c}$ and standard deviation σ is denoted by $\rho_{\sigma,\mathbf{c}}(\mathbf{x}) = \exp(-\frac{\|\mathbf{x}-\mathbf{c}\|^2}{2\sigma^2})$. The discrete Gaussian distribution over Λ of center $\mathbf{c}$ and standard deviation σ is denoted by $D_{\Lambda,\sigma,\mathbf{c}}(\mathbf{z}) = \rho_{\sigma,\mathbf{c}}(\mathbf{z}) / (\sum_{\mathbf{x} \in \Lambda} \rho_{\sigma,\mathbf{c}}(\mathbf{x}))$ for all $\mathbf{z} \in \Lambda$.

SIS Problem. Let $\mathbf{A} \in \mathbb{Z}_q^{m \times n}$ be an $m \times n$ matrix with entries in $\mathbb{Z}_q$ that consists of m uniformly random vectors $\mathbf{a_i} \in \mathbb{Z}_q^n$: $\mathbf{A} = [\mathbf{a_1} | \cdots | \mathbf{a_m}]$. The Short Integer Solution (SIS) problem asks to find a non-zero vector $\mathbf{x} \in \mathbb{Z}^m$ such that: $\|\mathbf{x}\| \leq \beta$ for some $\beta \in \mathbb{R}$ and $\mathbf{x}\mathbf{A} = \mathbf{0} \in \mathbb{Z}_q^n$. In order to guarantee that such a non-trivial, short solution exists, it is required that $\beta \geq \sqrt{n \log q}$, and $m \geq n \log q$. The R-SIS problem is a special case of the SIS problem, where the matrix $\mathbf{A}$ is restricted to negacyclic blocks $\mathbf{A} = [\mathsf{rot}(\mathbf{a_1}) | \cdots | \mathsf{rot}(\mathbf{a_m})]$

GPV Framework. The GPV framework is a way to construct lattice-based signatures based on the hash-and-sign paradigm. At the high-level, the GPV framework works as follows: The public key is a matrix $\mathbf{A} \in \mathbb{Z}_q^{n \times m}$, where $n > m$, and the private key is a matrix $\mathbf{B} \in \mathbb{Z}_q^{m \times m}$. Public and private key are linked by the relation $\Lambda_q(\mathbf{A})^\perp = \Lambda_q(\mathbf{B})$. A signature for a given a message $\mathtt{m} \in \{0,1\}^*$ is a short vector $\mathbf{s} \in \mathbb{Z}_q^m$ such that $\mathbf{s}\mathbf{A}^* = \mathcal{H}(\mathtt{m})$, where $\mathcal{H} : \{0,1\}^* \to \mathbb{Z}_q^n$ is a hash function. The function which finds $\mathbf{s}$ is called a pre-image trapdoor sampler. Given the public-key $\mathbf{A}$, it is straightforward to verify that $\mathbf{s}$ is a valid signature: it is sufficient to check that the signature $\mathbf{s}$ is indeed short and verify that $\mathbf{s}\mathbf{A}^* = \mathcal{H}(\mathtt{m})$. The GPV framework is proven secure in the (quantum) random oracle model assuming the hardness of the SIS problem.

NTRU Lattices. Let $\phi = x^n + 1$ for $n = 2^\kappa$ and $q \in \mathbb{N}^\star$. A set of NTRU secrets consists of four polynomials $f, g, F, G \in \mathbb{Z}[x]/(\phi)$ which verify the NTRU equation $fG - gF = q \bmod \phi$. Additionally, it is required that f is invertible modulo q. Then the public polynomial h is defined by $h \leftarrow gf^{-1} \bmod q$. The matrices $\mathbf{A} = \left[\begin{array}{c|c} 1 & h \\ \hline 0 & q \end{array}\right]$ and $\mathbf{B} = \left[\begin{array}{c|c} f & g \\ \hline F & G \end{array}\right]$ generate the same lattice, but the matrix $\mathbf{A}$ contains large polynomials, whereas the $\mathbf{B}$ matrix contains only small polynomials. Recovering the secret polynomials from the knowledge of h corresponds to breaking the NTRU problem.

Trapdoor Sampling. A trapdoor sampler takes a matrix $\mathbf{A}$, some additional information T (the 'trapdoor') and a target vector $\mathbf{c}$ as inputs and returns a short vector $\mathbf{s}$ as output, such that $\mathbf{s}\mathbf{A}^* = \mathbf{c} \bmod q$. Note that, using basic linear algebra, it is easy to find a vector $\mathbf{c_0}$ such that $\mathbf{c_0}\mathbf{A}^* = \mathbf{c} \bmod q$. Finding a short vector $\mathbf{s}$ is then equivalent to finding a vector $\mathbf{v} = \mathbf{s} - \mathbf{c_0} \in \Lambda_q^\perp(\mathbf{A})$ close to $\mathbf{c_0}$. In the GPV framework, the secret basis $\mathbf{B}$ serves as the trapdoor to achieve this.

Gram-Schmidt Orthogonalization (GSO). Let n, m be integers. Let $\mathbf{B} \in \mathbb{R}^{n \times m}$ be a full-rank matrix. The Gram-Schmidt Orthogonalization (GSO) of $\mathbf{B}$ is the unique matrix $\tilde{\mathbf{B}} = (\tilde{\mathbf{b}}_0, ..., \tilde{\mathbf{b}}_{n-1}) \in \mathbb{R}^{n \times m}$ such that $\mathbf{B} = \mathbf{L}\tilde{\mathbf{B}} \in \mathbb{R}^{n \times m}$, where $\mathbf{L}$ is a lower triangular with 1's on the diagonal and the vectors $\tilde{\mathbf{b}}_i$ are pairwise orthogonal. The value $\|\mathbf{B}\|_{\mathsf{GS}} = \max_{\tilde{\mathbf{b}}_i \in \tilde{\mathbf{B}}} \|\tilde{\mathbf{b}}_i\|$ is called the Gram-Schmidt norm of $\mathbf{B}$.

3 FALCON

FALCON stands for "Fast Fourier lattice-based compact signatures over NTRU" and instantiates the theoretical signature framework of [14] using NTRU lattices and fast Fourier trapdoor sampling. For brevity, we will refer to the FALCON version v1.2 (as submitted to NIST [27]) simply as 'FALCON' and to the version defined in [18] as 'deterministic FALCON'. In the following, we provide a brief description of those parts of FALCON which are required to make this paper self-contained. We refer the interested reader to the full FALCON specification [27] for those parts that are omitted here.

Parameters. The degrees of the reduction polynomial $\phi = x^n + 1$ are 512 and 1024 for NIST security level I (128-bit) and security level V (256-bit), respectively. Hence, there are two variants of FALCON which are called FALCON-512 and FALCON-1024, respectively. The integer modulus is fixed to $q = 12289$ for both variants.

Key Generation. The private key is generated by drawing two private polynomials $f, g \xleftarrow{\$} \mathcal{R}$ with small coefficients[1], where f is invertible modulo q, and by computing two unique private polynomials $F, G \in \mathcal{R}$ such that $fG - gF = q \bmod (x^n + 1)$. The public key is computed by computing a public polynomial $h \in \mathcal{R}$ such that $h = gf^{-1} \bmod q$. From these polynomials, the public basis matrix $\mathbf{A} \in \mathcal{R}^2$ and the private basis matrix $\mathbf{B} \in \mathcal{R}^{2\times 2}$ are defined as follows:

$$\mathbf{A} = \begin{pmatrix} 1 & h^* \end{pmatrix}, \quad \mathbf{B} = \begin{pmatrix} g & -f \\ G & -F \end{pmatrix}, \tag{1}$$

where the corresponding lattices are defined as follows:

$$\Lambda^{\perp}(\mathbf{A}) = \{\mathbf{y} \in \mathcal{R}^2 : \mathbf{y}\mathbf{A} = \mathbf{0}\}, \quad \Lambda(\mathbf{B}) = \{\mathbf{y} \in \mathcal{R}^2 : \mathbf{y} = \mathbf{x}\mathbf{B} \text{ for some } \mathbf{x} \in \mathcal{R}^2\}. \tag{2}$$

Note that these bases, and thus their associated lattices, are orthogonal modulo q, i.e., $\mathbf{B}\mathbf{A}^* = \mathbf{0} \bmod q$ and $\Lambda^{\perp}(\mathbf{A}) = \Lambda(\mathbf{B})$. Also note that the secret key is not directly stored as $\mathbf{B}$, rather as $\mathsf{sk} = (\hat{\mathbf{B}}, \mathsf{T})$, where $\hat{\mathbf{B}}$ is the entry-wise Fourier representation of the private basis $\mathbf{B}$, and T is a binary tree representing the GSO of $\mathbf{B}$. The public key of FALCON is defined as $\mathsf{pk} = h$.

Signature Generation. Signature generation consists in hashing a message $\mathtt{m}$, along with a random nonce r, into a polynomial c modulo ϕ, whose coefficients are uniformly mapped to integers in the 0 to $q - 1$ range. Then, a pair of short polynomials (s_1, s_2) such that $s_1 = c - s_2 h \bmod \phi \bmod q$ is generated using the secret lattice basis (f, g, F, G). Note that it is sufficient to transmit s_2 as the signature, because s_1 can be computed from (s_2, c, h). The signature algorithm of FALCON is illustrated in Algorithm 1. Note that we have simplified the presentation by leaving out the signature compression step in [27], because it is not relevant for our attack.

[1] In practice, the coefficients of the polynomials f and g are generated following a discrete Gaussian distribution with center 0 and standard deviation $\sigma = 1.17\sqrt{q/2n}$.

Algorithm 1. FALCON.SIGN(m, sk)

Require: A message m, a secret key $\mathsf{sk} = (\hat{\mathbf{B}}, T)$
Ensure: A signature sig
1: **procedure** FALCON-SIGN(m,sk)
2: $r \xleftarrow{\$} \{0,1\}^{320}$
3: $c \leftarrow$ HASHTOPOINT$(r\|\mathtt{m}, q, n)$ ▷ $c \in \mathcal{R}$
4: $\hat{\mathbf{t}} \leftarrow (\hat{c}, 0)\hat{\mathbf{B}}^{-1}$ ▷ pre-image $\hat{\mathbf{t}}$
5: **do**
6: $\hat{\mathbf{z}} \leftarrow$ FFSAMPLING$(\hat{\mathbf{t}}, \mathsf{T})$ ▷ trapdoor sampler
7: $\hat{\mathbf{s}} \leftarrow (\hat{\mathbf{t}} - \hat{\mathbf{z}})\hat{\mathbf{B}}$
8: **while** $\|\mathbf{s}\|^2 > \lfloor 2.42 n\sigma^2 \rfloor$
9: $(s_1, s_2) \leftarrow$ INVFFT$(\mathbf{s})$
10: **return** sig $= (r, s_2)$

Trapdoor Sampling. The trapdoor sampler of FALCON is implemented by the function FFSAMPLING, which takes a binary tree T (representing the GSO of the private basis) and a pre-image $\hat{\mathbf{t}}$ as input and returns an output vector $\hat{\mathbf{z}}$ such that $\mathbf{z} \in \Lambda$ and $\mathbf{s} = (\mathbf{t} - \mathbf{z})\mathbf{B} \sim D_{(c,0)+\Lambda(\mathbf{B}),\sigma,0}$, as illustrated in Algorithm 2. The trapdoor sampler of FALCON samples signatures of norm proportional to $\|\mathbf{B}\|_{GS}$ using a discrete Gaussian sampler SAMPLERZ with varying mean μ and standard deviation σ', as shown in Algorithm 3.

The discrete Gaussian sampler SAMPLERZ calls another random sampler called BASESAMPLER, whose distribution is statistically close to a half Gaussian distribution centred in 0 with a fixed standard deviation.

The BASESAMPLER function is shown in Algorithm 4, where $[\![P]\!]$ is a function that, for any logical proposition P, returns 1 if P is true or 0 otherwise, and RCDT is a reverse cumulative distribution table.

Algorithm 2. FFSAMPLING$_n(\mathbf{t}, \mathsf{T})$

Require: $\hat{\mathbf{t}} = (t_0, t_1) \in \mathsf{FFT}(\mathbb{Q}[x]/(x^n+1))^2$, a FALCON-tree T
Ensure: $\hat{\mathbf{z}} = (z_0, z_1) \in \mathsf{FFT}(\mathbb{Z}[x]/(x^n+1))^2$
1: **procedure** FFSAMPLING$_n(\mathbf{t}, \mathsf{T})$
2: **if** $n = 1$ **then**
3: $\sigma' \leftarrow \mathsf{T.value}$
4: $z_0 \leftarrow$ SAMPLERZ(t_0, σ')
5: $z_1 \leftarrow$ SAMPLERZ(t_1, σ')
6: **return** $\hat{\mathbf{z}} = (z_0, z_1)$
7: $(\ell, \mathsf{T}_0, \mathsf{T}_1) \leftarrow (\mathsf{T.value}, \mathsf{T.leftchild}, \mathsf{T.rightchild})$
8: $\hat{\mathbf{t}}_1 \leftarrow$ SPLITFFT(t_1)
9: $z_1 \leftarrow$ FFSAMPLING$_{n/2}(\hat{\mathbf{t}}_1, \mathsf{T}_1)$
10: $z_1 \leftarrow$ MERGEFFT$(\hat{\mathbf{z}}_1)$
11: $t_0' \leftarrow t_0 + (t_1 - z_1) \cdot \ell$
12: $\hat{\mathbf{t}}_0 \leftarrow$ SPLITFFT(t_0')
13: $\hat{\mathbf{z}}_0 \leftarrow$ FFSAMPLING$_{n/2}(\hat{\mathbf{t}}_0, \mathsf{T}_0)$
14: $z_0 \leftarrow$ MERGEFFT$(\hat{\mathbf{z}}_0)$
15: **return** $\hat{\mathbf{z}} = (z_0, z_1)$

Algorithm 3. SAMPLERZ(-)

Require: $\mu, \sigma' \in \mathbb{R}$ with $\sigma_{\min} \leq \sigma' \leq \sigma_{\max}$
Ensure: $z \in \mathbb{Z}$ is sampled from a distribution close to $D_{\mathbb{Z},\mu,\sigma'}$

```
1: procedure SAMPLERZ(μ, σ')
2:     r ← μ − ⌊μ⌋
3:     c ← σ_min/σ'
4:     while True do
5:         z0 ← BASESAMPLER()
6:         b ← UNIFORMBITS(8)&1
7:         z ← b + (2b − 2)z0
8:         x ← (z−r)^2/(2σ'^2) − z0^2/(2σ_max^2)
9:         if BEREXP(x, c) = 1 then          ▷ See [27] for the definition of BEREXP.
10:            return z + ⌊μ⌋
```

Algorithm 4. BASESAMPLER(-)

Require: -
Ensure: An integer $z^+ \sim D_{\mathbb{Z}^+,\sigma_{\max},0}$

```
1: procedure BASESAMPLER(-)
2:     u ← UNIFORMBITS(72)
3:     z+ ← 0
4:     for i ← 0 → 17 do
5:         z+ ← z+⟦u < RCDT[i]⟧
6:     return z+
```

Deterministic Signature Generation. In deterministic FALCON, signature generation uses a fixed (versioned) salt value instead of generating a random value r as in Line 2 of Algorithm 1. The fixed salt value consists of a byte representing the value of $\ell = \log_2(n)$ and an ASCII representation of the string FALCON_DET followed by all-zero padding bytes. The fixed salt is omitted from the signatures themselves. We refer the interested reader to the full specification of deterministic FALCON given in [18].

Signature Verification. Given a message $\mathtt{m}$, a signature s_2 and a public key h, the verifier first computes the hash of the message $c = \mathcal{H}(\mathtt{m})$, then computes the first signature component $s_1 = hs_2$, and finally verifies that (s_1, s_2) is sufficiently short. If so, the signature is accepted, otherwise the verifier rejects the signature.

4 The Fault Attack Against Deterministic FALCON

In this section, we first explain the general idea of the attack against deterministic FALCON and then go into the details and describe how to make it work in practice.

Suppose we can obtain two different signatures $\mathbf{s}, \mathbf{s}'$ for the same message hash c as in Algorithm 3, but with different outputs of the trapdoor sampler FFSAMPLING. We define the vector $\mathbf{v}$ as the difference between the two signatures:

$$\mathbf{v} = (v_1, v_2) = (s_1 - s_1', s_2 - s_2') = \mathbf{s} - \mathbf{s}'. \tag{3}$$

Because

$$\mathbf{v}\mathbf{A}^* = s\mathbf{A}^* - s'\mathbf{A}^* = (0, c) - (0, c) = 0, \tag{4}$$

we have found a vector $\mathbf{v}$ in the lattice generated by $\mathbf{B}$:

$$\mathbf{v} \in \Lambda^{\perp}(\mathbf{A}) = \Lambda(\mathbf{B}). \tag{5}$$

Moreover, because both signatures $\mathbf{s}$ and $\mathbf{s}'$ are relatively short, so it is also their difference $\mathbf{v}$ short. More precisely, we have:

$$\|\mathbf{v}\| = \|\mathbf{s} - \mathbf{s}'\| \leq \|\mathbf{s}\| + \|\mathbf{s}'\| \leq 2\sqrt{\lfloor 2.42n\sigma^2 \rfloor}. \tag{6}$$

Also note that, due to the definition of $\mathbf{B}$ in Eq. (1), there exists a vector $\mathbf{u} = (u_1, u_2) \in \mathcal{R}^2$ such that

$$(v_1, v_2) = \mathbf{v} = \mathbf{u}\mathbf{B} = (u_1, u_2) \begin{pmatrix} g & -f \\ G & -F \end{pmatrix} = (u_1 g + u_2 G, -u_1 f - u_2 F), \tag{7}$$

so v_2 lies in the sublattice of $\mathcal{R}$ generated by the secret key elements f, F.

Now, if we inject faults into the pseudo-random number generator called by FFSAMPLING, while repeatedly signing the same message with a deterministic variant of FALCON, we are in exactly the following situation: we obtain different valid signatures for the same message hash and subtracting different faulty signatures from each other we obtain several v_2 as just described above. If we have enough v_2 we can try to apply lattice reduction techniques to obtain f, F or an equivalent representation of the private key.

To summarize the main idea of our attack, by injecting faults into subroutines of FFSAMPLING, we want to obtain a set of different signatures $\{\mathbf{s}^{(0)}, \mathbf{s}^{(1)}, \ldots \mathbf{s}^{(m-1)}\}$ for the same input message. Subtracting pairs of different signatures from each other, gives us a set of "somewhat short" vectors

$$\{s_2^{(j)} - s_2^{(k)} \,|\, 0 \leq j < m, 0 \leq k < m, j \neq k\} \tag{8}$$

in a sublattice generated by f, F.

The obvious approach to obtain f or F from these vectors is lattice reduction. However, since the rank of the lattice generated by f, F is 512 for FALCON-512 and 1024 for FALCON-1024, reducing a "somewhat short" basis of vectors like v_2 to find a "really short" f can be still computationally hard in practice. To make our attack feasible in practice, we show a simple way to work in a suitable sublattice of smaller rank in the next subsections.

4.1 Fault Attack in a Suitable Sublattice

In order to obtain a practically feasible attack, we would like to work in a lower-rank sublattice that still contains the private key f. Injecting a fault in the FFSAMPLING routine will result in an incorrect value $\hat{\mathbf{z}}$ of the signing procedure (cf. Algorithm 1, line 6). Looking at Algorithm 1, line 7 we see that

$$\hat{\mathbf{s}} - \hat{\mathbf{s}}' = (\hat{\mathbf{t}} - \hat{\mathbf{z}})\hat{\mathbf{B}} - (\hat{\mathbf{t}} - \hat{\mathbf{z}}')\hat{\mathbf{B}} = (\hat{\mathbf{z}}' - \hat{\mathbf{z}})\hat{\mathbf{B}}. \tag{9}$$

Substituting the definition of private basis matrix $\mathbf{B}$ from Eq. (1) we obtain

$$\hat{s}_2 - \hat{s}_2' = (\hat{z}_0 - \hat{z}_0') \odot \hat{f} + (\hat{z}_1 - \hat{z}_1') \odot \hat{F}, \tag{10}$$

where $\odot$ denotes component-wise multiplication.

Now let us have a closer look at Algorithm 2 which performs trapdoor sampling in a recursive way. The output z_1 does not depend on the recursive call on line 13. If we track the call-tree of recursions in FFSAMPLING, we see therefore that z_1 does not depend on the later calls to SAMPLERZ on lines 4 and 5. These later calls to SAMPLERZ in turn result in calls of BASESAMPLER and UNIFORMBITS. So faults injected in the later calls to UNIFORMBITS will only affect the output z_0 of the FFSAMPLING call at the top of the recursion call tree. Hence, if the fault is injected in this way, then $z_1 = z_1'$ and Eq. (10) becomes

$$\hat{s}_2 - \hat{s}_2' = (\hat{z}_0 - \hat{z}_0') \odot \hat{f}. \tag{11}$$

Because the Fourier inverse is a linear operation, we obtain

$$s_2 - s_2' = (z_0 - z_0')f. \tag{12}$$

Note that the same calculations hold if we assume that $\mathbf{s}$ is also a faulty signature (but different from $\mathbf{s}'$). In this case, $z_0 - z_0'$ is the difference between the two injected faults.

Injecting faults repeatedly and taking the difference between two faulty signatures (or a faulty and a correct one) we can produce a set

$$\Delta = \{\delta_1 f, \delta_2 f, \ldots, \delta_m f\}, \tag{13}$$

where $\delta_i = z_i - z_i'$. We consider the lattice generated by Δ:

$$\Lambda(\Delta) \subset \Lambda(f) \subset \mathcal{R}. \tag{14}$$

If Δ is large enough, then there is a possibility that $f \in \Lambda(\Delta)$. Additionally, if the rank of $\Lambda(\Delta)$ is small enough, e.g. 40, we can apply lattice reduction techniques to find f.

Now the question is, what are the possible values of the δ_i, because this determines the rank of the sublattice generated by Δ. For example, if $\delta_i(x) = a_i + b_i(x)$ for all i, then the rank of the lattice $\Lambda(\Delta)$ is at most 2. In order to get a better understanding of the range of possible values for the δ_i, we have to have a closer look at the effect of a fault injection in concrete instance of the pseudo-random number generator (PRNG) used by UNIFORMBITS.

4.2 The PRNG of Deterministic FALCON

This paragraph is specific to the ChaCha20-based PRNG implementation used in the pre-image sampler of the deterministic FALCON reference implementation provided in [18], which is actually the the same as in the reference implementation of standard FALCON submitted to NIST as [27]. However, note that similar considerations are transferable to other PRNG implementations whenever necessary.

Essentially, the PRNG uses the key stream generated by the ChaCha20 stream cipher as pseudo-random output. The PRNG maintains a 256 byte buffer. So a fresh ChaCha20 pseudo-random output is only produced, whenever the buffer has been exhausted.

It seems entirely possible that a fault can be injected even during the fetching from this buffer. However, it seems far easier to inject a fault into a ChaCha20 computation, because this takes a lot longer than just fetching data from a buffer. The fault can have any of several well-known effects like instruction skipping or data corruption to cause a faulty output. Note that the structure of a ChaCha20 computation is normally easy to identify in a power trace. Hence, it is generally not difficult to identify trigger points for injecting a suitable fault.

We have to distinguish between two types of faults. The first type is a fault that just affects the output u of a single ChaCha20 computation underlying the call of UNIFORMBITS in Algorithm 4, line 2. Because the output of ChaCha20 is buffered, a single fault can lead to up to 256 faulty output bytes. This will affect several calls to UNIFORMBITS and hence affect several z_0, z_1 in Algorithm 2, lines 4 and 5.

The second type of fault changes the seed of the PRNG in UNIFORMBITS. The reference implementation uses ChaCha20 to implement UNIFORMBITS. Because ChaCha20 adds its output to its internal state, this type of fault is quite likely. Note that this second type of fault affects the current and all future outputs of BASESAMPLER and hence potentially more z_0, z_1 in Algorithm 2, lines 4 and 5 than the first type of fault.

Note that a fault may not necessarily result in a different signature. For example, different values of u can result in the same value of z^+ in Algorithm 4.

Because of the 256 byte buffer and because SAMPLERZ needs 10 pseudo-random bytes at a time, we expect a single fault of the first type to generate values δ from a roughly 25 dimensional subspace of $\mathsf{FFT}(\mathbb{Q}[x]/(x^n+1))$. As Fourier transformation is a linear operation, we can hence assume that the resulting differences v between faulty signatures will lie in a sublattice of rank ≈ 25. This is well within reach of current lattice reduction algorithms. See, for example, [2] for a discussion of modern sieving algorithms. We only need a basis of a sublattice of rank about 25, so we expect that with around 25 suitable faults an attacker should have a good chance of finding the secret key.

4.3 Combining the Fault Attack with Exhaustive Search

Recall that, according to Eq. (14), $\Lambda(\Delta)$ is only a sublattice of $\Lambda(f)$. So there is no guarantee that lattice reduction with Δ finds the secret private key f. In general, it will only output cf, for some polynomial $c \in \mathcal{R}$.

If $c = x^k$, then cf is just a negacyclic rotation of the original f. This is not a problem for the attacker, as all negacyclic rotations of f are equivalent private keys. Indeed, if (f, g, F, G) is a set of secret NTRU-polynomials, then $(x^k f, x^k g, -x^{n-k} F, -x^{n-k} G)$ also satisfies the NTRU equation

$$(x^k f)(-x^{n-k} G) - (x^k g)(-x^n - kF) = -x^n fG + x^n gF = fG - gF \\ = q \bmod \phi. \quad (15)$$

The public polynomial $h = g/f = (x^k g)/(x^k f)$ is the same for both sets of NTRU secrets. Hence, both NTRU secrets are equivalent in the sense that messages signed with either of them can be verified by the same public key.

If $c = c_k x^k$ with $c_k \neq \pm 1$, and cf is relatively short, then the coefficients of cf will have a greatest common divisor $|c_k| \neq 1$. This situation is easy to recognize for an attacker. The attacker just has to divide the candidate private key component cf by the greatest common divisor of its coefficients.

If c is not simply a monomial, but cf is relatively short (say up to about four times the length of f), then c is most likely sparse and its few non-zero coefficients are small. In these cases, an attacker can try to find c by exhaustive search and hence still determine the private key f. Note that, because rotated keys are equivalent, as we have just seen, the attacker can assume that one of the non-zero coefficients of c is simply c_0.

4.4 Extending the Attack to FALCON

The fault attack described so far works only for a deterministic version of FALCON, for example, as described in [18]. The standard non-deterministic version of FALCON described in [27] randomizes the function HASHTOPOINT, which hashes the input to a polynomial in $\mathcal{R}$, as illustrated in Algorithm 1. Hence, if the same message is signed twice with the standard version of FALCON, then FFSAMPLING in line 6 is called for two different values of $\hat{\mathbf{t}}$ with overwhelming probability. Thus, taking the difference of these two signatures is very unlikely to return a short vector, and so it becomes computationally infeasible to find f by lattice reduction techniques.

To apply our attack to standard FALCON, a second fault injection is necessary to make standard FALCON behave in a deterministic way. For example, if the random value r in Algorithm 1, line 3 can be forced to a fixed value by a fault injection or other means, then it will be possible to apply our fault attack as described above. Note, that the fault to force r to a fixed value does not have to be necessarily perfectly reproducible. Faulty signatures with the "wrong" r can be easily recognized by looking the difference with another signature, then this difference vector would be noticeably larger than expected. Note also that, therefore, it is not necessary to force r into one particular value. It is sufficient to force it to one of a small set $\{r_0, r_1, \ldots, r_{m-1}\}$ of possible values. The resulting faulty signatures can easily be grouped into sets coming from r_1, r_2 etc. by looking at the size of pairwise differences. This highlights the importance of using a fault attack resistant random number generator in Algorithm 1, line 3, e.g., by using a fault-resistant implementation of ChaCha20, e.g., [31].

5 Practical Evaluation

In order to validate the feasibility of our attack and our hypotheses on the number of faults required to a successful attack, we conducted a number of experiments with the FALCON implementations provided in [26] and [18].

5.1 Simulations on a PC

We modified the code in [26] slightly to make it deterministic as described in [18] and added instrumentation for simulating the injection of faults of the two types described above.

We simulated the attack for 10 randomly chosen keys for each of FALCON-512 and FALCON-1024. For each key we chose 10 random inputs. These inputs were hashed and signed by the implementation we have just described. As mentioned in Sect. 4.2, we distinguish between two types of faults: A fault may either only affect one block of output of the PRNG or it may affect the state of the PRNG and hence all future outputs. We call the former type of fault 'transient' and the latter 'persistent'. We simulated each type of fault with 25 and 50 attacks each. Key and input were fixed for each attack, as this is a requirement for the attack to work. We injected exactly one fault in each experiment. We attacked each of the final five calls of the PRNG. We repeated each of these experiments 10 times. To give an idea of the remaining effort to find the key by exhaustive search after the lattice reduction step as described in Sect. 4.3, we also give the number w of non-zero coefficients of s, when sf is the key candidate found by the lattice reduction step and f is the actual private key the attacker is looking for.

Summaries of the results of these experiments are illustrated in Fig. 1, 2, 3, 4, 5, 6, 7 and 8. The sum of the slices with $w = 1, 2, 3$ and $w \geq 4$ is not necessarily 1, because there are cases in which all signatures are the same, despite injected faults. This can happen, for example, because of rounding. Interestingly, all figures look quite similar. It is not surprising to see that the probability for an (almost) immediate success, i.e., a result with $w = 1$, is higher with 50 faults than with 25. In every experiment, for each of the 10 keys, there was one attack that revealed a rotation of the key immediately without exhaustive search. This is not obvious from the figures, but it also illustrates the effectiveness of the attack. To summarize, we see that the attack has a high probability of success, even with a very moderate number of faults.

5.2 Simulations on ARM Cortex-M4

We simulated our attack also on a STMF407G-DISC1 board featuring an a 32-bit ARM Cortex-M4 (`STMF407VGT6`) high-performance microcontroller with FPU core, 1-MB Flash memory, and 192-Kbyte RAM. We ran our attack using the C implementation of deterministic FALCON-1024 available at [18][2] compiled with `gcc version 12.2.0` and optimization flags `-O3 -mfloat-abi=hard`

[2] `commit 02a2a64c44147775e6870b2d957f2cfda1437895`.

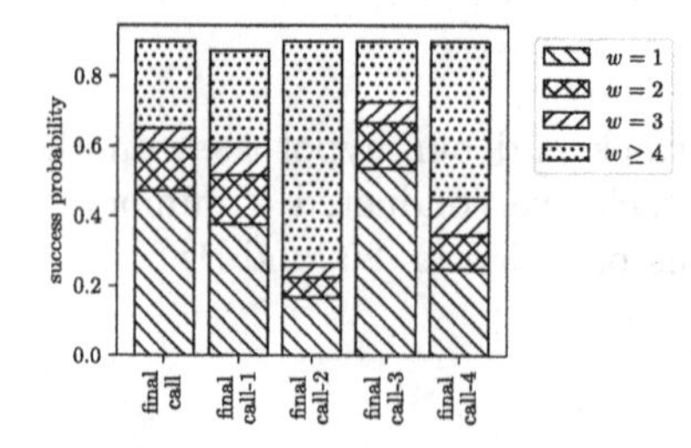

Fig. 1. 25 attacks against FALCON-512 with persistent faults. See text for further details.

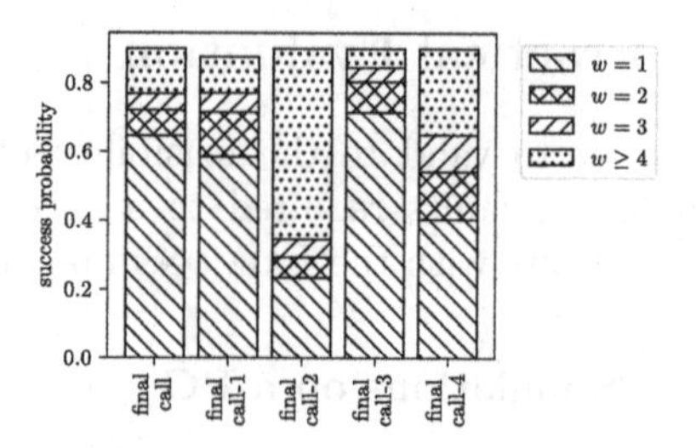

Fig. 2. 50 attacks against FALCON-512 with persistent faults. See text for further details.

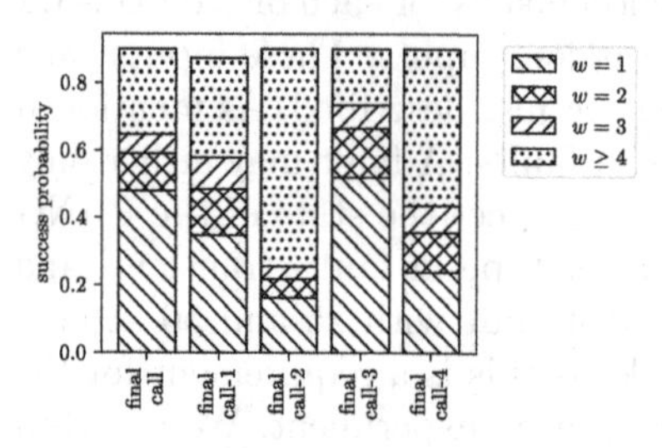

Fig. 3. 25 attacks against FALCON-512 with transient faults. See text for further details.

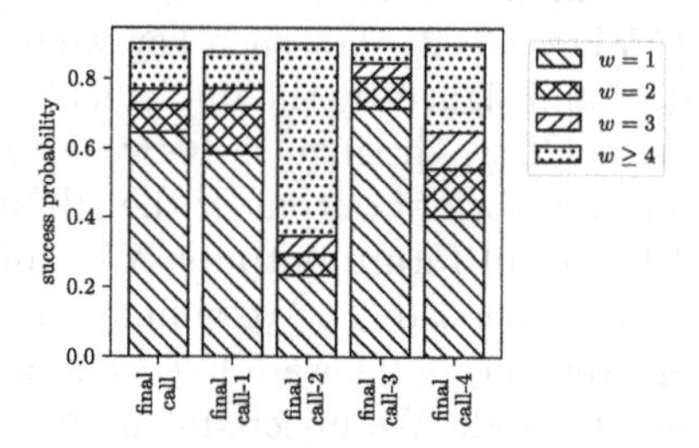

Fig. 4. 50 attacks against FALCON-512 with transient faults. See text for further details.

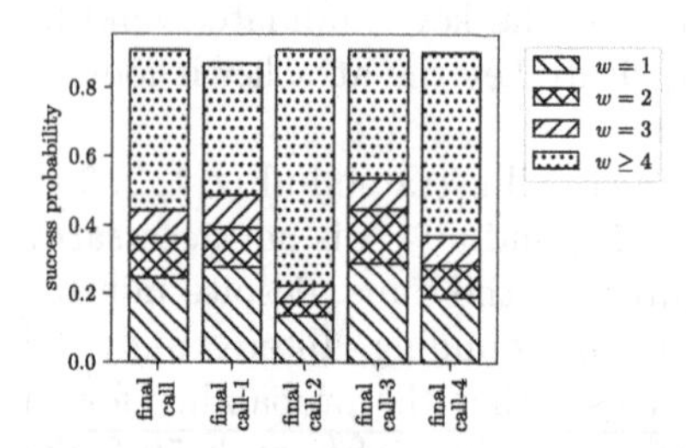

Fig. 5. 25 attacks against FALCON-1024 with persistent faults. See text for further details.

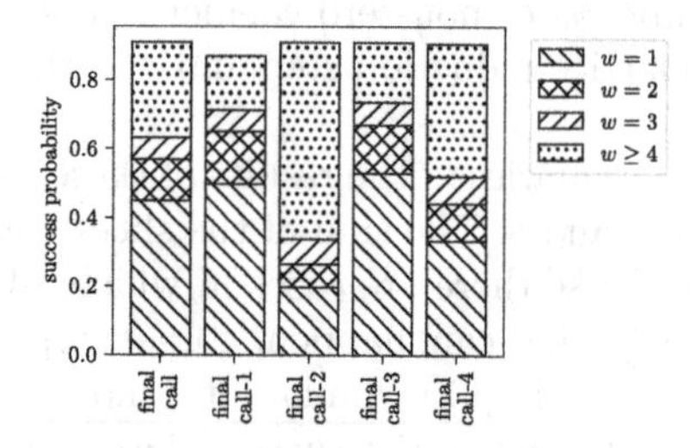

Fig. 6. 50 attacks against FALCON-1024 with persistent faults. See text for further details.

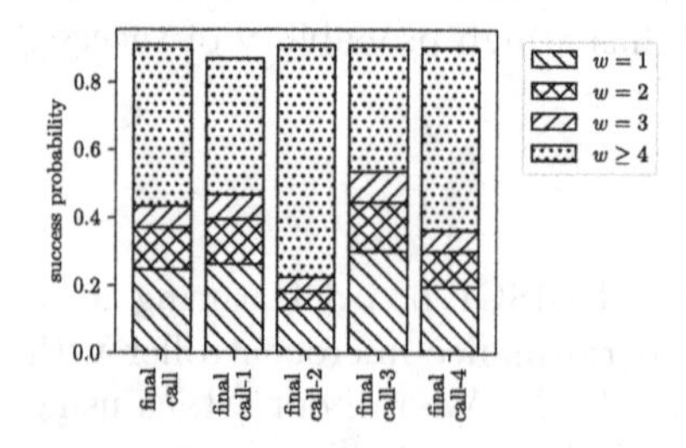

Fig. 7. 25 attacks against FALCON-1024 with transient faults. See text for further details.

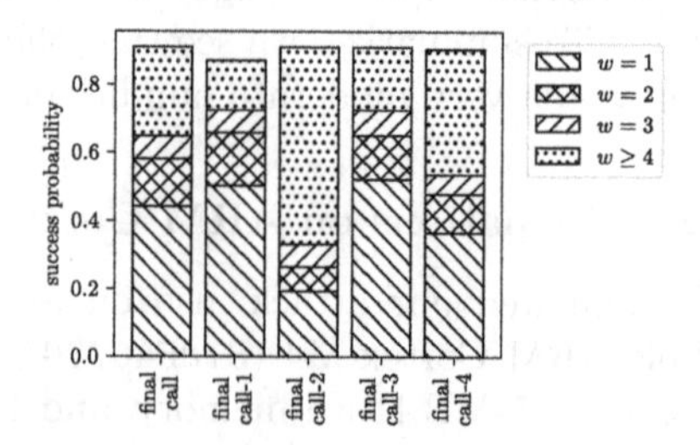

Fig. 8. 50 attacks against FALCON-1024 with transient faults. See text for further details.

-mfpu=fpv4-sp-d16. We simulated the injection of random faults in the register r2 right before the execution of the assembly instruction eors r3, r2 (cf. Listing 1.2 on line 4) with the aid of a debugger (GNU gdb 13.1). In practice, this corresponds in randomly changing the content of lower part of the 64-bit variable cc at line 287 of the file rng.c (cf. Listing 1.1 on line 4).

Listing 1.1. C code

```
1  uint64_t cc;
2  uint32_t state[16];
3  ...
4  state[14] ^= (uint32_t)cc;
```

Listing 1.2. Assembly code

```
1  ldr    r2, [sp, #32]
2  ldr    r3, [sp, #136]
3  ...
4  eors   r3, r2
```

In this way, we are able to change the state of ChaCha20, hence the generation of random numbers operated by the function falcon_inner_prng_refill(), thus leading to the generation of different signatures for the same input message. We were able to fully recover (a rotation) of the secret key f using 100 faulty signatures, when a random fault was injected in the fourth to last execution of falcon_inner_prng_refill() for a fixed input message with 78 executions, hence leading us to be able to forge legitimate signatures for it. When the faults were injected in the third to last execution of falcon_inner_prng_refill(), it was possible to recover the secret key f up to an additional (yet practically doable) brute-force effort. The attack recovered cf, as in Sect. 4.3, where c had only two non-zero coefficients; these were both from the set $\{-1, 1\}$.

We repeated the attack injecting random instruction skips (i.e., we randomly skipped up to three instructions in the portion of code comprised between falcon_inner_prng_refill()+70 and falcon_inner_prng_refill()+600 in order to simulate the effect of timing jitter during fault injection) during the third to last execution of falcon_inner_prng_refill(). We injected 100 faults, 20 of which lead to an unresponsive behaviour, and 8 of which were unsuitable because lead to the same output. In this case, we were able to recover the secret key f up to a practicable final brute-force effort. We stress the fact these attacks were run the attack only for a fixed key and input message to verify their applicability in practice. By using different inputs results may change as shown by the statistical simulations of Sect. 5.1.

Finally, note that, although the attack was carried out without using a real fault injection setup, the simulation results on real hardware clearly indicate that the attack is practically feasible as long as random instructions skip or random faults in a 32-bit register, memory, or during the computation of an XOR-operation are possible during the execution of the falcon_inner_prng_refill() function. All these kind of fault injections are typically not considered difficult to achieve with a low-cost fault injection setup, e.g., [15].

6 Countermeasures

The typical defence against fault attacks in signature generation is to verify the signature before returning it as output. However, this does not work in our case, because the faulty signatures generated by our attack are all perfectly valid signatures. (An attacker can actually exploit this property to filter the faulty signatures generated during an attack: A fault attack may sometimes produce random output or a fault of the wrong type. To avoid these 'faulty faults' breaking the lattice reduction step, an attacker can check whether the faulty signatures are valid and throw away those that are not. An obvious countermeasure is a complete re-calculation of each signature or at least re-compute FFSAMPLING. Of course this incurs a significant performance penalty. Also, it does not prevent the attack if the attacker is able to inject the same fault in both signature calculations.

Randomly shuffling the order of the computation of the sampled coefficients in FFSAMPLING may increase significantly the effort of an attacker. In fact, it is impossible for an attacker to guarantee that an injected fault only affects the value z_0 returned by FFSAMPLING in this case. The attacker is still able to find $\mathbf{s} - \mathbf{s}' \in \Lambda(\mathbf{B})$ as in Eq. (9), but $s_2' - s_2$ is not necessarily in the sublattice just generated by f, so the simplification from Eq. (10) to Eq. (11) does not apply any more. Instead, the attacker obtains vectors in the larger lattice generated by both f and F. This makes the lattice reduction harder and may make it infeasible to obtain f or at least pf for some sparse polynomial p with small coefficients.

However, note that this is not possible with the procedure illustrated in Algorithm 2, because the input to the second recursive call in line 13 depends on the output of the first recursive call in line 9. Also note that the ChaCha20-based PRNG that is currently used by FALCON does not allow to compute an arbitrary coefficient z_j in Algorithm 2, lines 4 and 5 without computing all previous (in the standard order) z_i first. The reason is, that Algorithm 3 does not consume a fixed number of pseudo-random bits. Hence, the same number of pseudo-random bits generated for the computation of each z_i may be different for different i.

So the introduction of shuffling in FFSAMPLING would require replacing the sampling algorithm by one with parallelizable steps and changing the PRNG in such a way that the index of the coefficient becomes a parameter. Such a change would allow to generate the bitstreams each coefficient independently and would hence make it possible to shuffle the order in which the coefficients are calculated. Note that signature generation would still be deterministic in this case.

If changes to the PRNG are undesirable in a concrete instantiation of a deterministic FALCON signature scheme, then the PRNG can be re-run and the output (or a checksum over the output) and its final state compared against the values for the original run when the signature was produced. However, this offers only incomplete protection, because a fault does not necessarily have to be injected into the PRNG for our attack.

7 Conclusion and Future Work

In this work, we have described the first fault attack of deterministic FALCON that exploits specific algorithmic properties of FALCON. The attack works under a very generic and realistic fault model using random faults injected during the execution of the pre-image trapdoor sampler. The attack can be naturally transferred to non-deterministic standard FALCON by suppressing the entropy in the call to HASHTOPOINT in standard FALCON with another fault injection (or, by other attack means).

It would be interesting to transfer the result of this work to deterministic variants of other signature schemes based on the GPV framework, such as Mitaka and ModFalcon (see [13] and [9], respectively). Another possible line of research is the practical investigation of the countermeasures we have suggested in Sect. 6.

References

1. Akleylek, S., Bindel, N., Buchmann, J.A., Krämer, J., Marson, G.A.: An efficient lattice-based signature scheme with provably secure instantiation. In: Pointcheval, D., Nitaj, A., Rachidi, T. (eds.) AFRICACRYPT 2016. LNCS, vol. 9646, pp. 44–60. Springer, Heidelberg (2016). https://doi.org/10.1007/978-3-319-31517-1_3
2. Albrecht, M.R., Ducas, L., Herold, G., Kirshanova, E., Postlethwaite, E.W., Stevens, M.: The general sieve kernel and new records in lattice reduction. In: Ishai, Y., Rijmen, V. (eds.) EUROCRYPT 2019, Part II. LNCS, vol. 11477, pp. 717–746. Springer, Heidelberg (2019). https://doi.org/10.1007/978-3-030-17656-3_25
3. Ambrose, C., Bos, J.W., Fay, B., Joye, M., Lochter, M., Murray, B.: Differential attacks on deterministic signatures. In: Smart, N.P. (ed.) CT-RSA 2018. LNCS, vol. 10808, pp. 339–353. Springer, Heidelberg (2018). https://doi.org/10.1007/978-3-319-76953-0_18
4. Barenghi, A., Pelosi, G.: A note on fault attacks against deterministic signature schemes. In: Ogawa, K., Yoshioka, K. (eds.) IWSEC 16. LNCS, vol. 9836, pp. 182–192. Springer, Heidelberg (2016). https://doi.org/10.1007/978-3-319-44524-3_11
5. Bindel, N., et al.: qTESLA. Technical report, National Institute of Standards and Technology (2017). https://csrc.nist.gov/projects/post-quantum-cryptography/round-1-submissions
6. Bindel, N., Buchmann, J., Krämer, J.: Lattice-based signature schemes and their sensitivity to fault attacks. In: 2016 Workshop on Fault Diagnosis and Tolerance in Cryptography, FDTC 2016, Santa Barbara, CA, USA, 16 August 2016, pp. 63–77. IEEE Computer Society (2016). https://doi.org/10.1109/FDTC.2016.11
7. Bruinderink, L.G., Pessl, P.: Differential fault attacks on deterministic lattice signatures. IACR TCHES **2018**(3), 21–43 (2018). https://doi.org/10.13154/tches.v2018.i3.21-43. https://tches.iacr.org/index.php/TCHES/article/view/7267
8. Cao, W., Shi, H., Chen, H., Chen, J., Fan, L., Wu, W.: Lattice-based fault attacks on deterministic signature schemes of ECDSA and EDDSA. In: Galbraith, S.D. (ed.) Topics in Cryptology - CT-RSA 2022. Lecture Notes in Computer Science, vol. 13161, pp. 169–195. Springer, Cham (2022). https://doi.org/10.1007/978-3-030-95312-6_8

9. Chuengsatiansup, C., Prest, T., Stehlé, D., Wallet, A., Xagawa, K.: ModFalcon: compact signatures based on module NTRU lattices. Cryptology ePrint Archive, Report 2019/1456 (2019). https://eprint.iacr.org/2019/1456
10. Ducas, L., Durmus, A., Lepoint, T., Lyubashevsky, V.: Lattice signatures and bimodal Gaussians. In: Canetti, R., Garay, J.A. (eds.) CRYPTO 2013, Part I. LNCS, vol. 8042, pp. 40–56. Springer, Heidelberg (2013). https://doi.org/10.1007/978-3-642-40041-4_3
11. Ducas, L., Lyubashevsky, V., Prest, T.: Efficient identity-based encryption over NTRU lattices. In: Sarkar, P., Iwata, T. (eds.) ASIACRYPT 2014, Part II. LNCS, vol. 8874, pp. 22–41. Springer, Heidelberg (2014). https://doi.org/10.1007/978-3-662-45608-8_2
12. Espitau, T., Fouque, P.A., Gérard, B., Tibouchi, M.: Loop-abort faults on lattice-based Fiat-Shamir and hash-and-sign signatures. In: Avanzi, R., Heys, H.M. (eds.) SAC 2016. LNCS, vol. 10532, pp. 140–158. Springer, Heidelberg (2016). https://doi.org/10.1007/978-3-319-69453-5_8
13. Espitau, T., et al.: Mitaka: a simpler, parallelizable, maskable variant of falcon. Cryptology ePrint Archive, Report 2021/1486 (2021). https://eprint.iacr.org/2021/1486
14. Gentry, C., Peikert, C., Vaikuntanathan, V.: Trapdoors for hard lattices and new cryptographic constructions. Cryptology ePrint Archive, Report 2007/432 (2007). https://eprint.iacr.org/2007/432
15. Guillen, O.M., Gruber, M., De Santis, F.: Low-cost setup for localized semi-invasive optical fault injection attacks - how low can we go? In: Guilley, S. (ed.) COSADE 2017. LNCS, vol. 10348, pp. 207–222. Springer, Heidelberg (2017). https://doi.org/10.1007/978-3-319-64647-3_13
16. Güneysu, T., Lyubashevsky, V., Pöppelmann, T.: Practical lattice-based cryptography: a signature scheme for embedded systems. In: Prouff, E., Schaumont, P. (eds.) CHES 2012. LNCS, vol. 7428, pp. 530–547. Springer, Heidelberg (2012). https://doi.org/10.1007/978-3-642-33027-8_31
17. Hulsing, A., et al.: SPHINCS+. Technical report, National Institute of Standards and Technology (2022). https://csrc.nist.gov/Projects/post-quantum-cryptography/selected-algorithms-2022
18. Lazar, D., Peikert, C., algoidan: Deterministic falcon implementation. https://github.com/algorand/falcon. Accessed 17 Nov 2022
19. Lu, X., Zhang, Z., Au, M.H.: Practical signatures from the partial Fourier recovery problem revisited: a provably-secure and Gaussian-distributed construction. In: Susilo, W., Yang, G. (eds.) ACISP 2018. LNCS, vol. 10946, pp. 813–820. Springer, Heidelberg (2018). https://doi.org/10.1007/978-3-319-93638-3_50
20. Lyubashevsky, V., et al.: CRYSTALS-DILITHIUM. Technical report, National Institute of Standards and Technology (2017). https://csrc.nist.gov/projects/post-quantum-cryptography/round-1-submissions
21. Lyubashevsky, V., et al.: CRYSTALS-DILITHIUM. Technical report, National Institute of Standards and Technology (2022). https://csrc.nist.gov/Projects/post-quantum-cryptography/selected-algorithms-2022
22. McCarthy, S., Howe, J., Smyth, N., Brannigan, S., O'Neill, M.: BEARZ attack FALCON: Implementation attacks with countermeasures on the FALCON signature scheme. Cryptology ePrint Archive, Report 2019/478 (2019). https://eprint.iacr.org/2019/478
23. NIST: NIST announces first four quantum-resistant cryptographic algorithms. https://www.nist.gov/news-events/news/2022/07/nist-announces-first-four-quantum-resistant-cryptographic-algorithms (2022). Accessed 21 Dec 2022

24. Poddebniak, D., Somorovsky, J., Schinzel, S., Lochter, M., Rösler, P.: Attacking deterministic signature schemes using fault attacks. Cryptology ePrint Archive, Report 2017/1014 (2017). https://eprint.iacr.org/2017/1014
25. Poddebniak, D., Somorovsky, J., Schinzel, S., Lochter, M., Rösler, P.: Attacking deterministic signature schemes using fault attacks. In: 2018 IEEE European Symposium on Security and Privacy, EuroS&P 2018, London, United Kingdom, 24–26 April 2018, pp. 338–352. IEEE (2018). https://doi.org/10.1109/EuroSP.2018.00031
26. Prest, T., 'Dan': falcon.py. https://github.com/tprest/falcon.py. Accessed 31 Dec 2022
27. Prest, T., et al.: FALCON. Technical report, National Institute of Standards and Technology (2022). https://csrc.nist.gov/Projects/post-quantum-cryptography/selected-algorithms-2022
28. Romailler, Y., Pelissier, S.: Practical fault attack against the ED25519 and EDDSA signature schemes. In: 2017 Workshop on Fault Diagnosis and Tolerance in Cryptography, FDTC 2017, Taipei, Taiwan, 25 September 2017, pp. 17–24. IEEE Computer Society (2017). https://doi.org/10.1109/FDTC.2017.12
29. Samwel, N., Batina, L.: Practical fault injection on deterministic signatures: the case of EdDSA. In: Joux, A., Nitaj, A., Rachidi, T. (eds.) AFRICACRYPT 2018. LNCS, vol. 10831, pp. 306–321. Springer, Heidelberg (2018). https://doi.org/10.1007/978-3-319-89339-6_17
30. Schwabe, P., et al.: CRYSTALS-KYBER. Technical report, National Institute of Standards and Technology (2022). https://csrc.nist.gov/Projects/post-quantum-cryptography/selected-algorithms-2022
31. Zeh, A., Meier, M., Rieger, V.: Parity-based concurrent error detection schemes for the ChaCha stream cipher. In: 2019 IEEE International Symposium on Defect and Fault Tolerance in VLSI and Nanotechnology Systems, DFT 2019, Noordwijk, Netherlands, 2–4 October 2019, pp. 1–4. IEEE (2019). https://doi.org/10.1109/DFT.2019.8875478

Fault Attacks Sensitivity of Public Parameters in the Dilithium Verification

Andersson Calle Viera[1,2](✉), Alexandre Berzati[1], and Karine Heydemann[1,2]

[1] Thales DIS, Meyreuil, France
{andersson.calle-viera,alexandre.berzati, karine.heydemann}@thalesgroup.com
[2] Sorbonne Université, CNRS, Inria, LIP6, 75005 Paris, France

Abstract. This paper presents a comprehensive analysis of the verification algorithm of the CRYSTALS-Dilithium, focusing on a C reference implementation. Limited research has been conducted on its susceptibility to fault attacks, despite its critical role in ensuring the scheme's security. To fill this gap, we investigate three distinct fault models - randomizing faults, zeroizing faults, and skipping faults - to identify vulnerabilities within the verification process. Based on our analysis, we propose a methodology for forging CRYSTALS-Dilithium signatures without knowledge of the secret key. Instead, we leverage specific types of faults during the verification phase and some properties about public parameters to make these signatures accepted. Additionally, we compared different attack scenarios after identifying sensitive operations within the verification algorithm. The most effective requires potentially fewer fault injections than targeting the verification check itself. Finally, we introduce a set of countermeasures designed to thwart all the identified scenarios rendering the verification algorithm intrinsically resistant to the presented attacks.

Keywords: Dilithium · Digital signature · Fault Attacks · Side-channel attacks · Post-quantum cryptography · Lattice-based cryptography

1 Introduction

Shor's algorithm [34], capable of breaking current cryptosystems [12,32], has underscored the urgency for post-quantum cryptography (PQC). As the third round of the National Institute of Standards and Technology (NIST) has concluded [1], four new post-quantum public key schemes are set to be standardized by 2024. Although estimates on the availability of sufficiently large quantum computers remain uncertain, there is a concerted effort by academia and industry to be ready when these algorithms are standardized. It is noteworthy that of the three digital signature schemes chosen, two are based on hard problems over structured lattices. This work focuses on CRYSTALS-Dilithium [3], hereafter referred to as Dilithium, which is the NIST recommended standard for quantum-safe digital signatures [1].

S. Bhasin and T. Roche (Eds.): CARDIS 2023, LNCS 14530, pp. 62–83, 2024.
https://doi.org/10.1007/978-3-031-54409-5_4

The effective and secure implementation of cryptographic algorithms on current hardware platforms poses a challenge, as it might be vulnerable to fault attacks (FA) and side-channel attacks (SCA). Securing embedded cryptographic applications against such attacks is essential but complex. It requires not only to consider a large set of attacks but also the potential impact on performances of the deployed protections. Although Dilithium has been designed to be resistant to timing attacks, recent work showed that its implementations are likely to leak secret informations [2,15,21,23,25,29,30]. Fault attacks significantly threaten the security of cryptographic systems, potentially undermining the integrity and confidentiality of sensitive data. In this paper, we focus on the analysis of Dilithium's verification algorithm - a crucial component of signature schemes - with a particular emphasis on its sensitivity to fault attacks. Unlike well-established signatures, like RSA or DSA and their variants, the verification algorithm of Dilithium has yet to be precisely analyzed [7,27,33]. By exploring theoretical attack vectors on the verification algorithm, we aim to provide a comprehensive understanding of the vulnerabilities associated with this procedure.

Our Contributions. In this work, we present three properties allowing the generation of forged Dilithium signatures that, when combined with appropriate faults, can bypass the verification without requiring the knowledge of the secret key. To facilitate a deeper understanding of the verification algorithm's sensitivity to fault attacks, we meticulously investigate each identified operation and assess its susceptibility to four common theoretical fault models. From this investigation, we detail three scenarios, the most realistic ones, but we also discuss other potential locations. We comprehensively summarize the analyzed locations and the required corresponding fault models in Table 1. In addition, we present a set of relevant dedicated countermeasures aiming at mitigating the different scenarios.

2 Preliminaries

In this section, we present the essential background on Dilithium to better understand the various attack paths presented. We also give a short summary of existing fault attacks and signature forgery methods relative to Dilithium.

Notations: Let us note $\mathbb{Z}_q$ the ring of integers modulo q and $\mathbb{Z}_q[X] = (\mathbb{Z}/q\mathbb{Z})[X]$ the set of polynomials with integer coefficients modulo q. We define $R = \mathbb{Z}[X]/(X^n + 1)$ the ring of polynomials with integer coefficients, reduced by the cyclotomic polynomial $X^n + 1$ and $R_q = \mathbb{Z}_q[X]/(X^n + 1)$ when considering integer coefficients modulo q. Elements in R_q are denoted by lowercase letters, while elements in R_q^k or R_q^l are denoted by bold lowercase letters. Matrices with coefficients in R_q are denoted by bold uppercase letters. In this context, implementations such as [13,17] represent $a \in R_q$ as a structure of n integers, often named `poly`, while an element $\mathbf{a} \in R_q^k$ (resp. $\mathbf{b} \in R_q^l$) is represented as a structure of k (resp. l) `poly` and is commonly named `polyveck` (resp. `polyvecl`).

In the remainder, we perform polynomial operations in R_q unless otherwise noted.

For an even (resp. odd) positive integer α, we define $r_0 = r \bmod^{\pm} \alpha$ to be the unique element r' in the range $-\frac{\alpha}{2} < r' \leq \frac{\alpha}{2}$ (resp. $-\frac{\alpha-1}{2} < r' \leq \frac{\alpha-1}{2}$) such that $r \equiv r \bmod \alpha$. For $\alpha \in \mathbb{N}$, we define $r' = r \bmod^{+} \alpha$ to be the unique element r' in the range $0 \leq r' < \alpha$.

For an element $w \in \mathbb{Z}_q$, we define $\|w\|_\infty$ as $|w \bmod^{\pm} q|$. For an element $\mathbf{w} \in R$, i.e., $\mathbf{w} = w_0 + w_1 X + \ldots + w_{n-1} X^{n-1}$, we define $\|\mathbf{w}\|_\infty$ as $\max\limits_i \|w_i\|_\infty$ and we define $\|\mathbf{w}\| = \sqrt{\|w_0\|_\infty^2 + \|w_1\|_\infty^2 + \ldots + \|w_{n-1}\|_\infty^2}$.

Let $S_\eta = \{\mathbf{w} \in R : \|\mathbf{w}\|_\infty \leq \eta\}$ and $\tilde{S}_\eta$ the set $\{\mathbf{w} \bmod^{\pm} 2\eta : \mathbf{w} \in R\}$.

For $\lambda \in \mathbb{Z}_q$ and an element $\mathbf{h}$ of k vectors of n coefficients. We define $|\mathbf{h}|_{h_j=\lambda}$ as the total number of coefficients of $\mathbf{h}$ equal to λ.

$[\![\mathtt{op}]\!]$ represents the boolean evaluation of the operation op.

2.1 Presentation of Dilithium

In 2022 Dilithium [3] was selected alongside Falcon [35] and SPHINCS$^+$ [4], yet it is the recommended PQC signature scheme for most use cases. Dilithium is a lattice-based signature scheme based on the Fiat-Shamir with aborts principle [22] and proposed by the "Cryptographic Suite for Algebraic Lattices" (CRYSTALS) team. Its security is derived from the hardness of solving the module learning with errors (M-LWE) [3], and SelfTargetMSIS [18] problems. For a given $(k, l) \in \mathbb{N}$, it operates over the module $R_q^{k \times l}$, with fixed $q = 8380417$, a 23-bit integer and $n = 256$. There are three security levels, from NIST level 1 to level 5, with changes essentially in the (k, l) chosen for the module $R_q^{k \times l}$. There are also two variants of the signing algorithm, one deterministic and one randomized, the only difference being how the randomness is sampled. For efficiency, the scheme makes use of rounding sub-functions such as `Power2Round`, `Decompose`, `HighBits`, `LowBits`, `MakeHint`, `UseHint` whose latest specification can be found in [3]. To reduce memory storage, polynomials are stored as a byte stream, using packing and unpacking functions specified in [3].

Algorithm 1. KeyGen

Output: $pk = (\rho, \mathbf{t}_1)$, $sk = (\rho, K, tr, \mathbf{s}_1, \mathbf{s}_2, \mathbf{t}_0)$

1 $\zeta \leftarrow \{0,1\}^{256}$
2 $(\rho, \rho', \mathcal{K}) \in \{0,1\}^{256} \times \{0,1\}^{512} \times \{0,1\}^{256} := \mathrm{H}(\zeta)$ ▷ H instantiated as SHAKE-256
3 $\mathbf{A} \in R_q^{k \times l} := \mathtt{ExpandA}(\rho)$ ▷ $\mathbf{A}$ is generated and stored in NTT Representation as $\hat{\mathbf{A}}$
4 $(\mathbf{s}_1, \mathbf{s}_2) \in S_\eta^l \times S_\eta^k := \mathtt{ExpandS}(\rho')$
5 $\mathbf{t} := \mathbf{A}\,\mathbf{s}_1 + \mathbf{s}_2$ ▷ Compute $\mathbf{A}\mathbf{s}_1$ as $\mathtt{NTT}^{-1}(\hat{\mathbf{A}}\,\mathtt{NTT}(\mathbf{s}_1))$
6 $(\mathbf{t}_1, \mathbf{t}_0) := \mathtt{Power2Round}_q(t, d)$
7 $tr \in \{0,1\}^{256} := \mathrm{H}(\rho \,\|\, \mathbf{t}_1)$
8 **return** $pk = (\rho, \mathbf{t}_1)$, $sk = (\rho, \mathcal{K}, tr, \mathbf{s}_1, \mathbf{s}_2, \mathbf{t}_0)$

Key Generation. The key generation in Algorithm 1 expands a matrix $\mathbf{A} \in R_q^{k \times l}$ from a public seed ρ via a function `ExpandA` [3]. It then samples random secret vectors $\mathbf{s}_1$ and $\mathbf{s}_2$. Elements of these vectors belong to R_q with small coefficients of size at most η, a small integer. The second part of the public key, denoted pk, is computed as $\mathbf{t} = \mathbf{A}\,\mathbf{s}_1 + \mathbf{s}_2$ but, for efficiency, only the higher bits $\mathbf{t}_1$ are made public while the lower part $\mathbf{t}_0$ is kept secret. Finally, tr is the hash of the pk and is added to the secret key, sk.

Algorithm 2. Sign

Input : $sk = (\rho, \mathcal{K}, tr, \mathbf{s}_1, \mathbf{s}_2, \mathbf{t}_0)$
Output: $\sigma = (\tilde{c}, \mathbf{z}, \mathbf{h})$

1 $\mathbf{A} \in R_q^{k \times l} := \mathtt{ExpandA}(\rho)$ ▷ $\mathbf{A}$ is generated and stored in NTT representation as $\hat{\mathbf{A}}$
2 $\mu \in \{0,1\}^{512} := \mathrm{H}(tr \,||\, M)$
3 $\kappa := 0$, $(\mathbf{z}, \mathbf{h}) := \perp$
4 $\rho' \in \{0,1\}^{512} := \mathrm{H}(\mathcal{K} \,||\, \mu)$ (or $\rho' \leftarrow \{0,1\}^{512}$ for randomized signing)
5 **while** $(\mathbf{z}, \mathbf{h}) = \perp$ **do** ▷ Pre-compute $\hat{s}_1 := \mathtt{NTT}(\mathbf{s}_1)$, $\hat{s}_2 := \mathtt{NTT}(\mathbf{s}_2)$ and $\hat{\mathbf{t}}_0 := \mathtt{NTT}(\mathbf{t}_0)$

6 $\quad \mathbf{y} \in \tilde{S}_{\gamma_1}^l := \mathtt{ExpandMask}(\rho', \kappa)$ ▷ κ is increased by 1 at each call
7 $\quad \mathbf{w} := \mathbf{A}\,\mathbf{y}$ ▷ $\mathbf{w} := \mathtt{NTT}^{-1}(\hat{\mathbf{A}} \cdot \mathtt{NTT}(\mathbf{y}))$
8 $\quad \mathbf{w}_1 = \mathtt{HighBits}_q(\mathbf{w}, 2\gamma_2)$
9 $\quad \tilde{c} \in \{0,1\}^{256} := \mathrm{H}(\mu \,||\, \mathbf{w}_1)$
10 $\quad c \in B_\tau := \mathtt{SampleInBall}(\tilde{c})$ ▷ Store c in NTT representation as $\hat{c} = \mathtt{NTT}(c)$
11 $\quad \mathbf{z} := \mathbf{y} + c\,\mathbf{s}_1$ ▷ Compute $c\mathbf{s}_1$ as $\mathtt{NTT}^{-1}(\hat{c} \cdot \hat{s}_1)$
12 $\quad \mathbf{r}_0 := \mathtt{LowBits}_q(\mathbf{w} - c\mathbf{s}_2, 2\gamma_2)$ ▷ Compute $c\mathbf{s}_2$ as $\mathtt{NTT}^{-1}(\hat{c} \cdot \hat{s}_2)$
13 $\quad$ **if** $\|\mathbf{z}\|_\infty \geq \gamma_1 - \beta$ or $\|\mathbf{r}_0\|_\infty \geq \gamma_2 - \beta$ **then**
14 $\quad\quad (\mathbf{z}, \mathbf{h}) := \perp$
15 $\quad$ **else**
16 $\quad\quad \mathbf{h} := \mathtt{MakeHint}_q(-c\,\mathbf{t}_0, \mathbf{w} - c\mathbf{s}_2 + c\,\mathbf{t}_0, 2\gamma_2)$ ▷ Compute $c\,\mathbf{t}_0$ as $\mathtt{NTT}^{-1}(\hat{c} \cdot \hat{\mathbf{t}}_0)$
17 $\quad\quad$ **if** $\|c\,\mathbf{t}_0\|_\infty \geq \gamma_2$ or $|\mathbf{h}|_{\mathbf{h}_j=1} > \omega$ **then**
18 $\quad\quad\quad (\mathbf{z}, \mathbf{h}) := \perp$
19 **return** $\sigma = (\tilde{c}, \mathbf{z}, \mathbf{h})$

Signature. The signature, described in Algorithm 2, consists of a rejection sampling loop, generating a new signature until it satisfies some security and correctness properties. The rejection loop starts by generating a masking vector $\mathbf{y}$[1] with coefficients below γ_1. Then, the signer computes $\mathbf{w} = \mathbf{A}\mathbf{y}$ and compresses it into high-order bits $\mathbf{w}_1$ and low-order bits $\mathbf{w}_0$. The message is hashed with $\mathbf{w}_1$ to sample a specific ternary challenge c with fixed weight τ and the rest 0's. The potential signature is computed, using $\mathbf{s}_1$ as $\mathbf{z} = \mathbf{y} + c\,\mathbf{s}_1$. Because the verifier does not know $\mathbf{t}_0$, the signature includes a vector $\mathbf{h}$ that keeps track of the coefficients that overflow onto the high part of $\mathbf{w}_1$. Checks are performed in

[1] This vector is essentially used as a random mask of the secret $\mathbf{s}_1$ line 11 in Algorithm 1.

lines 13 and 17 to ensure that no information about the secret key leaks and for correctness.

If any of these checks fails, a new signature candidate is generated.

Signature Verification. The verification algorithm, described in Algorithm 3, involves computing the high-order bits of $\mathbf{A}\,\mathbf{z} - c\,\mathbf{t}_1\,2^d$ using the signature and the public key, pk. The result is then corrected by the hint vector $\mathbf{h}$. If the signature is correct, this is equal to $\mathbf{w}_1$, which allows to recompute the challenge c. To verify a signature, a final check ensures that all the coefficients of $\mathbf{z}$ are less than $\gamma_1 - \beta$ and that the number of hints in $\mathbf{h}$ are less than ω.

Algorithm 3. Verify

Input : $pk = (\rho, \mathbf{t}_1)$, $\sigma = (\tilde{c}, \mathbf{z}, \mathbf{h})$
Output: $True$ or $False$

1 $\mathbf{A} \in R_q^{k \times l} :=$ `ExpandA`(ρ)
2 $\mu \in \{0,1\}^{512} := \mathrm{H}(\mathrm{H}(\rho \,||\, \mathbf{t}_1) \,||\, M)$
3 $c :=$ `SampleInBall`$(\tilde{c})$
4 $\mathbf{w}_1' :=$ `UseHint`$_q(\mathbf{h}, \mathbf{Az} - c\mathbf{t}_1\,2^d, 2\gamma_2)$
5 **return** $[\![||\mathbf{z}||_\infty < \gamma_1 - \beta]\!]$ and $[\![\tilde{c} = \mathrm{H}(\mu \,||\, \mathbf{w}_1')]\!]$ and $[\![|\mathbf{h}|_{\mathbf{h}_j = 1} \leq \omega]\!]$

2.2 Fault Models

Over the past two decades, fault injection attacks have emerged as a powerful method of compromising devices, even secured ones [39]. These attacks use a variety of techniques, including laser beam and electromagnetic (EM) pulse, which allow precise control over space and time. Extensive research has focused on characterizing the effects of fault injection in order to identify fault models at a given abstraction (circuit level, hardware logical level, assembly code, source code). Such models serve as a framework for categorizing and studying the potential fault attacks on systems and sub-systems such as cryptographic ones.

A fault model at hardware logical level includes the width of the fault (mono-bit, multi-bits, byte, or word) and the induced change (bit set, bit reset, bit-flip, random changes). The feasibility and cost of a fault model are related to the required equipment, the time, and the level of expertise needed. At software level, faults impact the data and computations, the control flow and executed instructions, or both. At this level, common fault models cover effects such as instruction skipping or conditional branch inversion. A threat (or attacker) model defines both the fault models and the number of faults needed to study the security of a system. Note that achieving a precise fault multiple times generally requires a strong attacker and means increased difficulty in real-life scenarios.

In this paper, we consider four fault models at C level, two on instruction flow and two on data, namely skipping faults, test inversion faults, randomization faults, and zeroizing faults. For each we explain how they can be achieved.

Skipping fault involves deliberately skipping selected lines of code within the execution of a program. It amounts to not executing specific program instructions or intentionally manipulating the program counter. Skipping faults can be achieved with many fault injection techniques such as CPU clock or voltage glitching [19,36,40], EM pulse injection [24] or laser beam [14]. Skipping faults can have severe consequences, such as bypassing critical security checks or essential cryptographic steps. The practical demonstration of higher-order skipping faults has recently underscored the significance of this type of fault attack [9], highlighting the potential security risks they pose.

Test inversion fault corresponds to the inversion of a conditional branch outcome (if-then or if-then-else constructs). It can be achieved by inverting the condition in the corresponding conditional jump instruction or by corrupting an instruction involved in the condition's computation, all of which can be achieved by several injection means [11,20,26]. It can also be the result of skipping the branch instruction. Test inversion enables bypassing security verification.

Zeroizing fault assumes that the attacker can set a variable, a constant or a portion thereof, to zero. While the realism of this attack scenario has been questioned, instances of zeroizing faults have been successfully executed in practice as it amounts to resetting or flipping some bits [6]. State-of-the-art fault injection allows controlling the fault affecting up to a dozen bits [10]. Therefore, zeroing less than a dozen bits can be considered as realistic. Zeroizing faults can also be a consequence of skipping faults, or instruction or operand corruption [26,37].

Randomization fault introduces random changes to data or computations within a targeted program, causing unexpected behavior. This means that after the injection, the attacker remains unaware of the exact result of the computation, but gains an advantage by knowing that it has been altered within a specific range. Randomization faults can lead to incorrect output, bypassing security checks, or compromising the integrity of the cryptographic operation.

In the remainder, we only consider the type and number of faults for each scenarios from which one can deduce the corresponding attacker model.

2.3 Related Works

In this section, we first give an overview of the state of the art in fault attacks on Dilithium. Then, we explain how one can forge signatures with partial information about the secret key.

Fault Attacks on Dilithium. For its PQC competition, NIST put an emphasis on security against side-channel and fault attacks. In this regard, Dilithium already has some constant-time properties to an extent. Still, several practical fault injection attacks leading to the key recovery have already been published. Among them, Bruinderink and Pessl [8] demonstrated the applicability of differential fault attacks on the deterministic Dilithium through multiple paths.

In contrast to the extensive research on fault injection attacks on the signature algorithm, the verification process has received less attention. The main reason is that only public information are handled during the verification process. Fault attacks on the verification procedure primarily target the comparison in line 5 of Algorithm 3, which is usually carefully implemented on secure devices to prevent acceptance of corrupted signatures. The algebraic parts of the verification process are often considered less sensitive, given the difficulty of forging a signature. Nonetheless, exploitable vulnerabilities can make these algebraic parts an attractive attack surface for fault injection. It is the case for the RSA signature scheme where manipulating the modulus N, which is sensitive to faults, allows an attacker to pass the verification with false signatures [33].

Although there have been efforts to explore fault injection attacks in the context of the verification procedure, such research remains scarce.

One notable study conducted by Bindel et al. [5] highlighted the potential consequences of zeroizing the challenge c within the verification process of other lattice-based signature schemes. They demonstrated that such zeroization could enable successful verification of invalid signatures for any message, all without needing the secret key. Furthermore, they showed that skipping the correctness check or the size check on $\mathbf{z}$ in line 5 of Algorithm 3 could have the same effect.

Achieving zeroization in practice is not a trivial task, and concerns have been raised regarding the practicality of this specific theoretical attack scenario. However, recently, Ravi et al. [31] presented the first practical zeroization fault attack on an implementation of the Dilithium signature verification. They showed that zeroizing the twiddle constants, a fixed table of coefficients, in the `NTT` reduces its entropy, thus achieving the same effect as Bindel et al. who zeroized the challenge c. They practically demonstrated this by noticing that zeroizing the starting address of the the twiddle constant's table is sufficient to set them all to zero.

Given the critical role of the verification algorithm in upholding the security of digital systems, it is essential to dedicate more attention to comprehensively evaluating its susceptibility to fault injection attacks.

Dilithium Signature Forgery. In the following, we present how to forge Dilithium signatures, assuming that only the $\mathbf{s}_1$ part of the secret key is known.

The verification algorithm essentially recomputes the value of $\mathbf{w}_1$ using only public information to accept a signature. To sign a message, the sk used is composed of (ρ, $\mathcal{K}$, tr, $\mathbf{s}_1$, $\mathbf{s}_2$, $\mathbf{t}_0$). The seed ρ needed to expand the matrix $\mathbf{A}$ is also part of the public key and tr is the hash of pk, so they can be both retrieved from public information. The nonce $\mathcal{K}$ is used to generate the vector $\mathbf{y}$, but no check on the verification allows to determine if this particular value was used. Thus, it can be replaced by a random value. The $\mathbf{s}_2$ part of the secret key is only used for rejection checks and in intermediate values. Regarding $\mathbf{t}_0$, the security proof considers it as public [3], but in practice it remains secret.

In the signing algorithm, if we assume that $\mathbf{y}$ is randomly chosen, as in the randomized version of Dilithium, then, with the knowledge of the public key and

$\mathbf{s}_1$, an attacker can proceed up to the computation of $\mathbf{z}$, in line 11 of Algorithm 2. From this step, there are basically two different methods to forge a signature.

Bruinderink and Pessl [8] presented a modified signing procedure to perform signature forgery with only public information and the knowledge of $\mathbf{s}_1$. They first compute with known values the value $\mathbf{u} := \mathbf{A}\,\mathbf{s}_1 - \mathbf{t}_1\,2^d = \mathbf{t}_0 - \mathbf{s}_2$. Given $\mathbf{s}_2$'s small coefficients, the quantity $\mathbf{u}$ approximates $\mathbf{t}_0$. It allows them to compute an alternative $\mathbf{h}$ with $\mathbf{u}$. They cannot check the rejections based on $\mathbf{s}_2$ and $\mathbf{t}_0$, so they remove them because this will not impact the correctness of the signature with high probability.

Ravi et al. [30] also showed an alternative signature forgery procedure. They start in the same way as Bruinderink and Pessl up to line 11 of Algorithm 2. They showed that the `UseHint` procedure can be inverted and used to compute the high bits of $\mathbf{w} - c\,\mathbf{s}_2$. But, because $\|\texttt{LowBits}_q(\mathbf{A}\,\mathbf{z} - c\,\mathbf{s}_2,\, 2\gamma_2)\|_\infty < \gamma_2 - \beta$ with probability very close to 1, they can be sure to recompute the correct $\mathbf{w}_1$.

3 Public Parameters Sensitivity Analysis of Verify

In this section, we present the main idea allowing the acceptance of random signatures by Algorithm 3 through the exploitation of specific faults. Then, we conduct a comprehensive analysis of one implementation of the verification algorithm. The goal is to identify sensitive operations and explain how to forge signatures that would be accepted, in the presence of the corresponding fault. To our knowledge, this is the first extensive study of the verification algorithm. Finally, we summarize the sensitivity of each location regarding our fault models.

3.1 Main Idea

The attacker's goal, here, is to produce a message-signature pair that will be accepted by Algorithm 3. The main idea behind the signature verification of Dilithium is that computing the value $\mathbf{A}\,\mathbf{z} - c\,\mathbf{t}_1\,2^d$ will be equal to the high bits of $\mathbf{A}\,\mathbf{y}$ plus some bounded small values. These small values can sometimes slightly overflow onto the high part $\mathbf{w}_1$, so the signing algorithm computes a specific vector $\mathbf{h}$ of hints that will be used to compensate for this effect. Given this hint, one can retrieve the same $\mathbf{w}_1$ as in the signing procedure but with only public values.

Let us remember that a Dilithium signature is composed of $(c, \mathbf{z}, \mathbf{h})$, given that $\mathbf{z} = \mathbf{y} + c\,\mathbf{s}_1$ and $\mathbf{t} = \mathbf{A}\,\mathbf{s}_1 + \mathbf{s}_2$, we have

$$\mathbf{A}\,\mathbf{z} - c\,\mathbf{t} = \mathbf{A}\,\mathbf{y} - c\,\mathbf{s}_2. \tag{1}$$

By replacing $\mathbf{w} = \mathbf{A}\,\mathbf{y}$ and $\mathbf{t} = \mathbf{t}_1\,2^d + \mathbf{t}_0$[2] in Eq. 1, we get

$$\mathbf{A}\,\mathbf{z} - c\,\mathbf{t}_1\,2^d - c\,\mathbf{t}_0 = \mathbf{w} - c\,\mathbf{s}_2,$$

[2] The vector $\mathbf{w}$ is computed line 7 in Algorithm 2 and $\mathbf{t}$ is computed line 5 in Algorithm 1.

and by rewriting this equation, we obtain

$$\mathbf{A}\,\mathbf{z} - c\,\mathbf{t}_1\,2^d = \mathbf{w} - c\,\mathbf{s}_2 + c\,\mathbf{t}_0. \tag{2}$$

Equation 2 is exactly the quantity computed for the verification line 4 of Algorithm 3.

Remember that $\mathbf{h} = \texttt{MakeHint}_q(-c\,\mathbf{t}_0,\, \mathbf{w} - c\,\mathbf{s}_2 + c\,\mathbf{t}_0,\, 2\,\gamma_2)$.

Then, from Lemma 1.1 in [3], we know that

$$\texttt{UseHint}_q(\mathbf{h},\, \mathbf{w} - c\,\mathbf{s}_2 + c\,\mathbf{t}_0,\, 2\gamma_2) = \texttt{HighBits}_q(\mathbf{w} - c\mathbf{s}_2,\, 2\gamma_2). \tag{3}$$

Since $\|c\,\mathbf{s}_2\|_\infty \leq \beta$ and $\|\texttt{LowBits}_q(\mathbf{w} - c\mathbf{s}_2,\, 2\gamma_2)\|_\infty < \gamma_2 - \beta$[3], according to Lemma 2 [3] we have

$$\texttt{HighBits}_q(\mathbf{w} - c\mathbf{s}_2,\, 2\gamma_2) = \texttt{HighBits}_q(\mathbf{w}, 2\gamma_2) = \mathbf{w}_1, \tag{4}$$

which shows how to retrieve the value $\mathbf{w}_1$ from the public key and the signature.

From line 4 in Algorithm 3 and the equations above, we can see that, at the top level, $\mathbf{w}_1$ is only dependant on $\mathbf{A}$, $\mathbf{z}$, c, $\mathbf{t}_1$ and $\mathbf{h}$ which are known. The matrix $\mathbf{A}$ can be considered as a fixed value, like the constant d. The values $\mathbf{z}$ and $\mathbf{h}$ are essentially random values on their respective intervals, and an attacker can choose them freely.

Given this information, we show how to bound the value $c\,\mathbf{t}_1\,2^d$ so that it doesn't impact too much the high bits of $\mathbf{A}\,\mathbf{z}$.

Proposition 1. *Let $\boldsymbol{z} \in \tilde{S}^l_{\gamma_1-\beta}$ be a random vector. If at least one of the following conditions is satisfied:*

P1. $c\,\mathbf{t}_1\,2^d = 0$

P2. $\|c\,\mathbf{t}_1\,2^d\|_\infty \leq \beta$ *and* $\|\mathit{LowBits}_q(\boldsymbol{A}\,\boldsymbol{z} - c\,\mathbf{t}_1\,2^d,\, 2\gamma_2)\|_\infty < \gamma_2 - \beta$

P3. $\|c\,\mathbf{t}_1\,2^d\|_\infty \leq \gamma_2$ *and* $\boldsymbol{h}' = \mathit{MakeHint}_q(c\,\mathbf{t}_1\,2^d,\, \boldsymbol{A}\,\boldsymbol{z} - c\,\mathbf{t}_1\,2^d,\, 2\,\gamma_2)$

Then, $\mathit{HighBits}_q(\boldsymbol{A}\,\boldsymbol{z} - c\,\mathbf{t}_1\,2^d,\, 2\gamma_2) = \mathit{HighBits}_q(\boldsymbol{A}\,\boldsymbol{z}, 2\gamma_2)$.

Proof.

P1. If $c\,\mathbf{t}_1\,2^d = 0$, the result is straightforward.

P2. Direct application of Lemma 2 in [3].

P3. If $\|c\,\mathbf{t}_1\,2^d\|_\infty \leq \gamma_2$ then from Lemma 1.1 in [3], we know that

$$\begin{aligned}&\texttt{UseHint}_q\big(\texttt{MakeHint}_q(c\,\mathbf{t}_1\,2^d,\, \mathbf{A}\,\mathbf{z} - c\,\mathbf{t}_1\,2^d,\, 2\,\gamma_2),\, \mathbf{A}\,\mathbf{z} - c\,\mathbf{t}_1\,2^d,\, 2\,\gamma_2\big)\\ &= \texttt{HighBits}_q(\mathbf{A}\,\mathbf{z} - c\,\mathbf{t}_1\,2^d + c\,\mathbf{t}_1\,2^d,\, 2\gamma_2).\end{aligned}$$

□

[3] If σ is a valid signature then we know that this condition is fulfilled thanks to the check on $\mathbf{r}_0$ on line 13 of Algorithm 2.

If we can fault some operations of Algorithm 3 and have one of these three conditions, then we can carefully construct signatures that will pass the verification.

- Even though at first glance P1 seems like a strong hypothesis to realize, Ravi et al. [31] recently showed the practical realization of a fault attack involving the challenge c that has the same effect.
- P2 is perhaps the hardest hypothesis to use because we need both conditions for the fault to have the desired effect.
- P3 seems convenient because the hint vector $\mathbf{h}$ is part of the signature, and γ_2 is not too small.

To illustrate the sensitivity analysis, we use the C implementation of Dilithium from the PQclean library [17], which is identical to the reference one [13] but more portable. The code structure is also reused in other implementations [16]. The function `PQCLEAN_DILITHIUM2_CLEAN_crypto_sign_verify` will be referred to as `verify` to simplify notations and is given in Fig. 1. In the following, we describe three relevant scenarios resulting from the analysis of the C code. For each scenario, we identify which fault model, as presented in Sect. 2.2, allows us to exploit propositions P1 and P3 to forge a signature. We provide two algorithms Algorithm 4 for P1 and Algorithm 5 for P3, to forge signatures given the corresponding faults. Each algorithm has been implemented in SageMath. We have verified that such carefully forged signatures using these algorithms, paired with specific fault effects, enable us to pass the verification.

3.2 Preliminary Analysis

A natural target is the corruption of the value returned by the verification process. An attacker must then force the return value to 0, corresponding to a valid signature. However, zeroizing 32 bits may be relatively hard for an attacker to accomplish in practice. Alternatively, the attacker can try to pass all of the three checks lines 14, 16, and 54 of `verify` in Fig. 1, necessitating three test inversion faults at minimum. These sensitive tests are typically hardened in secure applications [38], making such fault effects potentially hard to achieve. The analysis below focuses on arithmetic parts that might be less carefully implemented since they do not handle secure parameters.

```
int verify(const uint8_t *sig, size_t siglen, const uint8_t *m, size_t mlen,
           const uint8_t *pk) {
    unsigned int i;
    uint8_t buf[K * POLYW1_PACKEDBYTES], rho[SEEDBYTES], mu[CRHBYTES];
    uint8_t c[SEEDBYTES], c2[SEEDBYTES];
    poly cp;
    polyvecl mat[K], z;
    polyveck t1, w1, h;
    shake256incctx state;
    if (siglen != CRYPTO_BYTES)
        return -1;

    unpack_pk(rho, &t1, pk);
    if (unpack_sig(c, &z, &h, sig))
        return -1;
    if (polyvecl_chknorm(&z, GAMMA1 - BETA))
        return -1;

    /* Compute CRH(H(rho, t1), msg) */
    shake256(mu, SEEDBYTES, pk, CRYPTO_PUBLICKEYBYTES);
    shake256_init(&state);
    shake256_absorb(&state, mu, SEEDBYTES);
    shake256_absorb(&state, m, mlen);
    shake256_finalize(&state);
    shake256_squeeze(mu, CRHBYTES, &state);

    /* Matrix-vector multiplication; compute Az - c2^dt1 */
    poly_challenge(&cp, c);
    polyvec_matrix_expand(mat, rho);

    polyvecl_ntt(&z);
    polyvec_matrix_pointwise_montgomery(&w1, mat, &z);

    poly_ntt(&cp);                    Scenario 1: Sampling of c̃
    polyveck_shiftl(&t1);             Scenario 2: Shift by d
    polyveck_ntt(&t1);
    polyveck_pointwise_poly_montgomery(&t1, &cp, &t1);

    polyveck_sub(&w1, &w1, &t1);  Scenario 3: Subtraction
    polyveck_reduce(&w1);
    polyveck_invntt_tomont(&w1);

    /* Reconstruct w1 */
    polyveck_caddq(&w1);
    polyveck_use_hint(&w1, &w1, &h);
    polyveck_pack_w1(buf, &w1);

    /* Call random oracle and verify challenge */
    shake256_init(&state);
    shake256_absorb(&state, mu, CRHBYTES);
    shake256_absorb(&state, buf, K * POLYW1_PACKEDBYTES);
    shake256_finalize(&state);
    shake256_squeeze(c2, SEEDBYTES, &state);
    for (i = 0; i < SEEDBYTES; ++i) {
        if (c[i] != c2[i]) {
            return -1;
        }
    }
    return 0;
}
```

Fig. 1. PQClean Dilithium `verify` code snippet.

```
1  void unpack_pk(uint8_t rho[SEEDBYTES], polyveck *t1,
2                 const uint8_t pk[CRYPTO_PUBLICKEYBYTES]) {
3    unsigned int i;
4    for (i = 0; i < SEEDBYTES; ++i) {
5      rho[i] = pk[i];
6    }
7    pk += SEEDBYTES;
8    for (i = 0; i < K; ++i) {
9      polyt1_unpack(&t1->vec[i], pk + i * POLYT1_PACKEDBYTES);
10   }
11 }
```

Fig. 2. PQClean unpack *pk* code snippet

```
1 void polyt1_unpack(poly *r, const uint8_t *a) {
2   unsigned int i;
3   for (i = 0; i < N / 4; ++i) {
4     r->coeffs[4*i + 0] = ((a[5*i + 0] >> 0) | ((uint32_t)a[5*i + 1] << 8)) & 0x3FF;
5     r->coeffs[4*i + 1] = ((a[5*i + 1] >> 2) | ((uint32_t)a[5*i + 2] << 6)) & 0x3FF;
6     r->coeffs[4*i + 2] = ((a[5*i + 2] >> 4) | ((uint32_t)a[5*i + 3] << 4)) & 0x3FF;
7     r->coeffs[4*i + 3] = ((a[5*i + 3] >> 6) | ((uint32_t)a[5*i + 4] << 2)) & 0x3FF;
8   }
9 }
```

Fig. 3. PQClean unpack $\mathbf{t}_1$ code snippet

For instance, the unpacking of $\mathbf{t}_1$ is a potential location for fault injection. To avoid affecting other public variables, such as $\mathbf{A}$, the only feasible target is the constant `0x3FF` lines 4 to 7 of the function `polyt1_unpack` Fig. 3. Zeroizing this constant sets every coefficient of $\mathbf{t}_1$ to zero and we can use P1 through Algorithm 4, detailed in Scenario 1. However, this approach requires a total of $K \times N$ repeated faults, which can be challenging in practice. Yet, it is worth noting that $\mathbf{t}_1$ could be sensitive if declared as a global variable. Then, as by default it is initialized to 0, faulting the call to the function `polyt1_unpack`, line 9 in Fig. 2, could set $\mathbf{t}_1$ to 0 with just K repeated faults. Alternatively, one test inversion fault, line 8 Fig. 2, can force zero iterations of the loop.

```
1  void invntt_tomont(int32_t a[N]) {
2    unsigned int start, len, j, k;
3    int32_t t, zeta;
4    const int32_t f = 41978; // mont^2/256
5    k = 256;
6    for (len = 1; len < N; len <<= 1) {
7      for (start = 0; start < N; start = j + len) {
8        zeta = -zetas[--k];
9        for (j = start; j < start + len; ++j) {
10         t = a[j];
11         a[j] = t + a[j + len];
12         a[j + len] = t - a[j + len];
13         a[j + len] = montgomery_reduce((int64_t)zeta * a[j + len]);
14       }
15     }
16   }
17   for (j = 0; j < N; ++j)
18     a[j] = montgomery_reduce((int64_t)f * a[j]);
19 }
```

Fig. 4. PQClean $\mathtt{NTT}^{-1}$ code snippet

Our attention also turns to lines 34, 36, and 41 of Fig. 1 involving the NTT and $\mathtt{NTT}^{-1}$ conversions, given in Fig. 4 . Notably, Ravi et al. [31] already cover the conversion of c in line 34. At the end of the inverse conversion of $\mathbf{A}\,\mathbf{z} - c\,\mathbf{t}_1\,2^d$ each coefficient undergoes multiplication by the squared Montgomery factor divided by 256 in a `for` loop, line 18 Fig. 4. This 32-bit integer constant plays a critical role. It is used at each of the N iterations so it can potentially be stored in a register. Zeroizing this value once can set all polynomial $\mathbf{A}\,\mathbf{z} - c\,\mathbf{t}_1\,2^d$ to 0. However, this fault must be repeated K times, once for each polynomial of the vector processed by the $\mathtt{NTT}^{-1}$. We can exploit this fault to sample the challenge c with $\mathbf{w}_1 = 0$ and forge valid signatures with Algorithm 4. We notice that even if we first perform the $\mathtt{NTT}^{-1}$ of $\mathbf{A}\,\mathbf{z}$ and $c\,\mathbf{t}_1\,2^d$ separately, and then subtract the two, it would also be vulnerable. This is because we can apply the same fault to the $\mathtt{NTT}^{-1}$ of $c\,\mathbf{t}_1\,2^d$ to zeroize the result, enabling the exploitation of P1.

3.3 Scenario 1: Sampling of $\tilde{c}$

```
1 void poly_challenge(poly *c, const uint8_t seed[SEEDBYTES]) {
2   unsigned int i, b, pos;
3   uint64_t signs;
4   uint8_t buf[SHAKE256_RATE];
5   shake256incctx state;
6   shake256_init(&state);
7   shake256_absorb(&state, seed, SEEDBYTES);
8   shake256_finalize(&state);
9   shake256_squeeze(buf, sizeof buf, &state);
10  signs = 0;
11  for (i = 0; i < 8; ++i)
12    signs |= (uint64_t)buf[i] << 8 * i;
13  pos = 8;
14  for (i = 0; i < N; ++i)
15    c->coeffs[i] = 0;
16  for (i = N - TAU; i < N; ++i) {
17    do {
18      if (pos >= SHAKE256_RATE) {
19        shake256_squeeze(buf, sizeof buf, &state);
20        pos = 0;
21      }
22      b = buf[pos++];
23    } while (b > i);
24    c->coeffs[i] = c->coeffs[b];
25    c->coeffs[b] = 1 - 2 * (signs & 1);
26    signs >>= 1;
27  }
28  shake256_release(&state);
29 }
```

Fig. 5. PQClean sampling of c code snippet

For efficiency, the verification algorithm only compares the recomputed seed $\tilde{c}$ with the one from the signature, line 54 Fig. 1. In our investigation, we identify the procedure, in Fig. 5, for sampling the challenge c from its seed $\tilde{c}$ as sensitive.

This process involves setting all N coefficients of the challenge to zero using a first `for` loop, followed by another `for` loop setting τ coefficients as 1 or -1.

By exploiting skipping or test-inversion faults, an attacker can target the `for` loop, line 16 Fig. 5, abort it prematurely, and zeroize all coefficients of c with just one correctly targeted fault.

Similarly, the same effect can be achieved by faulting the loop's termination condition, such as zeroizing the constant `TAU`.

Suppose the challenge c has been successfully manipulated to be zero. We present an algorithm enabling an attacker to exploit this effect, resulting in the acceptance of false signatures without needing the secret key.

Algorithm 4. Sign based on P1

Input : $pk = (\rho, \mathbf{t}_1)$
Output: $\sigma = (\tilde{c}, \mathbf{z}, \mathbf{h})$

1 $\mathbf{A} \in R_q^{k \times l} := \texttt{ExpandA}(\rho)$ ▷ $\mathbf{A}$ is generated and stored in NTT representation $\hat{\mathbf{A}}$
2 $\mu \in \{0,1\}^{512} := \mathrm{H}(\mathrm{H}(\rho \,||\, \mathbf{t}_1) \,||\, M)$
3 $\mathbf{z} \in \tilde{S}^l_{\gamma_1 - \beta}$
4 $\mathbf{w} := \mathbf{A}\,\mathbf{z}$
5 $\mathbf{h} := \texttt{SampleInBall}_{\omega}()$
6 $\mathbf{w}_1 = \texttt{UseHint}_q(\mathbf{h}, \mathbf{w}, 2\gamma_2)$
7 $\tilde{c} \in \{0,1\}^{256} := \mathrm{H}(\mu \,||\, \mathbf{w}_1)$
8 **return** $\sigma = (\tilde{c}, \mathbf{z}, \mathbf{h})$

Algorithm 4 utilizes the fact that if $c = 0$, then $c\,\mathbf{t}_1\,2^d = 0$, therefore leveraging P1. We begin by sampling the vector $\mathbf{z}$ within the appropriate range. Similar to [31], our algorithm generates a random $\mathbf{h}$ satisfying its corresponding condition. Using the `UseHint` function, we compute the corresponding $\mathbf{w}_1$ to sample the resulting $\tilde{c}$. As observed earlier, we exploit faults that set c to 0 in the verification algorithm, meaning that the same seed $\tilde{c}$ is sampled as in Algorithm 4. Unlike [31], we don't perform a rejection on the first coefficient of c because the fault in the verification does not use this condition.

As a variation of Algorithm 4, we can directly set $\mathbf{h}$ to zero and use only the high bits of $\mathbf{A}\,\mathbf{z}$ to derive the seed $\tilde{c}$. It is worth noting that while $\mathbf{h}$ being completely null is a situation that could arise in practice, its probability is negligible. In current versions of Dilithium, this check is neither specified nor implemented. A thorough analysis is required to determine if adding the $\mathbf{h} = 0$ check to the verification algorithm would reject valid signatures. Furthermore, this scenario relies on the ability to set all coefficients of c to zero. Whereas, the challenge c should have precisely τ coefficients equal to 1 or -1. However, there are no checks in place to verify this in practice.

3.4 Scenario 2: Shift by d

```
1 void polyveck_shiftl(polyveck *v) {
2   unsigned int i;
3   for (i = 0; i <K; ++i)
4     poly_shiftl(&v->vec[i]);
5 }
```

Fig. 6. PQClean polyvec shift code snippet

```
1 void poly_shiftl(poly *v) {
2   unsigned int i;
3   for (i = 0; i <N; ++i)
4     a->coeffs[i] <<= D;
5 }
```

Fig. 7. PQClean poly code snippet

In this scenario, we focus on line 35 of `verify` given in Fig. 1. At this point, $\mathbf{t}_1$ has been unpacked, and the challenge c has been sampled from the seed $\tilde{c}$. Faulting either the shift of $\mathbf{t}_1$ by d or the multiplication of c with $\mathbf{t}_1$ can influence the magnitude of the product $c\mathbf{t}_1 2^d$. It is important to note that the result of the multiplication of c with $\mathbf{t}_1 2^d$, stored in the same location as $\mathbf{t}_1 2^d$, already contains coefficients outside the exploitable range of Proposition 1. Thus, faulting this operation does not yield usable outcomes.

Now, let us analyze the multiplication of $\mathbf{t}_1$ by 2^d. By considering skipping faults, an attacker can target the call to the `polyveck_shiftl` function on line 35 of `verify` by skipping the corresponding jump instruction with one fault.

Another potential target is line 3 of Fig. 6, where faulting the loop counter terminates the function prematurely. Alternatively, the call to `poly_shiftl` on line 4 can be targeted during each of the K iterations. However, this approach requires K repeated faults and can be more challenging to achieve.

The loop line 3 of Fig. 7 can be a potential target for a single fault. Similarly, we can target line 4 but this approach also requires K repeated faults.

Regarding zeroization faults, the constant d can be targeted to zeroize a bit or a byte of its value. It is worth noting that, in practice, for all versions of Dilithium, $d = 13 =$ `0b1101`, which is 3 bits to set to zero.

Considering randomization faults on d, the difference is that this time there is no control over the value d' so most of the random faults are not usable.

Our aim is to determine the suitable d' such that $\|c\mathbf{t}_1 2^{d'}\|_\infty \leq \gamma_2$ which allows us to utilize P3. Let us compute such a d' by bounding the product

$$\|c\mathbf{t}_1 2^{d'}\|_\infty \leq 2^{d'} \|c\|_1 \|\mathbf{t}_1\|_\infty, \tag{5}$$

since $\|c\|_1 = \tau$ and $\|\mathbf{t}_1\|_\infty \leq 2^{10} - 1$[4]

$$\leq 2^{d'} \tau (2^{10} - 1).$$

We want $2^{d'} \tau (2^{10} - 1) \leq \gamma_2$. Therefore $d' \leq \log_2 \left(\frac{\gamma_2}{\tau (2^{10} - 1)} \right)$.

[4] We must have this condition fulfilled in **Sign** for a signature to be valid.

Example: For Dilithium-2 we have $d' = 1$, while for Dilithium-3 and 5 we have $d' = 2$. In practice, however, the maximum erroneous d' tolerated for any version is 3. This is explained by the fact that we have analyzed the worst possible case, and so in practice the bound can be tightened.

Algorithm 5. Sign based on P3

Input : $pk = (\rho, \mathbf{t}_1)$
Output: $\sigma = (\tilde{c}, \mathbf{z}, \mathbf{h})$

1 $\mathbf{A} \in R_q^{k\times l} :=$ `ExpandA`(ρ) ▷ $\mathbf{A}$ is generated and stored in NTT representation $\hat{\mathbf{A}}$
2 $\mu \in \{0,1\}^{512} := \mathrm{H}(\mathrm{H}(\rho \,||\, \mathbf{t}_1) \,||\, M)$, $(\mathbf{h}) := \perp$
3 **while** $(\mathbf{h}) = \perp$ **do**
4 $\quad \mathbf{z} \in \tilde{S}^l_{\gamma_1-\beta}$
5 $\quad \mathbf{w} := \mathbf{A}\,\mathbf{z}$
6 $\quad \mathbf{w}_1 = \mathtt{HighBits}_q(\mathbf{w}, 2\gamma_2)$
7 $\quad \tilde{c} \in \{0,1\}^{256} := \mathrm{H}(\mu \,||\, \mathbf{w}_1)$
8 $\quad c \in B_\tau :=$ `SampleInBall`$(\tilde{c})$
9 $\quad \mathbf{h} := \mathtt{MakeHint}_q(-c\,\mathbf{t}_1\,2^{d'}, \mathbf{w} + c\,\mathbf{t}_1\,2^{d'}, 2\gamma_2)$
10 $\quad$ **if** $|\mathbf{h}|_{\mathbf{h}_j=1} > \omega$ **then**
11 $\quad\quad (\mathbf{h}) := \perp$
12 **return** $\sigma = (\tilde{c}, \mathbf{z}, \mathbf{h})$

Assuming we have effectively manipulated $\mathbf{t}_1\,2^d$ so that $\|c\,\mathbf{t}_1\,2^{d'}\|_\infty \leq \gamma_2$, we present an algorithm, Algorithm 5 enabling an attacker to exploit this with P3 and achieve the acceptance of false signatures without requiring the secret key.

Algorithm 5 closely resembles the correct signing algorithm employed in Dilithium, although lacking some rejection checks that we can't verify. It operates with the vector $\mathbf{z}$ sampled within the appropriate range and leverages the hint vector computed using $c\,\mathbf{t}_1 2^d$. Using P3, supposing we managed to produce the corresponding fault, we can assure that $c\,\mathbf{t}_1\,2^d$ remains sufficiently small to prevent excessive overflow into the higher bits. However, we still need to keep the rejection criterion based on the maximum value of non-zero coefficients within $\mathbf{h}$ for successful verification of such signatures. Our practical implementation of this algorithm, using SageMath library, has demonstrated low rejection rate for every security level of Dilithium, with no more than 3 on average.

3.5 Scenario 3: Subtraction

```
1 void polyveck_sub(polyveck *w, const polyveck *u, const polyveck *v) {
2   unsigned int i;
3   for (i = 0; i <K; ++i)
4     poly_sub(&w->vec[i], &u->vec[i], &v->vec[i]);
5 }
```

Fig. 8. PQClean `polyveck_sub` code snippet

```
1 void poly_sub(poly *c, const poly *a, const poly *b) {
2   unsigned int i;
3   for (i = 0; i <N; ++i) {
4     c->coeffs[i] = a->coeffs[i] - b->coeffs[i];
5 }
```

Fig. 9. PQClean `poly_sub` code snippet

To conclude our analysis, we direct our attention to line 39 of `verify` in Fig. 1. Notably, in current implementations, the result of the subtraction of $\mathbf{A}\,\mathbf{z}$ by $c\,\mathbf{t}_1\,2^d$ is stored in the same variable as $\mathbf{A}\,\mathbf{z}$. Introducing a fault in the subtraction, allows us to exploit this observation and leverage P1.

First, one can skip the call to the function `polyveck_sub` on line 39 of `verify`, Fig. 1, to fault the subtraction. Similarly, line 3 of Fig. 8 can be targeted to exit the `for` loop early. Since the result is stored in the same location as the first operand, skipping the call to `poly_sub` on line 4 of Fig. 8 at each of the K iterations yields the same outcome. However, this approach necessitates K repeated faults, which can be harder to do.

Within the `poly_sub` function given in Fig. 9, we can focus on skipping the loop on line 3. Alternatively, we can target line 4 of Fig. 9, although this requires $K \times N$ repeated faults.

In this scenario, once we achieved to fault the subtraction, we leverage P1 and Algorithm 4 remains applicable. It allows an attacker to produce a valid message-signature pair for verification. It is important to note that targeting this location has the same outcome as zeroizing the $\mathbf{t}_1$or zeroizing the challenge c in Scenario 1.

3.6 Experimental Validation

Our primary objective is to evaluate the functionality of Algorithm 4 and Algorithm 5 under the conditions specified by P1 and P3, respectively. To achieve this, we have chosen to model faults exclusively at the algorithmic level. This decision is based on the following reasons:

- Within the C code, there are multiple potential locations and various types of exploitable faults that can lead to the three scenarios discussed in Sects. 3.3, 3.4, and 3.5.
- As outlined in Sect. 2.2, there are numerous ways to achieve the desired outcomes.
- The specific faults required will depend heavily on the target platform and binary code, which depends on the source code, and both the compiler and compilation options used.

Therefore, to cover a broad range of possible faults, we have developed three modified versions of Dilithium in Python that correspond to each scenario, and ensure the desired algorithmic effects.

- Version 1 for Scenario 1, where we arbitrarily set c to 0.
- Version 2 for Scenario 2, where we set d to match the value of d'.
- Version 3 for Scenario 3, where we removed the subtraction operation entirely.

We have validated that the signatures generated by Algorithm 4 are accepted when using the versions 1 and 3. Likewise, we have verified that the signatures generated by Algorithm 5 are accepted when using version 2.

4 Countermeasures

It is essential to implement the scheme thoughtfully, to minimize potential attacks, identifying and securing vulnerable operations within it. We outline several countermeasures to address the sensitive locations identified in this section.

For example, line 39 of `verify`, storing the result of $\mathbf{A}\,\mathbf{z}$ minus $c\,\mathbf{t}_1\,2^d$ in the same memory location as $c\,\mathbf{t}_1\,2^d$ prevents the exploitation of this subtraction in Scenario 3. Even if an attacker attempts to fault the subtraction, the subsequent computation of the high bits of $c\,\mathbf{t}_1\,2^d$ at line 45 of `verify` renders them unusable for accepting false signatures. Thus protecting this location with no extra cost.

Proposition P1 relies on the fact that all $K \times N$ coefficients of $c\,\mathbf{t}_1\,2^d$ are smaller than they should be. Therefore, if we can prevent even a single coefficient from being changed in size, the presented scenarios will not work.

A first set of commonly used countermeasures aims to make it more difficult for the attacker to induce faults or reproduce them [38]. There are also mechanisms that can detect and prevent fault injections targeting loops [28]. This can ensure data is handled correctly throughout the process. However, these countermeasures are fragile and complex to deploy, as we must ensure their presence in the final code.

Consequently, it is more advantageous to have a Dilithium verification algorithm that is intrinsically resistant to propositions P1 and P3. Let us introduce specific countermeasures tailored for the identified sensitive operations.

Distribution Check of the value $c\,\mathbf{t}_1\,2^d$ before the subtraction. By verifying if it is the expected one, we can effectively detect the faults used in Scenario 1 and 2. However, in practice, this means computing some statistical test on the values which can be computationally expensive

Verify d. Alternatively, we can check the correctness of the value d before using it. One way to do this verification is by first noticing that $(2^d)^{-1} = 1 - 2^{10} \mod q$, which can be computed easily and only with shift operations. Therefore, checking that $2^d \times (2^d)^{-1} = 1 \mod q$ before using the value d could ensure that it is the correct one used. However, this method only detects the faults of Scenario 2.

Split d. Another equivalent implementation would be to do the multiplication by 2^d in two times, with little overhead. If we set $d_1 > 3$ and $d_2 > 3$ such that $d = d_1 + d_2$, we can ensure that even if we fault one of the intermediate d, the result will be too big to use P3 in Scenario 2.

Alternative implementation. We can remark that by computing $\mathbf{z}' := \mathbf{z}\,(2^d)^{-1}$, at the beginning of the verification, we can write $\mathbf{A}\,\mathbf{z} - c\,\mathbf{t}_1\,2^d = \big(\mathbf{A}\,\mathbf{z}\,(2^d)^{-1} - c\,\mathbf{t}_1\big)\,2^d$. This time, the signatures will always be invalid if an attacker can skip the multiplication by $(2^d)^{-1}$ or by 2^d thus completely preventing Scenario 2. We give in Algorithm 6 a possible implementation of this countermeasure.

Algorithm 6. Verify Alternative

Input : $pk = (\rho, \mathbf{t}_1)$, $\sigma = (\tilde{c}, \mathbf{z}, \mathbf{h})$
Output: *True* or *False*

1 $\mathbf{A} \in R_q^{k \times l} := \texttt{ExpandA}(\rho)$
2 $\mu \in \{0,1\}^{512} := \mathrm{H}(\mathrm{H}(\rho \,||\, \mathbf{t}_1) \,||\, M)$
3 $c := \texttt{SampleInBall}(\tilde{c})$
4 $\mathbf{z}' := \mathbf{z}\,(2^d)^{-1}$
5 $temp_1 := \mathbf{A}\,\mathbf{z}'$
6 $temp_2 := -c\mathbf{t}_1$
7 $temp_2 := temp_2 + temp_1$
8 $\mathbf{w}_1' := \texttt{UseHint}_q(\mathbf{h}, temp_2\,2^d, 2\gamma_2)$
9 **return** $[\![||\mathbf{z}||_\infty < \gamma_1 - \beta]\!]$ and $[\![\tilde{c} = \mathrm{H}(\mu \,||\, \mathbf{w}_1')]\!]$ and $[\![|\mathbf{h}|_{\mathbf{h}_j=1} \leq \omega]\!]$

Norm Check. One last possible countermeasure would be to only accept a signature as valid if the check $\|c\,\mathbf{t}_1\,2^d\|_\infty > \gamma_2$ passes. The idea behind this check we introduce is that all three possibilities for Proposition 1 are based on the fact that $c\,\mathbf{t}_1\,2^d$ is smaller than it should be. By verifying if it is not too small, one can completely prevent its use. One thing to note is that the probability for every of the $K \times N$ coefficients to be naturally less than γ_2 is negligible. Thus, it should not change the verification algorithm of Dilithium. If this check doesn't affect the verification, it could prevent the faults used in Scenario 1 and 2.

Here, we give a summary of the previous two sections in the form of a table with the different scenarios, the type of fault that can be exploited for each, and the countermeasure associated.

Table 1. Summary of the vulnerable locations of the verification algorithm to the corresponding fault models. (✓: easy exploitation, ✓: possible exploitable, –: not applicable), together with the applicable countermeasures

Versions		Skipping	Test-Inv	Randomization	Zeroizing	Countermeasures
Scenario 1	for	✓	✓	–	✓	Distribution Check,
	`TAU`	–	–	✓	✓	Norm Check
Scenario 2	polyvec for	✓	✓	–	✓	Distribution Check,
	poly for	✓	✓	–	✓	Norm Check,
	d	✓	–	✓	✓	Verify d, Split d
Scenario 3	polyvec for	✓	✓	–	✓	Alternative
	poly for	✓	✓	–	✓	implementation
	function call	✓	–	–	✓	

5 Conclusion

This works aims at proving that, similarly to RSA, Dilithium verification shall be implemented carefully even if it does not handle secret data. Hence, we presented a comprehensive analysis of the verification algorithm of Dilithium, focusing on a common implementation in C and considering four common fault models: skipping faults, test inversion faults, randomization faults, and zeroizing faults. For each of them we establish a methodology for forging Dilithium signatures based on the specific type of fault employed during the verification process. Furthermore, our analysis provides valuable insights into the vulnerabilities and sensitive operations within the Dilithium verification algorithm. Building upon these findings, we propose a set of novel countermeasures covering the various scenarios introduced, and designed to mitigate the risks associated with these sensitive operations.

References

1. Alagic, G., et al.: Status report on the third round of the NIST post-quantum cryptography standardization process (2022)
2. Azouaoui, M., et al.: Protecting dilithium against leakage: revisited sensitivity analysis and improved implementations. In: CHES (2023)
3. Bai, S., et al.: CRYSTALS – Dilithium. National Institute of Standards and Technology (2022). https://csrc.nist.gov/Projects/post-quantum-cryptography/selected-algorithms-2022
4. Bernstein, D., Hülsing, A., Kölbl, S., Niederhagen, R., Rijneveld, J., Schwabe, P.: The SPHINCS+ signature framework. In: CCS (2019)
5. Bindel, N., Buchmann, J., Krämer, J.: Lattice-based signature schemes and their sensitivity to fault attacks. In: FDTC (2016)
6. Breier, J., Hou, X.: How practical are fault injection attacks, really? IEEE Access **10**, 113122–113130 (2022)
7. Brier, E., Chevallier-Mames, B., Ciet, M., Clavier, C.: Why one should also secure RSA public key elements. In: CHES (2006)

8. Bruinderink, L.G., Pessl, P.: Differential fault attacks on deterministic lattice signatures. CHES **2018**(3), 21–43 (2018)
9. Claudepierre, L., Péneau, P., Hardy, D., Rohou, E.: TRAITOR: a low-cost evaluation platform for multifault injection. In: ASSS (2021)
10. Colombier, B., et al.: Multi-spot laser fault injection setup: new possibilities for fault injection attacks. In: CARDIS (2021)
11. Colombier, B., Menu, A., Dutertre, J., Moëllic, P., Rigaud, J., Danger, J.: Laser-induced single-bit faults in flash memory: instructions corruption on a 32-bit microcontroller. In: IEEE HOST (2019)
12. Diffie, W., Hellman, M.: New directions in cryptography. IEEE Trans. Inf. Theory **22**(6), 644–654 (1976)
13. Ducas, L., et al.: PQ-CRYSTALS, Dilithium (2022). gitHub repository. Accessed 15 Dec 2022
14. Dutertre, J., Riom, T., Potin, O., Rigaud, J.: Experimental analysis of the laser-induced instruction skip fault model. In: NordSec (2019)
15. Islam, S., Mus, K., Singh, R., Schaumont, P., Sunar, B.: Signature correction attack on dilithium signature scheme. In: EuroS&P (2022)
16. Kannwischer, M., Petri, R., Rijneveld, J., Schwabe, P., Stoffelen, K.: PQM4: post-quantum crypto library for the ARM Cortex-M4. Accessed 15 Dec 2022
17. Kannwischer, M.J., Schwabe, P., Stebila, D., Wiggers, T.: PQClean (2022). https://github.com/PQClean/PQClean. GitHub repository Accessed 15 Sep 2023
18. Kiltz, E., Lyubashevsky, V., Schaffner, C.: A concrete treatment of Fiat-Shamir signatures in the quantum random-oracle model. In: Nielsen, J., Rijmen, V. (eds.) Advances in Cryptology – EUROCRYPT 2018. EUROCRYPT 2018. LNCS, vol. 10822, pp. 552–586. Springer, Cham (2018). https://doi.org/10.1007/978-3-319-78372-7_18
19. Korak, T., Hoefler, M.: On the effects of clock and power supply tampering on two microcontroller platforms. In: FDTC (2014)
20. Kumar, D., Beckers, A., Balasch, J., Gierlichs, B., Verbauwhede, I.: An in-depth and black-box characterization of the effects of laser pulses on atmega328p. In: CARDIS (2019)
21. Liu, Y., Zhou, Y., Sun, S., Wang, T., Zhang, R., Ming, J.: On the security of lattice-based Fiat-Shamir signatures in the presence of randomness leakage. IEEE Trans. Inf. Forensics Secur. **16** (2021)
22. Lyubashevsky, V.: Fiat-Shamir with aborts: applications to lattice and factoring-based signatures. In: Matsui, M. (eds.) Advances in Cryptology – ASIACRYPT 2009. ASIACRYPT 2009. LNCS, vol. 5912, pp. 598–616. Springer, Berlin, Heidelberg (2009). https://doi.org/10.1007/978-3-642-10366-7_35
23. Marzougui, S., Ulitzsch, V., Tibouchi, M., Seifert, J.: Profiling side-channel attacks on dilithium: a small bit-fiddling leak breaks it all. ePrint (2022)
24. Menu, A., Dutertre, J., Potin, O., Rigaud, J., Danger, J.: Experimental analysis of the electromagnetic instruction skip fault model. In: DTIS (2020)
25. Migliore, V., Gérard, B., Tibouchi, M., Fouque, P.A.: Masking dilithium. In: ACNS (2019)
26. Moro, N., Dehbaoui, A., Heydemann, K., Robisson, B., Encrenaz, E.: Electromagnetic fault injection: towards a fault model on a 32-bit microcontroller. In: FDTC (2013)
27. Muir, A.: Seifert's RSA fault attack: simplified analysis and generalizations. In: Ning, P., Qing, S., Li, N. (eds.) ICICS 2006. LNCS, vol. 4307, pp. 420–434. Springer, Heidelberg (2006). https://doi.org/10.1007/11935308_30

28. Proy, J., Heydemann, K., Berzati, A., Cohen, A.: Compiler-assisted loop hardening against fault attacks. ACM 2017 (2017)
29. Qiao, Z., Liu, Y., Zhou, Y., Ming, J., Jin, C., Li, H.: Practical public template attacks on CRYSTALS-dilithium with randomness leakages. IEEE Trans. Inf. Forensics Secur. **18**, 1–14 (2023). https://doi.org/10.1109/TIFS.2022.3215913
30. Ravi, P., Jhanwar, M.P., Howe, J., Chattopadhyay, A., Bhasin, S.: Side-channel assisted existential forgery attack on dilithium - a NIST PQC candidate. ePrint
31. Ravi, P., Yang, B., Bhasin, S., Zhang, F., Chattopadhyay, A.: Fiddling the twiddle constants - fault injection analysis of the number theoretic transform. CHES (2023)
32. Rivest, R., Shamir, A., Adleman, L.: A method for obtaining digital signatures and public-key cryptosystems. ACM Commun. (1978)
33. Seifert, J.P.: On authenticated computing and RSA-based authentication. In: CCS (2005)
34. Shor, P.: Algorithms for quantum computation: discrete logarithms and factoring. In: FOCS (1994)
35. Soni, D., Basu, K., Nabeel, M., Aaraj, N., Manzano, M., Karri, R.: FALCON, pp. 31–41. Springer, Cham (2021). https://doi.org/10.1007/978-3-030-57682-0_3
36. Timmers, N., Spruyt, A., Witteman, M.: Controlling pc on arm using fault injection. In: FDTC (2016)
37. Trouchkine, T., Bouffard, G., Clédière, J.: EM fault model characterization on SoCs: from different architectures to the same fault model. In: FDTC (2021)
38. Witteman, M.: Secure application programming in the presence of side channel attacks. https://www.riscure.com/publication/secure-application-programming-presence-side-channel-attacks/
39. Yuce, B., Schaumont, P., Witteman, M.: Fault attacks on secure embedded software: threats, design and evaluation. CoRR (2020)
40. Zussa, L., Dutertre, J.M., Clédière, J., Robisson, B., Tria, A.: Investigation of timing constraints violation as a fault injection means. In: DCIS (2012)

Side-Channel Analysis

Attacking at Non-harmonic Frequencies in Screaming-Channel Attacks

Jeremy Guillaume[1](✉), Maxime Pelcat[2], Amor Nafkha[1], and Rubén Salvador[3]

[1] CentraleSupélec, IETR UMR CNRS 6164, Gif-sur-Yvette, France
jeremy.guillaume@centralesupelec.fr
[2] Univ Rennes, INSA Rennes, CNRS, IETR - UMR 6164, 35000 Rennes, France
[3] CentraleSupélec, Inria, Univ Rennes, CNRS, IRISA, Rennes, France

Abstract. Screaming-channel attacks enable Electromagnetic (EM) Side-Channel Attacks (SCAs) at larger distances due to higher EM leakage energies than traditional SCAs, relaxing the requirement of close access to the victim. This attack can be mounted on devices integrating Radio Frequency (RF) modules on the same die as digital circuits, where the RF can unintentionally capture, modulate, amplify, and transmit the leakage along with legitimate signals. Leakage results from digital switching activity, so previous works hypothesized that this leakage would appear at multiples of the digital clock frequency, i.e., harmonics.

This work demonstrates that compromising signals appear not only at the harmonics and that leakage at non-harmonics can be exploited for successful attacks. Indeed, the transformations undergone by the leaked signal are complex due to propagation effects through the substrate and power and ground planes, so the leakage also appears at other frequencies. We first propose two methodologies to locate frequencies that contain leakage and demonstrate that it appears at non-harmonic frequencies. Then, our experimental results show that screaming-channel attacks at non-harmonic frequencies can be as successful as at harmonics when retrieving a 16-byte AES key. As the RF spectrum is polluted by interfering signals, we run experiments and show successful attacks in a more realistic, noisy environment where harmonic frequencies are contaminated by multi-path fading and interference. These attacks at non-harmonic frequencies increase the attack surface by providing attackers with more potential frequencies where attacks can succeed.

Keywords: Cybersecurity · hardware security · electromagnetic side channels · screaming-channel attacks

1 Introduction

Side-Channel Attacks (SCAs) [9,26] allow retrieving confidential information from computing devices by exploiting the correlation of internal data with the leakage produced while computing over these data. The term *side channel* is

S. Bhasin and T. Roche (Eds.): CARDIS 2023, LNCS 14530, pp. 87–106, 2024.
https://doi.org/10.1007/978-3-031-54409-5_5

therefore used to refer to physical leakage signals carrying confidential information. Side channels are general to CMOS computing devices and can take many forms, from runtime variations of system power consumption [18] to Electromagnetic (EM) emanations [12]. Screaming channels are a specific form of EM side channel that occurs on mixed-signal devices, where a Radio Frequency (RF) module is co-located on the same die as digital modules. In this context, the leakage of the digital part reaches the RF module, which can transmit it over a distance of several meters. This phenomenon allows attackers to mount side-channel attacks at distances from the victim. The seminal work of Camurati et al. [7] demonstrated how screaming-channel attacks can succeed at distances of up to 15 m.

Leakage is generated by the switching activity of the transistors from the digital part of the victim system, which operates at a clock frequency F_{clk}. When observed on a spectrum analyzer, the leakage power spectral density is shaped as peaks at the harmonics of F_{clk} (*i.e.* $n \times F_{clk}$ where $n \in \mathbb{Z}$). What makes screaming-channel attacks different from other SCAs is that the harmonics, after being modulated by the RF module, are visible around the carrier frequency F_{RF} of the legitimate RF signal (Sect. 2.2).

A limitation of this attack is that the harmonics of F_{clk} can be modulated at the same frequency as some interfering signals, such as WiFi signals. Since these interfering signals are transmitted voluntarily, they are stronger than the leakage signal, which, as a result, can be easily polluted and hence quickly become non-exploitable.

To overcome this limitation and further study the risk posed by screaming channels, this paper studies the attack's feasibility when capturing signals at frequencies other than the harmonics of the digital processing clock. Specifically, we seek to answer the following questions: **is exploitable leakage also present at frequencies other than the harmonics? In case it is, is the difficulty for a successful attack higher at non-harmonics than at harmonics frequencies?** If the first question is answered positively, attackers can have an extensive choice of potential frequencies to select from and find one not polluted by environmental noise during the attack. Such a property can also be an enabler to effectively extend the framework of multi-channel attacks [4], which attack by combining different side-channel sources and different frequencies in the context of modulated leakage signals.

To summarize, we investigate the presence of leakage over the spectrum at non-harmonic frequencies and demonstrate that this leakage can be used to build successful attacks. We propose the following contributions:

- **Two methodologies to search for exploitable leakage over the spectrum**. The first is based on a *fixed vs. fixed* t-test [11], and the second is an original contribution based on **Virtual Trigger (VT)** [15].
- With these methodologies, we **demonstrate that leakage in screaming-channel attacks is not only present at harmonic frequencies**, as explored in previous works [7,27], but it is also spread over a large share of the near-carrier spectrum.

- We compare both methods and demonstrate a **significant reduction in the exploration time** when looking for exploitable frequencies with the second methodology **based on pattern detection**.
- We evaluate the **effectiveness of attacks at non-harmonic frequencies** in a noiseless environment and show how this effectiveness can sometimes be higher at non-harmonics.
- We demonstrate **successful attacks in more realistic scenarios**. We apply the proposed methodologies, and the insights learned when attacking at non-harmonics and build attacks in a context where most harmonics are polluted by other standard signals typically found in the spectrum.

The rest of this paper is organized as follows: Sect. 2 introduces related and previous works on screaming-channel attacks. The attack scenario of our work and the setup are described in Sect. 3. The two methods we propose to search for leakage over the spectrum are presented in Sect. 4. Afterward, in Sect. 5, we demonstrate the attack feasibility by exploiting the leakages found at non-harmonic frequencies. Section 6 demonstrates the attack in a more challenging, and therefore more realistic, scenario. Lastly, Sect. 7 concludes the paper.

2 Related Works

2.1 Side-Channel Attacks

By capturing the leakage generated by computing devices, attackers can mount SCAs to jeopardize confidentiality and recover internal secret data. The most common methods used to build SCAs are Differential Power Analysis (DPA) [16], Correlation Power Analysis (CPA) [5], Mutual Information Analysis (MIA) [13], Template Attacks (TA) [8] and more recently Deep Learning (DL) [19]. The leakage signal of interest is generated by the digital part of the system from the switching activity of the transistors occurring when data is being computed or moved through the chip. The steep peaks of the signals resulting from the switching activity of transistors in digital devices produce leakage signals whose variations correlate with that switching activity and, therefore, with the data over which the processor computes. That leakage signal is most often found in the EM emanations [12] or the power consumption variations [18] of the victim device.

In a synchronous device, transistors switch at the pace of the digital clock. As a result, the leakage resulting from successions of transitions and non-transitions has a period equal to one clock cycle. The Fourier Transform of a signal having a period T corresponds to peaks at each harmonic of the frequency $F = 1/T$ [1] as formalized in Eq. (1), with A_n being the respective amplitudes of each harmonic. Then, theoretically, the leakage should appear at harmonics of $F_{clk} = 1/T_{clk}$ as formalized in Eq. (2) and illustrated in Fig. 1. This corresponds to what is empirically observed in works on side-channel attacks where the leakage is found at harmonics of the clock frequency [3].

$$H(f) = \Sigma_{n=-\infty}^{\infty} A_n \delta(f - n/T) \tag{1}$$

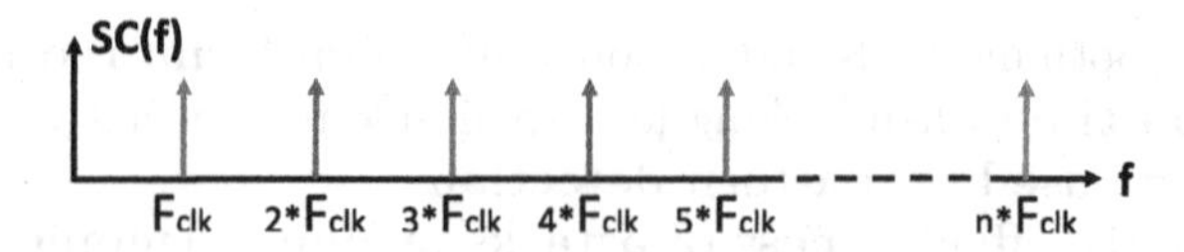

Fig. 1. Leakage presence over the spectrum: The Fourier transform of a signal having a period Tclk consists of peaks at both odd harmonics (green peaks) and even harmonics (blue peaks) of frequency $F_{clk} = 1/T_{clk}$. (Color figure online)

$$SC(f) = \Sigma_{n=-\infty}^{\infty} A_n \delta(f - nF_{clk}) \tag{2}$$

A limitation of traditional SCAs to capture clean leakage signals is that attackers must be in very close proximity to the victim device for a successful attack, usually only a few millimeters away. However, some specific scenarios allow attackers to take distance from the device [10,24]. One scenario for gaining distance from the device, which is the focus of this paper, is the so-called screaming-channel attack [7]. In this attack, the leakage is transmitted by an RF module that sits beside the digital part of a mixed-signal chip on the same die, allowing the attacker to capture it at a distance of several meters.

2.2 Screaming-Channel Attacks

Mixed-signal devices are heterogeneous platforms with digital and analog modules integrated into the same die. One of these analog modules can be the RF chain needed to build a System-on-a-Chip (SoC) with radio communications capabilities. This tight integration has the advantage of reducing the power consumption or the transmission delay between the digital and RF modules, as well as the cost of the final device. However, the very nature of this type of device has already been proven to be a hardware vulnerability [7].

Figure 2 illustrates a mixed-signal device. Compared to regular side-channels in digital devices, the leakage resulting from the switching activity in mixed-signal systems can travel through the substrate by the so-called *substrate coupling* effect [2,17,20,23]. This way, leakage signals can reach the radio transceivers of the RF part, which is very sensitive to noise and hence prone to capture these slight variations carrying information that correlates with secret data.

The leakage from the digital part, which is the one that would be used to build a traditional SCA is, as expressed in Eq. (3), modulated at the frequency of the legitimate RF signal F_{RF}, amplified and then transmitted by the RF module through the antenna. As a result, this amplification can bring, i.e., *scream*, the leakage signal at distances of several meters. In Camurati et al. use case [7], the transmitted Bluetooth signal is centered at 2.4 GHz, and the device clock frequency is 64 MHz. The leakage harmonics are therefore at 2.4 GHz + multiples of 64 GHz, i.e., 2.464 GHz, 2.528 GHz, 2.592 GHz, etc.

$$ScreamC(f) = SC(f) * F_{RF} = \Sigma_{n=-\infty}^{\infty} A_n \delta(f - nF_{clk} - F_{RF}) \tag{3}$$

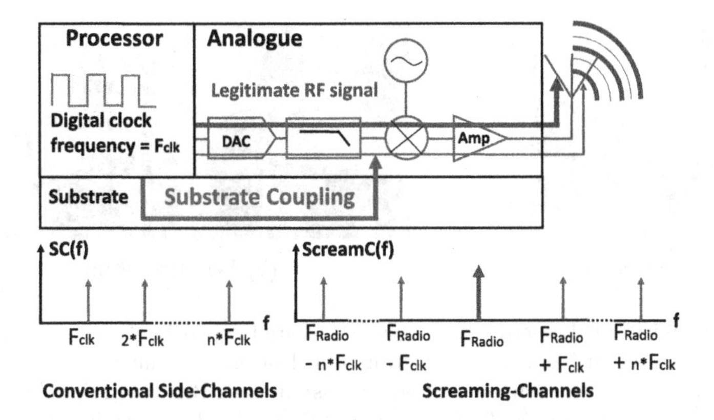

Fig. 2. Screaming-channel attacks: The conventional leakage of the digital part perturbates the RF module of the analog part, present on the same die. The radio transceivers here modulate the leakage around the frequency of the legitimate signal and transmit it at a longer distance (up to several meters).

In their seminal work, Camurati et al. [7] demonstrated that an attack using this modulated leakage is possible. Authors also conducted further works to understand better the properties of the leakage in this context [6]. In this original use case, the device transmits a Bluetooth signal from a commodity low-power SoC while the digital part (an Arm Cortex-M4 microcontroller) executes AES encryptions. The authors reported performing the screaming-channel attack at up to 15 m and observed some leakage at up to 60 m.

The limitation of this scenario is that the leakage carries low energy compared to other legitimate signals potentially present at the harmonic frequencies. As a result, attacks can be very difficult to conduct at those frequencies when they are polluted. This sets a high uncertainty on the feasibility of the attack in a noisy environment where all harmonics would be polluted. The following sections demonstrate that attackers do not have to limit themselves to attack at a very limited set of harmonic frequencies. On the contrary, a wide spectrum is at their disposal to compromise the system, which increases the threat that screaming-channel attacks represent. To the best of our knowledge, this is the first work to demonstrate that the attack is possible at non-harmonics. All previous works on screaming-channel attacks [6,7,27] used one harmonic (the second harmonic at 2.528 GHz) to perform the attack.

3 The Attack Scenario and Setup

This paper considers a scenario where the victim is a mixed-signal device that executes a Cryptographic Process (CP) while transmitting an RF signal in parallel. Since the leakage is emitted by the RF module, the attacker can capture it with a Software Defined Radio (SDR) device. The attacker's goal is to recover the secret key used by the CPs from this remotely captured leakage signal.

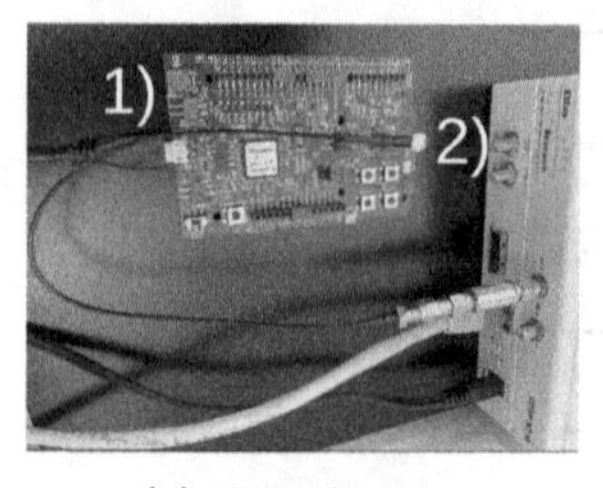

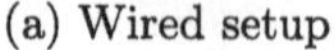

(a) Wired setup

(b) Wireless setup

Fig. 3. Experimental setup: (a) shows our wired setup for noiseless experiments and (b) an attacker at 7 m from the victim. This last one contains: 1) (blue arrow) the victim PCA10040 device running AES and transmitting a Bluetooth signal at 2.4 GHz.; and 2) (orange arrow) the SDR device that collects the data-correlated leakage from the victim. (Color figure online)

Figure 3 shows the attack setup used for all experiments. The victim device is an nRF52832 from Nordic instrument[1]. It contains an Arm Cortex M4 processor and an RF module. The attacker device is a USRP N210[2] using an SBX daughter board that can measure a signal between 400 MHz and 4.4 GHz and has a bandwidth of 40 MHz. To collect the leakage at a distance, a parabolic grid antenna with a gain of 26 dBi is used. The computer used for all our experiments has a 4-core Intel Xeon(R) CPU E3-1226 V3 @ 3.30 GHz, and 8 GB RAM.

The legitimate RF signal is a Bluetooth signal transmitted at 2.4 GHz without frequency hopping[3]. The attacked encryption algorithm is a software implementation of AES-128, whose encryption on the considered microcontroller takes 870μs. In the following, we describe the steps used in the experiments to collect traces. One *trace* corresponds to the collected leakage signal produced by one CP execution.

3.1 Leakage Collection

Figure 4 shows the steps undergone by the leakage signal. First, the USRP demodulates the RF signal at a given frequency, $Ftested$, that potentially carries leakage. Then, the USRP samples the baseband signal at 5 MHz. The choice of this sampling frequency is based on previous works on screaming-channel attacks [6,7,27], which we also use as it has so far provided sufficient resolution for successful attacks. Although a study of the impact of sampling frequency on screaming-channel attack success could be an interesting subject, it is out of the scope of this paper.

1 https://www.nordicsemi.com/Software-and-tools/Development-Kits/nRF52-DK..

2 https://www.ettus.com/all-products/un210-kit/.

3 Frequency-hopping is the repeated switching of the carrier frequency during radio transmission to reduce interference and avoid interception. In the case of Bluetooth transmissions, switching occurs among 81 channels, from 2.4 GHz to 2.48 GHz with 1 MHz wide bands.

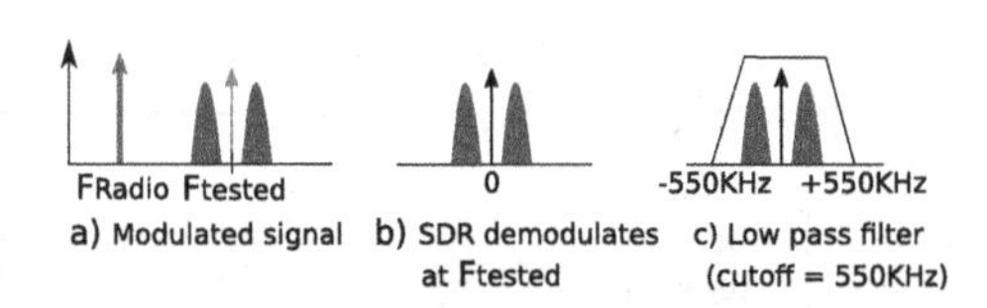

Fig. 4. Signal processing steps for leakage collection: a) The SDR device collects the leakage at the targeted frequency, potentially containing leakage. b) Trace segmentation is processed on the demodulated signal. c) After being segmented, the leakage is low-pass filtered at 550 KHz to reduce noise.

3.2 Trace Segmentation

To segment the obtained raw traces[4], i.e., to separate the segments corresponding to each individual AES encryption, pattern recognition is used. It consists in identifying the locations within the raw trace matching with the shape of the leakage produced by CPs. The steps applied for pattern recognition, chosen empirically for their good performance on the problem at hand, are the following:

1. Low-pass filter the raw trace with a cutoff frequency at the sampling frequency divided by 4: 5 MHz/4 = 1.125 MHz.
2. Compute the sliding correlation between the pattern and the filtered trace.
3. The peaks obtained during the sliding correlation are expected to correspond to the locations of the AES segments. Segment the raw trace by cutting it at these locations.
4. To reduce noise, low-pass filter the obtained segments with a cutoff frequency of 550 KHz.

To extract an initial pattern, we use VT [15] as the pattern extraction technique, illustrated in Fig. 5. This study shows that by knowing the precise time duration L_{cp} between 2 CPs, it is possible to segment a raw trace containing leakage produced by a series of CP executions. Averaging the obtained segments returns a representative pattern of the leakage produced by that device each time it executes a CP. The study also describes a procedure to find this precise length L_{cp}.

3.3 Time Diversity

As in previous works on screaming-channel attacks, time diversity is used to reduce noise from the leakage collection. It consists in running N encryptions with exactly the same data (plaintext and key) and averaging their leakage. Since the N encryptions are computing the same data, the leakage they produce would be very similar, and hence averaging tends to cancel the random noise contributions. While also typical in regular side-channel attacks, this is especially important in the case of screaming-channel attacks, as the leakage captured by the attacker contains additional noise due to the transmission channel.

[4] A *raw trace* corresponds to the collected signal, sampled and quantized by the SDR.

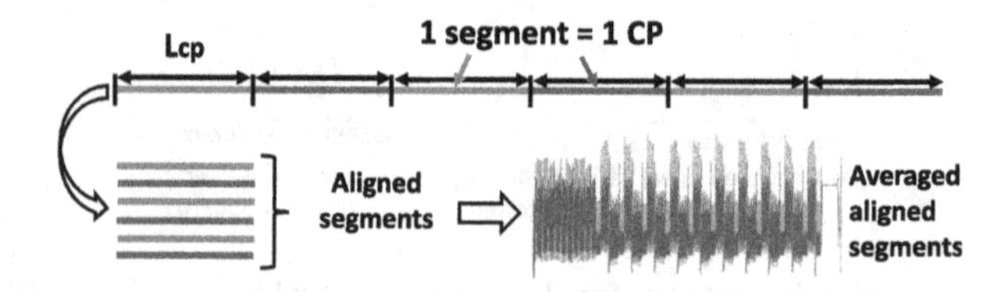

Fig. 5. Virtual Trigger (VT) [15]: Knowing the precise duration between 2 CP executions L_{cp} enables the segmentation of a raw trace. Splitting this raw trace in segments whose starts are separated by L_{cp} returns segments all containing leakage of one CP execution. Averaging these segments returns a reduced-noise segment, which can be used as a pattern

In the rest of this paper, this reduced-noise segment is called *a trace*. The number N is set to 10 in the experiments where the leakage is collected through a cable (Sects. 4.1 and 5). It is increased to 50 for the experiments where it is collected at a distance with an antenna (Sect. 6). These values significantly differ from the 500 traces used in [6] when attacking at a distance. While this brings additional difficulty in running a successful attack, it allowed us to run the experiments in a more reasonable time of 6 days instead of 5 weeks. For each experiment, the number of traces collected will be noted as $Nb_{Traces} \times N_{Time\_Diversity}$.

4 Searching for Leakage at Non-harmonic Frequencies

Figure 6 shows a part of the frequency spectrum at the output of the victim board while transmitting a Bluetooth signal at 2.4 GHz. Next to the legitimate signal peak, other peaks with lower energy appear. These correspond to the first 4 harmonics of the leakage from the digital part. They are present at frequencies equal to 2.4 GHz plus multiples of 64 MHz, which is the digital clock frequency. Therefore, one would intuitively assume that the leakage from the digital part transmitted by the victim board is stronger at these harmonics and, hence, that it would be more challenging to perform the attack at other frequencies. For this reason, previous works on screaming-channel attacks [6,27] use harmonic frequencies for the attack. Typically, the second harmonic is used, as it is often less polluted by interfering signals. In contrast, the first harmonic is located within a frequency band typically used by signals like Bluetooth or WiFi. Nevertheless, it can be seen on the spectrum that some variations in energy are also present between these harmonics, suggesting that leakage could potentially be distributed continuously along the spectrum.

The first question we want to answer is: **is the leakage from the digital part only present at the harmonic frequencies, or does it also appear at other frequencies?** To answer this question, we propose two methodologies to investigate at which frequencies the leakage exists in the spectrum. The first involves running a t-test at each tested frequency. The side-channel community

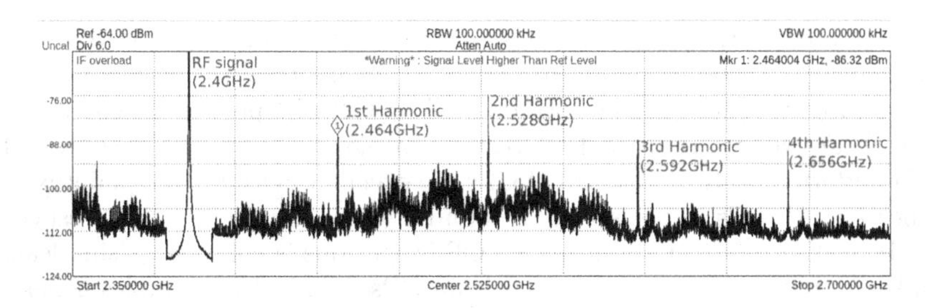

Fig. 6. Frequency spectrum of the victim output: The highest peak on the left (blue text) is the legitimate RF signal at 2.4 GHz. The following peaks (red text) are the first four leakage harmonics at 2.464 GHz, 2.528 GHz, 2.592 GHz, and 2.656 GHz. (Color figure online)

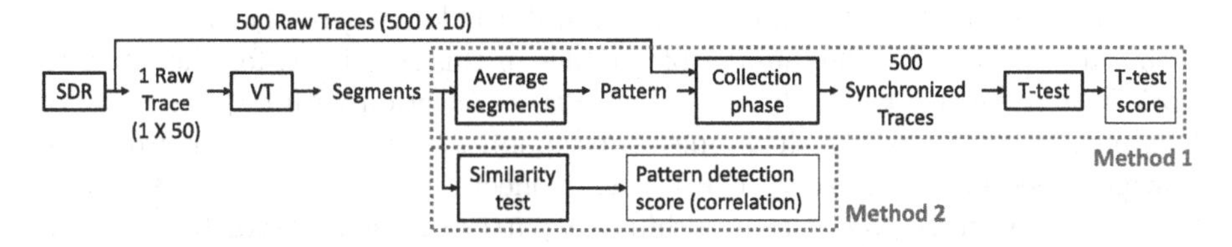

Fig. 7. Comparison of the steps of the two methods. Both first segment a raw trace using VT. Then, the first builds a pattern used for the collection phase and performs a t-test with the N collected traces (in the presented experiment, $N = 500$). The second uses the segments returned by VT to test if the raw trace contains leakage.

commonly uses this test to determine whether the internal data computed by the CPs impacts the leakage. Then, we propose a second method that reduces the implementation complexity while giving similar results. It is an adaptation of the method used in [15] to extract the CP leakage pattern; we refer to this second method as *pattern detection.* It consists of analyzing whether a signal collected at a given frequency is good enough to extract a CP pattern; if yes, this means that leakage is present. Figure 7 illustrates the difference between the two methods, described in the remainder of this section. Compared with the first methodology, the second method removes the most time-consuming phase: the collection of (500) synchronized traces needed for the t-test.

4.1 Leakage Localization Using T-Test Method

In this first investigation, a fixed vs. fixed t-test is performed [11] at each tested frequency. This test indicates whether or not some leakage samples depend on the internal data computed by the Cryptographic Processes (CPs). Therefore, if the t-test is conclusive at a given frequency, i.e., the score is over 4.5[5], this means

[5] This score means that there is information leakage with confidence > 0.99999 [11, 14, 25].

leakage is present there, as it is a necessary condition for the t-test to detect a dependency. We chose to perform a fixed vs. fixed t-test as Durvaux et al. [11] demonstrated that it needs fewer traces to detect data dependencies compared to the classical fixed vs. random Test Vector Leakage Assesment (TVLA) test [14].

To perform this test, two sets of CPs are executed with unique plaintext and key per set. All CPs in one set are computed with the values of their respective set. The leakage generated during the CP executions is collected. Using leakage samples from a common time point within CPs, i.e., leakage samples generated by the same operations, we compute a t-test as indicated in the following Eq. (4):

$$\text{t-test} = \frac{\hat{u}_1 - \hat{u}_2}{\sqrt{\frac{\sigma_1^2}{N_1} + \frac{\sigma_2^2}{N_2}}} \quad (4)$$

where $\hat{u}_i$ represents the average of the samples belonging to the i_{th} set, σ_i^2 is their variance, and N_i the number of samples in this set. Thus, for the t-test to work, it is necessary to synchronize the leakage collection with the execution of CPs perfectly to know which samples correspond to which time point from one leakage collection to another. The leakage collection method in Sect. 3 is used for this. For pattern recognition, a different pattern must be extracted for each frequency, as the shape of the CP leakage differs among frequencies. The VT [15] technique is used to automate pattern extraction.

In the experiment we collect 500 traces at each frequency using the wired setup from Fig. 3a. This methodology is used over the frequency range 1.4 GHz to 3.4 GHz (2.4 +/− 1 GHz), with a resolution of 1 MHz. These parameters are application-dependent and can be adjusted according to each particular use case to test a wider frequency band or to change the resolution.

Experimental Results. Figure 8 shows the results of the t-test. At each frequency, the maximum absolute value of the t-test is kept as a score. Frequencies corresponding to the harmonics are highlighted in blue. If the leakage had only been present at these frequencies, the score would have been higher than this threshold only at these locations. Still, it can be observed that the score is also above 4.5 at other frequencies, suggesting that leakage is also present at non-harmonics. In fact, even if the highest peaks are at the first 2 harmonics, the score is very high around the first 3 harmonics and is above the threshold over a band of more than 500 MHz of width (almost until the $6th$ harmonic). This is true on both the right and left sides of the spectrum.

The advantage of employing this t-test method to localize leakage in the spectrum is that it gives a result that we know how to interpret, as the t-test is a well-known tool in the side-channel community. The downside of this method is that the collection phase takes 27 h. A way to reduce this time would be to collect fewer traces at each frequency. As introduced, these results were obtained using 500 traces at each frequency sub-band (250 traces per set). If we repeat the experiment for 300 traces, the time is reduced to 15 h. However, 16.89% of the frequencies previously identified as carrying leakage with the experiment

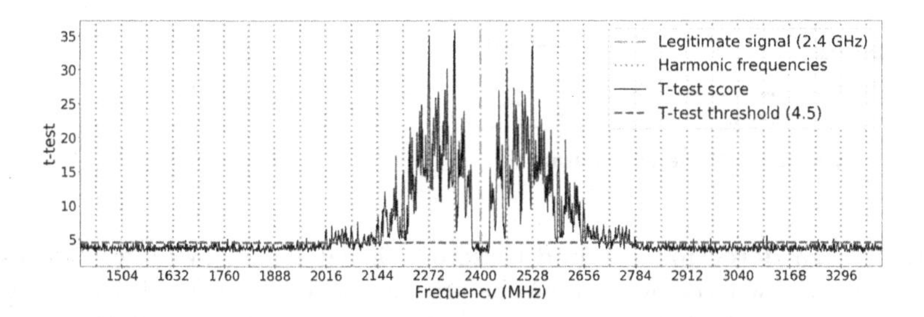

Fig. 8. Results of t-test (black curve): frequencies with a score higher than 4.5 (red horizontal dashed line) are considered as the ones where the leakage has been detected. Green central dash-dotted line: frequency of the legitimate signal (2.4 GHz). Blue vertical dotted lines: position of harmonic frequencies. (Color figure online)

using 500 traces are not detected anymore. Then if this method is used with an insufficient number of traces, there is a risk of not detecting frequencies that actually carry leakage. For example, in our experiments, frequencies between the $5th$ and $6th$ harmonics, which already have a low score with 500 traces per frequency, are not detected anymore by the t-test when it is performed with only 300 traces. It is, in fact, very difficult to determine how many traces are enough because the exploitable frequencies are unknown, and hence we cannot know if some are missing. We limit the number of traces to 500 to maintain the experiment time tractable. However, there is no guarantee that some frequencies, that could in fact contain exploitable leakage, remain undetected.

Therefore, in the next section, we propose a methodology that is more adapted to our requirements: detecting the presence of leakage in as many frequencies as fast as possible. The objective of this original method is to reduce the test complexity and, thus, its processing time. We apply this methodology to the same experiment and demonstrate equivalent results.

4.2 Leakage Localization Using Pattern Detection Method

This method consists of testing the similarity of segments returned by VT (c.f. Sect. 3). If leakage is present in the raw trace collected at the tested frequency F, then these segments should all correspond to the leakage of one Cryptographic Process (CP) execution and have the same shape, so the similarity test should be conclusive. We evaluate segment similarity with an adaptation of Virtual Trigger (VT) [15] as illustrated in Fig. 9.

We refer to this test as *pattern detection*. The algorithm of the pattern detection method is formalized in Algorithm 1. First, a raw trace is collected at the tested frequency f (line 6). This trace is segmented using VT (line 7), it is cut in $Nsegs = 50$ segments, each segment with a size L_{cp} equal to the sampling frequency$\times$the time duration between 2 CPs. The similarity test between the resulting segments is applied (lines 8 to 10). This test can be repeated

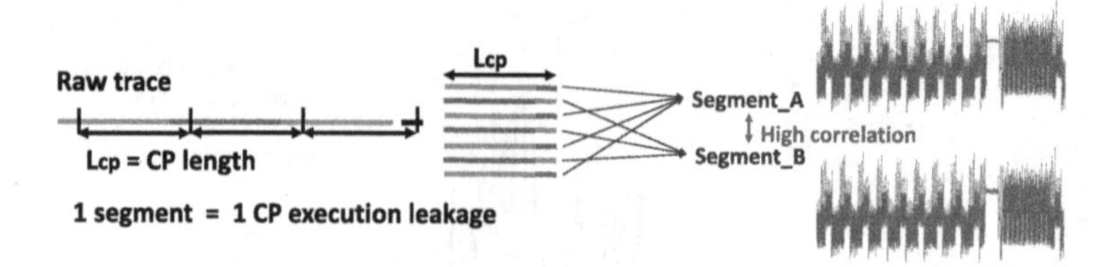

Fig. 9. Pattern detection method based on Virtual Trigger (VT), which consists in segmenting a raw trace by knowing the precise time duration L_{cp} between 2 CPs. If the raw trace is collected at a frequency actually containing leakage, the obtained segments should correspond to leakage coming from the CP and then be similar to each other, i.e., have the same shape.

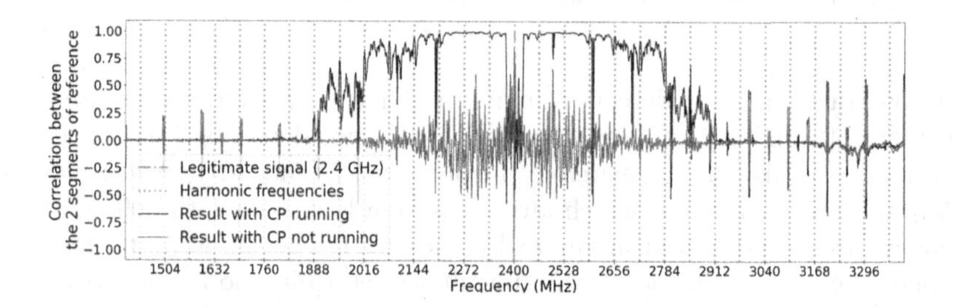

Fig. 10. Presence of leakage over the spectrum: The graph shows the correlation between the segment of references according to the frequency where the signal was collected when CPs are running (Black curve) and not running (Red curve). Green Central dash-dotted line: frequency of the legitimate signal (2.4 GHz). Blue vertical dotted lines: position of harmonic frequencies. (Color figure online)

N_{tests} times (loop from lines 4 to 11) and the results averaged (line 12). In our experiments, N_{tests} is initially set to 10. This method is applied over the same frequency range as in the first method.

Experimental Results and Discussions. The black curve in Fig. 10 corresponds to the results $L_{presence}(f)$ in Algorithm 1. They are expressed as a correlation, which is representative of the similarity between segments. The higher the correlation, the more likely the CP pattern is present. To make sure that, as expected, the correlation is high only at certain frequencies due to the presence of CP leakage, we repeat the same experiment, but now the victim does not perform any CPs. The red curve shows the results of this second test. It can be seen that when CPs are not executed, there are still peaks around the carrier frequency (2.4 GHz) due to digital activity. However, the correlations are much weaker because when segmenting the raw traces the segments obtained are not similar as they did not correspond to CP leakage.

Algorithm 1. Pattern detection for leakage evaluation on a frequency range

Input:
L_{cp} : The precise CP length
N_{segs} : The number of CPs to cut in one Raw Trace
N_{tests} : The number of tests per frequency
F_{start}, F_{stop}, δ_F : Start, Stop and Step Frequencies

Output:
$L_{presence}$: Estimation of the leakage presence at each tested frequency

1: $F \leftarrow F_{start}$
2: **repeat**
3: $S \leftarrow [\,]$ ▷ S: Similarity
4: **for** $i \leftarrow 0$ to N_{tests} **do**
5: The victim device starts executing a series of CPs
6: $RawTrace \leftarrow$ Signal collected by SDR at the tested frequency F
7: $Segs \leftarrow Segmentation$(RawTrace) using VT
8: $Seg_A \leftarrow average(pair\ Segs)$
9: $Seg_B \leftarrow average(odd\ Segs)$
10: $S.append(\rho(Seg_A, Seg_B))$ ▷ Compute correlation ρ
11: **end for**
12: $L_{presence}(F) \leftarrow Mean(S)$
13: $F \leftarrow F + \delta_F$
14: **until** $f >= F_{stop}$
15: **return** $L_{presence}$

To compare both methods, we show their results side by side in Fig. 11. Similarly to the threshold of 4.5 used in the t-test method, we need to set another condition for this second method to consider the detection as positive. The comparison of both methods for a threshold of minimum correlation larger or equal to 0.75 is shown in Fig. 11. We chose this threshold as it corresponds to the highest score of the test in the absence of leakage (red curve). Therefore, any score below this threshold cannot be taken as an indication of the presence of leakage. For both methods, the frequencies where their respective condition is met are highlighted in green. For this selected threshold of 0.75, we found after analyzing the results for each frequency that both methods yield the same results for 93.95% of the tested frequencies.

For this second method, it took 50 min instead of the 27 h in method 1 of Sect. 4.1. It is possible to further reduce this time by reducing the number of similarity tests performed at each frequency. As introduced, N_{tests} was initially set to 10 tests per frequency. We repeated the experiment reducing N_{tests} to only 1. In this case, leakage localization took only 15 min without significantly altering the results, and 94.78% of the detected frequencies with N_{tests}=10 were still detected with N_{tests}=1.

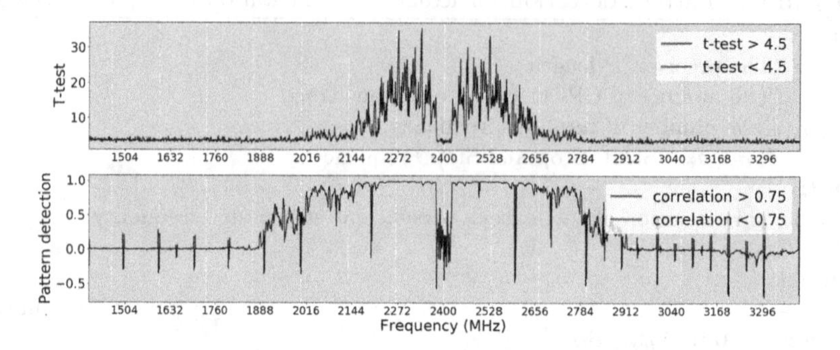

Fig. 11. Comparison of the results of the two methods: The upper graph shows frequencies where *t-test score* > 4.5 (green underlined), hence those carrying leakage as introduced in Sect. 4.1. The lower graph shows the results for the pattern detection method where *correlation threshold* > 0.75 . The results obtained using the pattern detection method with this threshold are the same for 93.95% of the tested frequencies. (Color figure online)

5 Attacking at Non-harmonics

In Sect. 4, we demonstrated that the leakage is also present at non-harmonics frequencies. This brings another question: **how efficient are attacks at these non-harmonic frequencies?** To answer the question, the attack is performed on a part of the spectrum where the leakage localization methods gave the best results. To keep the experimentation phase tractable, experiments only cover the right-hand side of the spectrum (i.e., positive) with respect to the legitimate signal. Indeed, the objective of the experiment is not to evaluate all possible frequencies on this particular type of device but to determine if an attack is possible at non-harmonic frequencies and to compare their performance to attacks at harmonic frequencies. The experiment covers a range of frequencies from 2.45 GHz to 2.6 GHz, and attacks are centered at 150 different frequencies, with 1 MHz steps. One may note that the attack would probably also be possible at other frequencies, including the one on the left part of the spectrum. However, as indicated, we focus only on the right half side to keep the experimentation phase in a reasonable time.

5.1 The Attack and Score

In our experiments, we run profiled correlation attacks [11], where the attacker has access to a similar device as the victim. This enables the attacker to build a profile for this type of device and learn the leakage behavior. During the attack phase, for each key byte, an assumption (i.e., hypothesis) is made on their value. The 256 possible hypotheses are tested, each getting a probability to be the correct one based on the correlation between the estimated leakage using the profile on one side, and the real leakage produced by the victim on the other. The hypothesis giving the highest probability is assumed to be the correct one.

In many cases, some bytes are incorrectly guessed, but it is still possible to brute-force the correct key. A brute-force approach [22] consists in testing the ranked keys from the most probable, according to the probabilities computed during the attack, until finding the right one. The number of keys tested by the brute-force algorithm before reaching the correct one is the *Key Rank* and is representative of the complexity of recovering the key. The lower the key rank is, the better the attack performs, as the brute-force attack needs less time to reach the correct key. When the key rank is lower than 2^{32}, it takes about 5 minutes on the experimental computer to brute-force the key. When lower than 2^{35}, the brute-force takes about 1 h. In the remainder of this paper, this key rank is kept as the criteria to evaluate the efficiency of an attack as in previous works [6].

5.2 Experiment and Results

In this first experiment, the victim and attacker are still connected by a cable (Fig. 3a). Two sets of 15000×10 traces are collected at each frequency, one to build the profile and the other to test the attack. The collection phase takes 4 days. For each attack, we compute the key rank and show the results in Fig. 12a. The experiment confirms that the attack is not only succeeding at the harmonics as expected but also at many other frequencies. Then, this finding increases the number of potential frequencies to use to succeed in the attack. The key rank is lower than 2^{32} at the 3 harmonics as they are not polluted. Among the 147 non-harmonics, 105 have a key rank lower than 2^{32} and 12 lower than 2^{35}.

As the experiment is fully automated, it is important to notice that a very high key rank at a given frequency does not ensure the attack is necessarily more difficult there. But it can be tried to make it work better by putting more effort into it, which is not our concern here.

6 Attacking in Challenging Conditions

After demonstrating attacks are also possible at frequencies other than the harmonics, we investigate how useful this finding is in a noisy environment when attacking at a distance. The questions targeted by this second experiment are the following: **In a noisy environment where harmonics are polluted, can non-harmonic frequencies keep the attack feasible? Can the attack be better at non-harmonic frequencies than at harmonic frequencies**

In the experiments presented in this section, the leakage is collected using the antenna and the setup shown in Fig. 3b. Compared to the first experiment we increase time diversity from 10 to 50, but collect the same number of traces at each frequency (15000×50). The patterns used for the collection phase and profiles used for the attack are the ones that were built in Sect. 5.2. The key rank is kept as the attack score.

6.1 Attacking at a Distance in a Noisy Environment

A first test is performed with the antenna at 2 m. In these conditions, the collection phase takes six days. Figure 12b shows the results. As expected, we can observe how the noisy environment reduces the number of exploitable frequencies. This is particularly visible around the first harmonic at 2.464 GHz, where WiFi and Bluetooth signals are present. Among the 150 frequencies, the rank is lower than 2^{32} only at 2 harmonics, as the first one is polluted. Among the 147 non-harmonics, 78 have a rank lower than 2^{32} and 12 lower than 2^{35}.

6.2 Attacking with Fewer Traces

A common goal of side-channel attacks is to succeed with as few traces as possible. We re-computed the attack with the traces collected at 2 m but reduced the number of traces per attack to 750 × 50 (from 15000 × 50). We were then able to set up (15000/750 = 20 attacks at each frequency). The experiment provides the results shown in Fig. 12c After sorting the 150 frequencies according to the average of their scores (average of the 20 log2 (key rank)), the harmonics ranked at the 3rd, 7th and 121st place. This proves that some non-harmonics can even get better scores than harmonic frequencies.

6.3 Attacking at a Further Distance

As screaming-channel attacks try to enable attacks where attackers are as far as possible from the victim, we run a new round of attack experiments with the antenna put at a distance of 7 meters. This time 50[6] attacks are performed under the exact same conditions at each frequency. To keep the experiments feasible in a reasonable time, we reduced the number of tested frequencies. The frequencies selected for this attack at 7 meters are chosen among the ones where the attack performed best in the 2-meters scenario using 750 traces. The results of the attacks at 7 meters are shown in Fig. 12d. In this case, there is **only one harmonic that still gets a key rank lower than 2^{35}, while most of the selected non-harmonics still work**. Among the 9 which were selected, 5 have a lower rank than 2^{32} and 2 than 2^{35}.

Figure 13 shows the results of the same attacks, but it focuses on the evolution of the key rank according to the number of traces used. Again, we keep the average of the 50 log2 (key rank) as the attack score for each frequency. When using the same harmonic as in previous works [6,7,27] (the second one at 2.528 GHz), the key rank decreases but very slowly. Then in our case, to get a rank lower than 2^{35} using this frequency, it is necessary to collect up to 30286 × 50 traces. These results show how allowing to search for leakage at frequencies other than the harmonics considerably reduces the number of traces needed to get the same results. The best non-harmonic frequency, at 2.484 GHz, needs only 65 × 50 traces to get this result, which is even better than the best harmonic at 2.592 GHz, where the number of traces required is 166 × 50.

[6] 50 is the minimal number usually considered by the side-channel community for statistically meaningful results.

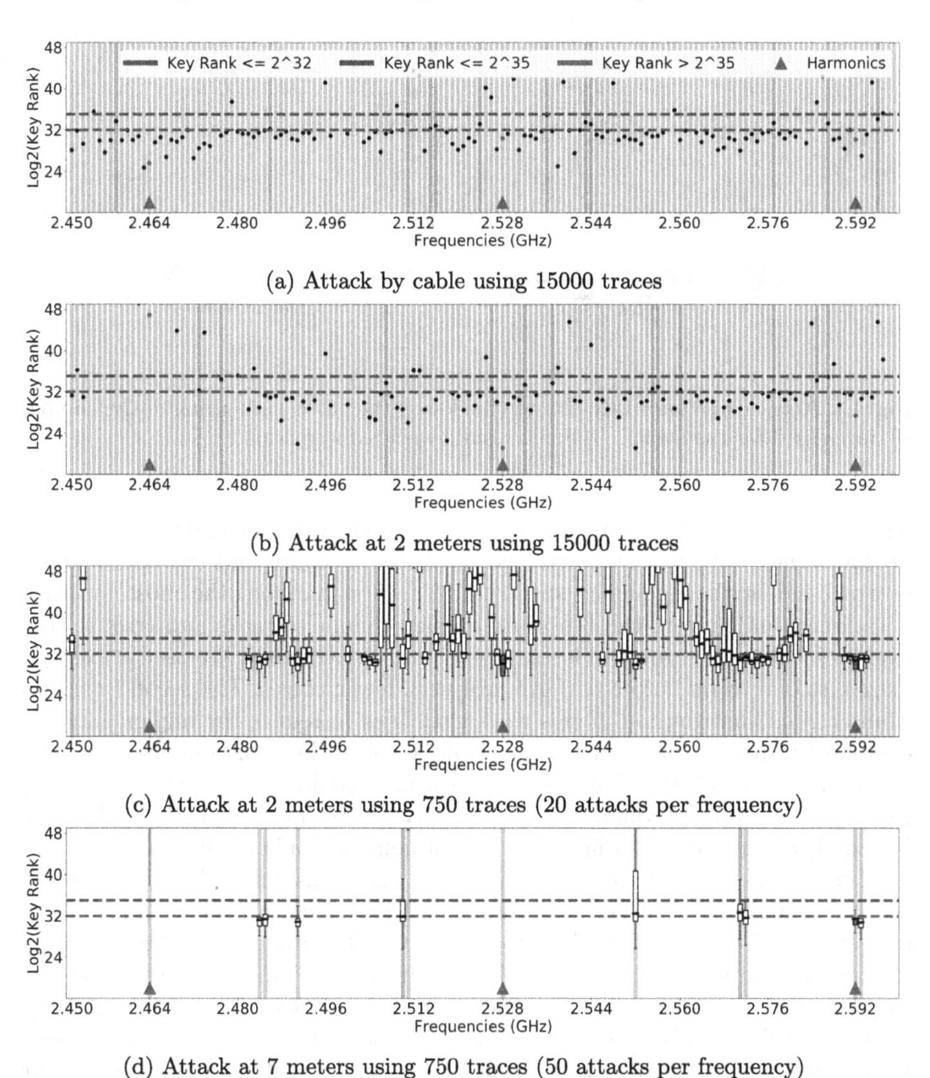

(a) Attack by cable using 15000 traces

(b) Attack at 2 meters using 15000 traces

(c) Attack at 2 meters using 750 traces (20 attacks per frequency)

(d) Attack at 7 meters using 750 traces (50 attacks per frequency)

Fig. 12. Experimental results: Key Rank (the lower, the better). Underlying colors per frequency indicate key ranks of: green <= to 2^{32} (approx. 5 mins brute-force), blue <= to 2^{35} (approx. 1 hour brute-fore), red > to 2^{35}. For (c) and (d), the graph shows: (1) a central box with the distribution of scores from the first quartile to the third; (2) whiskers with the extension of the remaining scores by 1.5× the inter-quartile range (equal to the difference between the third and first quartile); and (3) a horizontal line with the median score of the attacks. (Color figure online)

7 Discussion and Conclusion

This work defied the assumption that screaming-channel attacks perform best (or only) at harmonics of the digital processing clock of the victim, frequencies

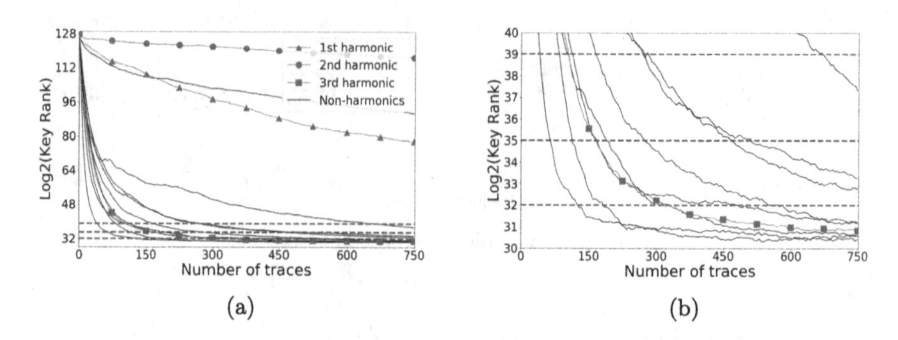

Fig. 13. Attack at 7 m: a) Key Rank (the lower, the better) according to the number of traces used for the attack. b) Same results with reduced scale.

where the leakage was so far supposed to be present with the highest amplitude. To investigate this, we proposed two methods to locate and evaluate leakage over a band of the frequency spectrum. The first method, the most intuitive and direct, builds from the literature on side-channel attacks and tries to find exploitable leakage through a t-test at each tested frequency. We used a fixed vs. fixed test due to its better performance with fewer traces. The second method is an original contribution of this work that tries to reduce the implementation complexity and the processing time while keeping the same quality of results as the first method, which uses a standard methodology accepted by the side-channel community. Exploiting these two methods, we demonstrate that the leakage is also present at a large amount of non-harmonic frequencies.

The presence of leakage at non-harmonics is consistent with previous studies [21], demonstrating that when the digital part creates noise at a given frequency, and as this noise travels through the CMOS substrate, the latter acts as a filter that spreads the noise over a wider frequency band. As a consequence, the noise can be found on the RF side at frequencies other than the harmonics.

We considered only one type of device, the same used by previous works on screaming-channel attacks, that is still available off-the-shelf at the time of our study. The present study does not prove that leakage will always be present at non-harmonics on any other device. However, it highlights the fact that leakage presence has to be checked there, too, as it is possible to find it at these frequencies, even if stronger peaks at harmonics give the intuitive idea that leakage would appear mainly there. This is exactly what we have proved in this work.

This study also demonstrates how this phenomenon can make attacks feasible in cases where all exploitable harmonics are polluted by interfering signals, as is the case in more realistic, real-life scenarios. The studied phenomenon can also reduce the number of traces needed for the attack. Compared with the performance of the attack at the best harmonic, using the best non-harmonic enables to reduce by 60% the number of traces needed to get a key rank under 2^{35}.

In future works, it could be interesting to detect the best frequencies first and then focus the efforts only on them. For example, by building better profiles: in our work, in order to build profiles at a large number (150) of frequencies, these profiles were built with only 150K traces (15000×10), which is relatively small compared with previous works (1 to 5M traces per profile). One can also extend the range of attacks to a distance where no harmonic gives a reasonable key rank (for example, superior to 2^{39}) with a given maximum number of traces and observe how many meters a non-harmonic attack can gain.

Acknowledgment. We want to acknowledge the reviewers of the current and previous versions of this paper, as well as Dr. Maria Méndez Real and Dr. Dennis Gnad for their constructive feedback.

References

1. Adamczyk, B.: Foundations of Electromagnetic Compatibility: With Practical Applications. Wiley, Hoboken (2017)
2. Afzali-Kusha, A., Nagata, M., Verghese, N.K., Allstot, D.J.: Substrate noise coupling in SoC design: modeling, avoidance, and validation. Proc. IEEE **94**(12), 2109–2138 (2006)
3. Agrawal, D., Archambeault, B., Rao, J.R., Rohatgi, P.: The EM Side—Channel(s). In: Kaliski, B.S., Koç, ç.K., Paar, C. (eds.) Cryptographic Hardware and Embedded Systems-CHES 2002. CHES 2022, LNCS, vol. 252, pp. 29–45. Springer, Heidelberg (2003). https://doi.org/10.1007/3-540-36400-5_4
4. Agrawal, D., Rao, J.R., Rohatgi, P.: Multi-channel attacks. In: Walter, C.D., Koç, Ç.K., Paar, C. (eds.) CHES 2003. LNCS, vol. 2779, pp. 2–16. Springer, Heidelberg (2003). https://doi.org/10.1007/978-3-540-45238-6_2
5. Brier, E., Clavier, C., Olivier, F.: Correlation power analysis with a leakage model. In: Joye, M., Quisquater, J.J. (eds.) CHES 2004. LNCS, vol. 3156, pp. 16–29. Springer, Heidelberg (2004). https://doi.org/10.1007/978-3-540-28632-5_2
6. Camurati, G., Francillon, A., Standaert, F.X.: Understanding screaming channels: from a detailed analysis to improved attacks. IACR Trans. Cryptograph. Hardware Embed. Syst. 358–401 (2020)
7. Camurati, G., Poeplau, S., Muench, M., Hayes, T., Francillon, A.: Screaming channels: when electromagnetic side channels meet radio transceivers. In: ACM Conference on Computer and Communications Security, pp. 163–177 (2018)
8. Chari, S., Rao, J.R., Rohatgi, P.: Template Attacks. In: Kaliski, B.S., Koç, Ç.K., Paar, C. (eds.) CHES 2002. LNCS, vol. 2523, pp. 13–28. Springer, Heidelberg (2003). https://doi.org/10.1007/3-540-36400-5_3
9. Choi, J., Yang, H.Y., Cho, D.H.: TEMPEST comeback: a realistic audio eavesdropping threat on mixed-signal SoCs. In: Proceedings of the 2020 ACM SIGSAC Conference on Computer and Communications Security (2020)
10. Dessouky, G., Sadeghi, A.R., Zeitouni, S.: SoK: secure FPGA multi-tenancy in the cloud: challenges and opportunities. In: IEEE EuroS&P, pp. 487–506 (2021)
11. Durvaux, F., Standaert, F.X.: From improved leakage detection to the detection of points of interests in leakage traces. In: Fischlin, M., Coron, J.S. (eds.) EUROCRYPT 2016. LNCS, vol. 9665, pp. 240–262. Springer, Heidelberg (2016). https://doi.org/10.1007/978-3-662-49890-3_10

12. Gandolfi, K., Mourtel, C., Olivier, F.: Electromagnetic analysis: concrete results. In: Koç, Ç.K., Naccache, D., Paar, C. (eds.) CHES 2001. LNCS, vol. 2162, pp. 251–261. Springer, Heidelberg (2001). https://doi.org/10.1007/3-540-44709-1_21
13. Gierlichs, B., Batina, L., Tuyls, P., Preneel, B.: Mutual information analysis: a generic side-channel distinguisher. In: Oswald, E., Rohatgi, P. (eds.) CHES 2008. LNCS, vol. 5154, pp. 426–442. Springer, Heidelberg (2008). https://doi.org/10.1007/978-3-540-85053-3_27
14. Gilbert Goodwill, B.J., Jaffe, J., Rohatgi, P.: A testing methodology for side-channel resistance validation. In: NIST Non-Invasive Attack Testing Workshop, vol. 7, pp. 115–136 (2011)
15. Guillaume, J., Pelcat, M., Nafkha, A., Salvador, R.: Virtual triggering: a technique to segment cryptographic processes in side-channel traces. In: 2022 IEEE Workshop on Signal Processing Systems (SiPS), pp. 1–6. IEEE (2022)
16. Kocher, P., Ja, J.: Differential power analysis. In: Wiener, M. (ed.) CRYPTO 1999. LNCS, vol. 1666, pp. 388–397. Springer, Heidelberg (1999). https://doi.org/10.1007/3-540-48405-1_25
17. Le, J., Hanken, C., Held, M., Hagedorn, M.S., Mayaram, K., Fiez, T.S.: Experimental characterization and analysis of an asynchronous approach for reduction of substrate noise in digital circuitry. IEEE Trans. Very Large Scale Integr. (VLSI) Syst. **20**(2), 344–356 (2011)
18. Mangard, S., Oswald, E., Popp, T.: Power Analysis Attacks: Revealing the Secrets of Smart Cards, vol. 31. Springer, New York (2008). https://doi.org/10.1007/978-0-387-38162-6
19. Masure, L., Dumas, C., Prouff, E.: A comprehensive study of deep learning for side-channel analysis. IACR Trans. Cryptograph. Hardware Embed. Syst. 3488–375 (2019)
20. Mohamed, C., Barelaud, B., Ngoya, E.: Physical analysis of substrate noise coupling in mixed circuits in SoC technology. In: The 5th European Microwave Integrated Circuits Conference, pp. 274–277. IEEE (2010)
21. Noulis, T., Baumgartner, P.: CMOS substrate coupling modeling and analysis flow for submicron SoC design. Analog Integr. Circ. Sig. Process **90**, 477–485 (2017)
22. Poussier, R., Standaert, F.X., Grosso, V.: Simple key enumeration (and rank estimation) using histograms: an integrated approach. In: Gierlichs, B., Poschmann, A. (eds.) CHES 2016. LNCS, vol. 9813, pp. 61–81. Springer, Heidelberg (2016). https://doi.org/10.1007/978-3-662-53140-2_4
23. Rhee, W., Jenkins, K.A., Liobe, J., Ainspan, H.: Experimental analysis of substrate noise effect on PLL performance. IEEE Trans. Circ. Syst. II Express Briefs **55**(7), 638–642 (2008)
24. Schellenberg, F., Gnad, D.R.E., Moradi, A., Tahoori, M.B.: An inside job: remote power analysis attacks on FPGAs. In: 2018 Design, Automation & Test in Europe Conference & Exhibition (DATE), p. 6 (2018)
25. Schneider, T., Moradi, A.: Leakage assessment methodology: extended version. J. Cryptogr. Eng. **6**, 85–99 (2016)
26. Standaert, F.X.: Introduction to side-channel attacks. Secure Integr. Circ. Syst. 27–42 (2010)
27. Wang, R., Wang, H., Dubrova, E.: Far field EM side-channel attack on AES using deep learning. In: 4th ACM Workshop on Attacks and Solutions in Hardware Security, pp. 35–44 (2020)

Bernoulli at the Root of Horizontal Side Channel Attacks

Gauthier Cler[1,2](✉), Sebastien Ordas[2], and Philippe Maurine[1]

[1] University of Montpellier LIRMM, Montpellier, France
{gcler,pmaurine}@lirmm.fr
[2] SERMA Safety & Security ITSEF, Pessac, France
{g.cler,s.ordas}@serma.com

Abstract. Vertical side-channel attacks represent a major threat to the confidentiality of enclosed secrets in hardware devices. Fortunately, countermeasures such as blinding or masking are nowadays quasi-systematically used to protect implementations of asymmetric cryptographic algorithms (RSA, ECDSA). Horizontal attacks thus constitute an interesting alternative for adversaries. They aim at recovering the secret exponent or scalar using a single trace, thus bypassing the blinding countermeasure. Several attacks have been proposed, based for instance on statistical distinguisher or clustering techniques. However, the success of these attacks relies heavily on the selection of Points of Interest (PoI) carrying leakage, extracted from acquired signals.

In this context, this work aims at providing a framework for the selection of PoI in the context of noisy traces. It is based on statistical tests applied to the distribution of each point; these tests have been derived from the analysis of noise impact on distributions. Experiments performed with this framework emphasize a gap reduction in terms of attack success rates between unsupervised and supervised attacks.

Keywords: Security · Side Channel Analysis · Unsupervised · Horizontal Attacks

1 Introduction

1.1 Background

Asymmetric (or public-key) cryptography algorithms are frequently used in smart cards and embedded systems to ensure secured transmissions. The most frequent ones are probably RSA [19] and ECDSA [11]. As with any cryptographic algorithm, they can be targeted by physical attacks like side-channel attacks (SCA) which aim at recovering the private exponent or the scalar used in the deciphering phase. Several attacks have been proposed in the literature, either profiled [5,18] or unprofiled [4,12,13]. However, blinding countermeasures of the exponent or scalar renders multiple traces aggregation-based attacks, also so called vertical attacks, ineffective.

S. Bhasin and T. Roche (Eds.): CARDIS 2023, LNCS 14530, pp. 107–126, 2024.
https://doi.org/10.1007/978-3-031-54409-5_6

As a consequence, Horizontal attacks have been proposed and have been demonstrated as a relevant alternative especially when dealing with low noise traces. They aim at recovering the secret exponent or scalar using a single trace. Several approaches have been proposed in the literature. Among them, one can identify those based on a statistical distinguisher such as the correlation [3,6], or those based on clustering techniques [10,15–17].

Because only one trace is used, the efficiency of Horizontal attacks is extremely sensitive to noise in traces. This is all the more true as the minimal bit recovery rate required to disclose the secret exponent using recovery methods such as [20] is very high.

Their efficiency also depends on the number of informative points in exponentiation or scalar multiplication patterns. This number is usually low in robust implementations and attacks fail in providing a high enough bit recovery rate of the randomized secret. To solve this issue two approaches are usually considered. Both aim at reducing the dimensionality of the problem either by considering only a few components from linear transformations like PCA [10] or by selecting temporal points of interest (PoI) from some heuristics [15–17] based on a priori leakage model. In the end, using only the selected points of interest, the goal is to classify in k (usually 2) classes all exponentiation/multiplication patterns often using a clustering algorithm like the k-means, the expectation-maximization algorithm, etc.

While the clustering methods based on [17] are still relevant, some issues can be addressed regarding the considered leakage model. Indeed, all the approaches mentioned above analyze all points, taking into account the fact that they derive from a two-class phenomenon. Although an ideal leaking point theoretically displays two components (two different behaviors according to the target implementation), it is not necessarily true in practice. Indeed, the observed distribution of a point can include additional components coming from different behaviors or from noise sources. Ignoring these additional behaviors or noise sources results in biased estimations when selecting PoI. Therefore, restricting the analysis of points to a two-class problem may lead to irrelevant results.

Furthermore, as acquired signals are the result of a physical phenomenon over time, neighboring points are more likely to share common information (if the sampling rate is high enough) about the processed data than distant points. This should also be considered during the analysis even if the initial analysis is done in a univariate manner.

1.2 Contributions

Within this context, we propose an alternative technique for the selection of PoI, based on a common leakage model from which a statistical test is progressively derived. The paper also explains how to adapt this test to real-life scenarios characterized by significant noise and outliers in the measurements. Eventually, it is shown how to reinforce a set of PoI, selected using the aforementioned test, by considering the eventual monotonic transformations of their statistical distribution. The resulting procedure for PoI selection is assessed on several sets of real traces including public datasets. Resulting attacks are compared to the state of the art methods and to a supervised PoI selection, considered as an upper bound of what can be achieved in an unsupervised way.

1.3 Paper Organization

The rest of the paper is organized as follows. Section 2 presents the related works on Horizontal attacks. It emphasizes the main issues with the state of the art implementations regarding the context and explains briefly the reasoning that led to the proposed contribution. Section 3 introduces the proposed methodology derived from a common leakage model. In Sect. 4, the latter is then adapted to real-life scenarios, this is to say to much noisier traces. Section 5 proposes a way to extend the set of identified PoI by integrating neighboring points with high Spearman rank correlation. The relevance of the proposed approach is then assessed through several experiments in Sect. 6. This includes the definition of targets and measurement conditions, as well as the chosen settings of parameters. The results of these experiments are then discussed and compared with other known approaches. Finally, Sect. 7 gives an overall conclusion about this work and suggests improvements and possible future work.

2 Related Works and Remarks

Horizontal attacks are an important topic in the context of black box side-channel analysis. Since the introduction of the first horizontal attack, namely the Single Power Analysis (SPA) [13], they have been gradually improved to cope with the countermeasures deployed by integrated circuit designers. To prevent timing attacks, regular and constant time exponentiation algorithms such as the one described in [7] have been adopted. Then, after the publication of the first vertical attacks, namely the DPA [12] and CPA [4], they enhanced their regular algorithm implementations with additional countermeasures to prevent these attacks. Regarding asymmetric algorithms implementations, the most common countermeasures might be the scalar/exponent blinding and the message blinding [8,13]. The combination of several of these countermeasures usually makes unsupervised side-channel attacks based on the statistical analysis of thousands of traces ineffective.

Therefore, adversaries have adapted their practice and more sophisticated Horizontal attacks have emerged. Such attacks try to exploit the leakage contained in a single trace to recover the secret exponent or scalar. Several approaches have been proposed. Among them, some are based on a statistical distinguisher like [6]. Others use clustering methods like [10,17]. However, patterns related to the processing of one bit of the exponent (or the scalar) are usually long in length and may contain very few informative points. They are thus usually not exploitable as they are. To cope with this problem, two approaches are typically used.

A first solution consists in reducing the dimension of the problem using for instance a principal component analysis (PCA) as in [2]. It allows to drastically reduce the size of patterns by projecting the actual points in a lower dimensional subspace. Such an approach however assumes that the leakage is characterized by a high variance. This is not always true, especially in the case of very noisy patterns. In that case, PCA often fails at extracting relevant components regarding the secret and rather highlights the measurement noise.

Alternatively, authors of [17] proposed a univariate and unsupervised PoI selection method. Each temporal point across all patterns is first classified using a clustering algorithm in two clusters. Then the difference of means (DoM) of each cluster is calculated to select an arbitrarily chosen number of points with the highest DoM. The attack carries on by clustering patterns by only considering the selected PoI. Eventually, a majority vote (or another statistical distinguisher) is applied to identify the secret bits. Following this publication, other works suggested to consider additional criteria to improve the attack. For instance, authors of [15] demonstrated the interest in considering the relative weighting of the clusters while [14] used multidimensional clustering algorithms.

Eventually, authors of [16] proposed an improvement of [17] by applying an error correction method after clustering, using a Deep Learning iterative framework.

In the end, frameworks based on [17] still represent as of today a pertinent approach in the context of Horizontal attacks. Still, in more realistic attack scenarios, where acquired signals are highly polluted, they could fail to apply properly.

Firstly, as mentioned above, the actual methods for the selection of PoI assume that the number of classes in point distribution is always the same (usually two) as expected from the targeted algorithm analysis. This is however not always true as in the presence of artifacts (that could be caused by charge pumps, noise, or outliers during the signal acquisition), the number of components in points distribution could be different from the theoretically awaited ones. This could thus potentially result in biased leakage estimations during the PoI selection. In the exploitation of highly noisy signals, these supplementary undesired behaviors should be taken into consideration for the analysis of PoI.

In addition, the performed analysis is carried out in a univariate manner (analysis of each temporal point across all patterns) and it would be nontrivial to extrapolate this multidimensional PoI analysis using a similar approach as the space complexity grows very fast. Still, after performing the univariate PoI selection, it would be relevant to expand this extracted set of PoI in order to make it more robust for the attack phase.

With these remarks in mind, we propose in the next sections, starting from the theoretical leakage model usually considered for the analysis of PoI, to derive a more realistic model taking into account additional behavior that could be present in acquired signals. From this adapted leakage model, we then propose a way to enrich the selected PoI set, considering neighboring temporal points if they display suitable properties.

3 A Bernoulli Distribution Based Leakage Model

3.1 Notation

Capital letters are used to denote random variables (for instance X) and lowercase letters x for their realization according to their distribution. $\Pr(Y|X)$ is the conditional probability of Y given X and $\Pr(X,Y)$ is the joint probability of X and Y. For statistical testing, α denotes the significance level.

A matrix of n rows and p columns is defined as $\mathbf{M} \in \mathbb{R}_{n \times p}$ such that $m_{i,*}$ denotes the ith row and $m_{*,i}$ the ith column, unless specified explicitly.

A dataset consisting of one or several signal traces acquisitions is denoted as matrix $\mathcal{D}$. Each row represents a single pattern of exponentiation (or scalar multiplication). $\mathcal{D}$ is associated with a vector of labels $\mathcal{Y} \in \mathbb{N}_n$ giving the class associated with each pattern. $\mathcal{D}$ can also be viewed as a vector of p random variables $[X_1, \ldots, X_p]$, each one following its own statistical distribution.

3.2 Leakage Model

In supervised SCA (knowing the classes beforehand), leakage assessment for the selection of points of interest (PoI) is usually done considering the univariate leakage of each temporal point individually. Many metrics can be used such as the Welch t-test (or the ANOVA/NICV for multi class probelems), mutual information or correlations. The goal is to select as PoI the points with the highest disparities between the different classes ectionfor instance points with the greatest difference of means or inter-class variances. Since the context is supervised, the weight of each class is not a concern.

However, in an unsupervised context, the problem is different since the adversary has no clear insight about the actual distribution of the data or the weights of its components. Thus, one or several assumptions should be made about the data distribution. These assumptions can usually be formulated from the knowledge of the attacked algorithm and potential countermeasures used to protect its implementation. This is usually sufficient to fix the number of classes to identify and discriminate as well as the weights (or repartition) $0 \leq \pi_i \leq 1$ of each component forming the overall distribution prior to any analysis.

In a context where an adversary attacks a regular exponentiation or scalar multiplication algorithm (or any regular algorithm) in which the number of classes is $k = 2$, the unknown vector of labels $\mathcal{Y}$ is expected (especially if the scalar or exponent is randomized) to be distributed as a succession of n Bernoulli trials:

$$Bern(\pi_0) = \pi_0^b (1 - \pi_0)^{1-b} \quad (1)$$

with $b \in \{0, 1\}$ and, in our case, $\pi_0 = 0.5$ the probability of an exponent or scalar bit to be ectionequal to 1. Thus $\mathcal{Y}$ follows a binomial distribution $B(\pi_0, n)$ with parameters π_0 and n which probability mass function ectionis:

$$pmf(b) = \binom{n}{h} \pi_0^b (1 - \pi_0)^{n-b} \quad (2)$$

with $b \in \{0, 1, \ldots, n\}$ the number of exponent or scalar bits equal to 1 and $\binom{n}{b}$ the binomial coefficient.

At that point, it is straightforward to state that an X_i of $\mathcal{D}$ probably carries information about an exponent or scalar bit, and is thus a PoI, if its observed distribution is binomial. The following statistical test can even be carried out to identify such X_i:

$$\begin{cases} H_0: & \mathcal{Y} \sim B(\pi_0, n) \\ H_1: & \mathcal{Y} \nsim B(\pi_0, n) \end{cases}$$

It allows assessing if a sample of n values in $\{0; 1\}$ actually follows a binomial distribution with parameter π_0. The statistic S of the test is the number of 1 in the sample and H_0 is accepted if $\alpha \leq pmf(S) \leq (1 - \alpha)$, with α the p-value of the test.

However, this test cannot be applied directly to X_i since they are continuous random variables (at least because of the Gaussian measurement noise) which are discretized into 8 to 16 bits according to the vertical resolution of the digital sampling oscilloscope used for measurements. ectionThus, such a procedure could only be applied after application of a clustering process with $k = 2$, transforming the discretized continuous random variables, X_i, into a binary random variable $\mathcal{Y}_i$. Alternatively, this can be also done by modeling the related distributions as Gaussian mixtures as explained below.

4 Identification of PoI

4.1 From Bernoulli Distribution to Gaussian Mixtures

Because we are dealing with continuous physical phenomenons discretized into 8 to 16 bits along with measurement Gaussian noise, the distribution of each X_i can be redefined as a Gaussian mixture. Define $K = \{1, 2\}$ the set of possible class values. Define $\boldsymbol{\theta} = \{\mu_i, \sigma_i^2, \pi_i\}_{i=1}^{k}$ the set of mixture parameters to estimate, with π_i the weight of each component i among the mixture s.t. $\sum_{i \in K} \pi_i = 1$, μ_i and σ_i^2 the corresponding mean and variance of the components. Define $\mathbf{x} = (x_1, \ldots, x_n)$ the set of observed data for each column following distribution X_i. Similarly, define $\mathbf{z} = (z_1, \ldots, z_n)$ as a vector of latent unobserved variables, indicating the membership of each observed data to its component. The probability density function of the mixture model according to parameters $\boldsymbol{\theta}$ is given then by:

$$\begin{aligned} X_i \sim f(x, \boldsymbol{\theta}) &= \sum_{i \in K} \pi_i \mathcal{N}(x|\mu_i, \sigma_i^2) \\ &= \sum_{i \in K} \pi_i \frac{1}{\sigma_i \sqrt{2\pi}} e^{-\frac{1}{2}\left(\frac{x - \mu_i}{\sigma_i}\right)^2} \end{aligned} \tag{3}$$

where $\mathcal{N}$ denotes the normal distribution function. The associated likelihood function is given by:

$$L(\mathbf{x}, \mathbf{z}; \boldsymbol{\theta}) = \Pr(\mathbf{x}, \mathbf{z}|\boldsymbol{\theta}) \tag{4}$$

The estimation of mixture parameters can be done following the Expectation Maximization (EM) process, as presented in previous works [14,15]. EM ection-aims at finding the Maximum Likelihood Estimate (MLE) by maximizing the marginal likelihood $L(\mathbf{x};\boldsymbol{\theta}) = \Pr(\mathbf{x}|\boldsymbol{\theta})$. This is done by iteratively applying two steps:

1. Compute the expectation using current estimated parameters $\boldsymbol{\theta}^{(t)}$ (initialized randomly at first) and actual conditional distribution $\mathbf{z}|\mathbf{x},\boldsymbol{\theta}^{(t)}$:

$$E(\boldsymbol{\theta}|\boldsymbol{\theta}^{(t)}) = \mathbb{E}_{\mathbf{z}|\mathbf{x},\boldsymbol{\theta}^{(t)}}\left[\log L(\mathbf{x},\mathbf{z};\boldsymbol{\theta})\right]$$

2. Find and update parameters that maximize the computed expectation:

$$\boldsymbol{\theta}^{(t+1)} = \arg\max_{\boldsymbol{\theta}} E(\boldsymbol{\theta}|\boldsymbol{\theta}^{(t)})$$

These subsequent steps are applied until convergence. This process reformulates the aim of the unsupervised PoI selection into the identification of points being Gaussian mixtures with parameters π_i consistent with the theoretical awaited distribution imposed by the Bernoulli process. In our case, this resumes in finding points being mixtures of two Gaussian distributions with parameter $\pi_1 = \pi_2$ passing the Binomial test described in Sect. 3.2.

4.2 Homogeneous Effect of Gaussian Noise

After the selection of PoI, done with the help of the Binomial test, it is possible to reject additional points by considering another criterion to further analyze the selected Gaussian mixtures. This step is not mandatory and can be viewed as a refinement to take into account the effect of Gaussian noise. This additional criterion is related to the variance of the mixture components. Indeed, because the Gaussian measurement noise affects both similarly and significantly both components, their variance is expected to be similar. As a result, a high disparity of variances should be perceived as a behavior not related to the exponent or scalar. The disparity of variance can be assessed using the F-test of equality of variances where the null hypothesis is given by $H_0 : \sigma_1^2 = \sigma_2^2$ and the associated statistic of the test is:

$$Z = \frac{\sigma_1^2}{\sigma_2^2} \tag{5}$$

H_0 is accepted if $q(\frac{\alpha}{2}) \leq Z \leq q(1-\frac{\alpha}{2})$, where quantiles q are calculated with corresponding F-distribution $F(n\pi_1 - 1, n\pi_2 - 1)$. This thus provides a refinement of the set of PoI by excluding points being relevant in terms of weights of their components (passing the Binomial test) but irrelevant in terms of relative variance.

4.3 Ranking PoI

After the PoI selection is done using the Binomial and F tests, one should want to rank them versus their interest or exploitability. This can be achieved by assessing the clustering ease of their components. To that aim, one can compute the probability of membership of each value $x \in \mathbf{x}$ to each component using estimated mixture model parameters by applying Bayes rule:

$$\forall x \in \mathbf{x}, \forall i \in K : \Pr(i|x) = \frac{\Pr(i)\Pr(x|i)}{\Pr(x)} = \frac{\Pr(i)\Pr(x|i)}{\sum_{j \in K}\Pr(j)\Pr(x|j)} \equiv \frac{\pi_i \mathcal{N}(x|\mu_i, \sigma_i^2)}{\sum_{j \in K}\pi_j \mathcal{N}(x|\mu_j, \sigma_j^2)} \tag{6}$$

This allows building the function $g(x) = \min_{i \in K}\{\Pr(i|\mathbf{x} = x)\}$ emphasizing the interval where the class ownership is uncertain. Finally, the area of uncertainty, or confusion probability, P_{conf}, can be calculated:

$$P_{conf} = \int_{-\infty}^{\infty} g(x)dx \tag{7}$$

This metric takes values in $[0, 1]$ where 0 denotes the perfect separation of the components and 1 their complete overlapping. It thus gives an insight about the ease of classification of each univariate distribution and allows to rank PoI. Obtained probability as defined in Eq. 6 for each possible value $x \in \mathbf{x}$ as well as deducted P_{conf} from Eq. 7 can be visualized for simulated univariate data on Fig. 1.

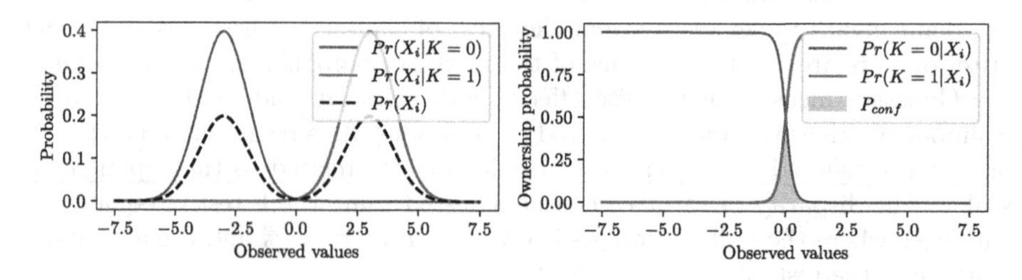

Fig. 1. Illustration of the probability density function of mixture components (left) and ownership probability along with P_{conf} area (right).

4.4 Additional Components Due to Outlier Behaviors and Noise

In a context where patterns, acquired with a digital sampling oscilloscope, are polluted by outliers or partially saturated because of a bad tuning of the vertical caliber, the previously described methodology fails to apply because such pollution results in the apparition of the parasitic components among the two searched ones. In this case, the analysis should be adapted to include these ℓ parasitic components, denoted as the set $L = \{k, \dots, k + \ell\}$. Otherwise, the estimate of components is biased. This yields $s = k + \ell$ components such that $S = K \cup L$. The mixture distribution can then be reformulated as:

$$X \sim f(x) = \sum_{i \in S} \pi_i \mathcal{N}(x|\mu_i, \sigma_i^2) \tag{8a}$$

$$\equiv \sum_{i \in K} \pi_i \mathcal{N}(x|\mu_i, \sigma_i^2) + \sum_{j \in L} \pi_j \mathcal{N}(x|\mu_j, \sigma_j^2) \tag{8b}$$

with $\sum_{i \in S} \pi_i = 1$.

Let us denote $\boldsymbol{\pi} = \{\pi_1, \ldots, \pi_s\}$, considering elements are sorted in decreasing order. Thus π_1 and π_2 represent weights of informative components (that should account for the majority of repartition among components). In this adapted leakage model, the previously introduced methodology to test for class repartition cannot be applied anymore. Indeed, vector of labels $\mathcal{Y}$ now follows a multinomial distribution and the binomial test should be replaced. The χ^2 test can be chosen to decide if points follow a multinomial distribution or not:

$$\chi^2_{test} = \sum_{i \in S} \frac{(O_i - E_i)^2}{E_i} \tag{9}$$

with $E_i = n\pi_i$ the number of expected (theoretical) occurrences for class $i \in S$, n the number of patterns observed and O_i the number of observed occurrences.

In our case, because the leakage originates from a polluted Bernoulli experiment s.t. a binomial distribution with additional components due to noise, this leads to consider the following test:

$$\begin{cases} H_0 : \pi_1 = \pi_2 = \frac{1 - \sum_{i \in L} \pi_i}{2} \\ H_1 : \pi_1 \neq \pi_2 \end{cases} \tag{10}$$

which allows verifying if a distribution is a multinomial one with two balanced components, or not. If the p-value of this test:

$$p_{val} = Pr_{\chi^2_{s-1}}(X \geq \chi^2_{test}) \tag{11}$$

where χ^2_{s-1} denotes the χ^2 distribution with $(s-1)$ degrees of freedom, is lower than α, the null hypothesis is rejected because the observed distribution is not a polluted binomial distribution (a multinomial distribution) with two balanced components.

Furthermore, as an additional constraint of the application, only points with multinomial distribution composed of two balanced (considering $\pi_0 = 0.5$) and dominating components should be selected as PoI. Reformulating Eq. 10, the above test, becomes:

$$\text{PoI Test:} \begin{cases} H_0 : \pi_1 = \pi_2 = \frac{1 - \sum_{i \in L} \pi_i}{2} \text{ and } \sum_{i \in K} \pi_i \geq 0.9 \\ H_1 : \pi_1 \neq \pi_2 \end{cases} \tag{12}$$

It is denoted as the PoI test in the rest of the paper since it allows to decide if a point is a PoI (if H_0 is accepted) or not (H_1). The setting of the threshold to 0.9 comes from [20] demonstrating that at least 90% of patterns must be correctly classified to deduce the remaining ones. Thus, below this threshold, full exponent/scalar recovery would be impossible even if all patterns are perfectly classified.

Eventually, as in the noise-free case, selected PoI are ranked by computing the probability of confusion as:

$$P_{conf} = \sum_{j \in L} \pi_j + \sum_{i \in K} \pi_i \int_{-\infty}^{\infty} g(x)dx \tag{13}$$

where the right term of Eq. 13 is computed with adjusted weights, as in Eq. 10.

4.5 Final PoI Selection Procedure

All in all, the above establishes the procedure described by the Algorithm 1 to identify and rank PoI. This is the procedure used in the section devoted to experimental results.

Data: $\mathcal{D}, w$
Result: $Q \in \mathbb{R}_p$
for $X_i \in \mathcal{D}^T$ **do**
  $P \leftarrow 0$
  for $i \in \{1, \ldots, w\}$ **do**
    $R \subseteq X_i : |R| = m, R \sim \mathcal{U}(X_i)$;
    $\boldsymbol{\theta} = \mathrm{GMM}(R, s)$ (Sect. 4.1);
    Apply PoI test (Eq. 12);
    Optionally apply F-test (Sect. 4.2);
    if $\mathbf{H}_0$ *(all tests) is accepted* **then**
      $P \leftarrow P + P_{conf}$ (Eq. 13), using estimated $\boldsymbol{\theta}$;
    else
      $P \leftarrow P + 1$;
    end
  end
  $Q_i = \frac{P}{w}$;
end
return Q

Algorithm 1: Mixture analysis procedure

As illustrated the process is applied independently to each X_i of $\mathcal{D}$. For each X_i a sub-routine is repeated w times in order to obtain averaged values of P_{conf} and limit any bias due to the random selection of the m realizations of X_i, as well as removing bias in the random initialization when applying EM optimization. This sub-routine consists first in randomly selecting m realizations of X_i. These realizations (denoted as R) are then used to build a Gaussian Mixture Model

(GMM) using the EM algorithm. The parameters of the obtained mixture are then used to perform the PoI test (Eq. 12) and the F-test of Sect. 4.2. If the null hypothesis of both tests is accepted then the accumulator of P is updated with the computed P_{conf} value, otherwise it is incremented by 1 (equal to 100% confusion). At the end of this sub-routine, the averaged P_{conf} value is computed and then stored at the right index Q_i. From there, one can be build the set $\mathcal{I}_{PoI}$ of PoI indexes s.t:

$$\mathcal{I}_{PoI} = \{i \mid i \in \{1, \dots, p\}, Q_i < 1\} \tag{14}$$

to be used as it for the attack phase or completed as explained in the next section.

5 Completing the Set of PoI

The set of PoI obtained by applying above procedure thus gathers points meeting drastic constraints deriving from a common leakage model about what must be an ideal leaking point. However, this set is not complete. Indeed, one can imagine transformations (at least monotonic) of multinomial distributions resulting in distributions very close to a Gaussian (for instance by simple reduction of the gap between the means of their components or in the case of high variances) failing the PoI test given by Eq. 12, or getting a very high P_{conf} value, while they carry nearly the same leakage.

To identify such points, we propose to use the Spearman correlation ρ and its associated test of significance. As we consider monotonic transformations of distributions, Spearman correlation is preferred as it is computed between variables ranks, over Pearson correlation which detects linear correlations by considering actual values. More precisely, the Spearman correlation coefficient between each PoI X_i and close points X_j is computed, and a significance test

$$\begin{cases} H_0: & \rho(X_i, X_j) = 0 \\ H_1: & |\rho(X_i, X_j)| > 0 \end{cases} \tag{15}$$

is applied with a p-value equal to $1 - \alpha$ (this is to say with a very high type I error).

All X_j getting a significant correlation with a PoI, X_i, this is to say rejecting H_0, are then included into the PoI set and Q is updated so that all the associated Q_j are equal to Q_i as they carry the same information than X_i.

The choice of such a high p-value could be surprising. However, it just guarantees that the monotonic link between X_i (already selected PoI) and its neighboring temporal points (X_j) is very strong and nearly perfect, albeit it fails the PoI test or gets a high P_{conf} value. We thus just limit the type II error consisting in pushing a point in the PoI set while it is not a leaking point.

This leads to the following procedure. $\forall i \in \mathcal{I}_{PoI}$, iterate through successive neighbors j of PoI i (with $j \in \{i-1, i+1, i-2, i+2, \dots\}$) and update

$$Q_j = \min(Q_j, Q_i)$$

as long as H_0 is rejected and $|\rho(X_i, X_j)| > \eta$, meaning that both temporal points carry the same information. The assessment of neighbors of PoI i ends when either condition does not hold anymore.

The application of this procedure yields an updated set of PoI $\mathcal{I}'_{PoI}$, containing previously identified PoI as well as their newly considered neighbors.

This augmentation step can also be seen as a multivariate generalization of the previously applied univariate PoI selection, as local leakage of each PoI is now supported by its neighbors, displaying monotonic transformation of said PoI (and conveying the same information), thus reinforcing the exploitability of initial identified leakages for the attack.

In the end, to ensure an optimal application of the presented method, an attacker would preferably choose to apply a strict PoI selection (Sect. 4), in order to build a set of most prominent PoI according to the leakage model, and then expand it by applying the aforementioned neighbors procedure.

6 Experimental Results

6.1 Datasets

To demonstrate the relevance of the proposed PoI selection methodology, it has been applied to three different datasets. Two of them are publicly available data sets [16] while the last one results from our measurements. These datasets are described as follows:

Cswap Pointer and Arith. It consists of two sets of traces and associated labels from [16]. The target is an ECC Curve25519 from μNaCl library. Traces come in two sets, from the two ECSM implementations, both using Montgomery ladder. They differ only in the implementation of the way the conditional swap is made: by arithmetic means or by pointers swapping. The targeted device is a STM32F4 microcontroller, with a cortex M4 running at 168 MHz. The acquisition has been made using an EM probe from Langer, a low-noise amplifier, and a low-pass filter. It consists of 300 traces of 255 patterns with a random scalar for each execution such that $\mathcal{D}_{Pointer} = (76500, 1000)$, $\mathcal{D}_{arith} = (76500, 8000)$. More details can be found in [16].

MbedTLS. RSA software 2048 bits from MbedTLS library [1]. The implementation is a sliding window exponentiation implementation but as the defined window size is set to 1, it behaves as a standard Square and Multiply. As the same function is used to compute the squaring and multiply operations results patterns display the same processing time. Considering that the first bit value should be always equal to 1, we target the third pattern from the exponentiation. The key is randomized for each exponentiation. The device on which run the exponentiation is an LPC55S69 from NXP with a Cortex M-33. It runs at 96MHz and is mounted on a CW308 ChipWhisperer UFO attack board from

NewAE. Power acquisitions were made using a Tektronik CT6 probe and a low-noise amplifier. The sampling rate of the scope was set to 5GS/s. The cutting and alignment of patterns were done following the methodology from [9]. This set is defined as $\mathcal{D}_{Mbedtls} = (80000, 4100)$, of square and multiply operations.

6.2 Supervised Approach

To assess the efficiency of the proposed PoI selection method, a reference is required. We thus first performed a clustering attack using a supervised PoI selection method, in order to fix this reference which is considered in the following as an upper bound of what could be achieved with an unsupervised approach. This means that the selection of PoI was done following a leakage analysis performed with the knowledge of the labels $\mathcal{Y}$, and by applying a Welch's t-test on each temporal point across all patterns from $\mathcal{D}$. Figure 2 reports the results of these leakage analyses. Considering the T-value of this test as a rough estimator quantifying the strength of the univariate leakage, it can be observed that the amount of leakage significantly varies from one dataset to the other. For instance, MbedTLS shows very high leakages uniformly spread over the pattern, whereas for Pointer dataset, there are only a few points with very high leakage. Univariate leakages observed for Arith dataset are significantly lower than that observed for Pointer but are more numerous and spread out all over the pattern.

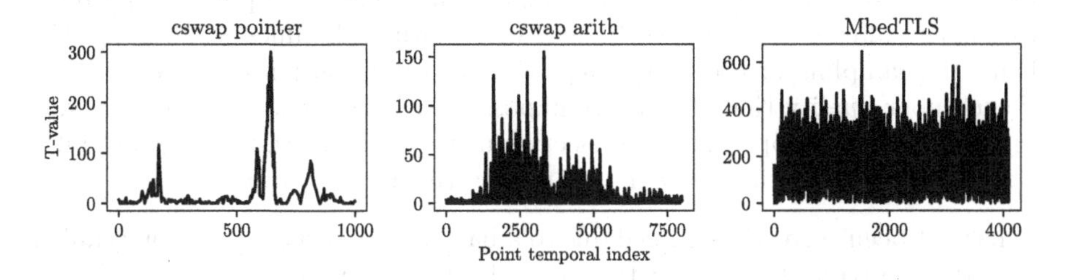

Fig. 2. Supervised leakage analysis: Absolute T-values for the three considered datasets, using all patterns from datasets $\mathcal{D}$

To define an upper bound for unsupervised attacks, the fuzzy k-means algorithm was successively applied to the subsets of points featuring at most the 200 points with the highest T-values. During this process, points were successively chosen in descending order of T-values. This process was repeated 100 times by randomly selecting 255 and 2048 patterns among the dataset respectively for the considered ECDSA algorithm and RSA algorithm to avoid any bias due to the random selection of the patterns or clustering initialization. In the end, the average bit recovery rate (BRR) was computed for each length of the subset.

The results are portrayed in Fig. 3 which gives the quantiles as well as the minimal, maximal and mean values of the BRR observed during the 100 trials. As shown, with 20 to 50 PoI, the BRR is usually higher than 90%. This demonstrates that the three datasets are exploitable by a clustering based horizontal attack, supposing the right PoI selection.

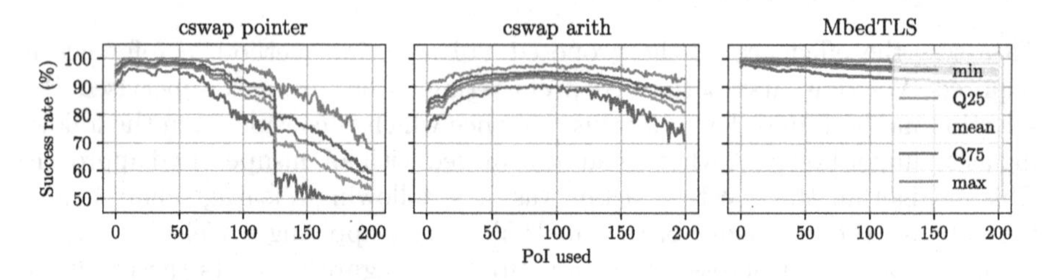

Fig. 3. Minimal value, maximal value, mean and quartiles of the BRR versus the number of PoI accumulated in the descending order of the T-values.

6.3 Unsupervised Approach

Before applying the same process as for the supervised approach, this time in an unsupervised way, this is to say with the proposed PoI selection method, patterns were first analyzed. This analysis has shown that some points of patterns are heavily affected by outliers (points with saturated values -128 or $+127$). Thus, the sampling of the underlying statistical distribution of these points is of such poor quality that it appears impossible to correctly estimate their distribution and thus apply any PoI selection method. It was thus decided to filter the patterns by rejecting all temporal points which distribution X_i:

- has not been correctly sampled due to a bad setting of the oscilloscope caliber leading to truncating one or the two tails of the distribution.
- is too affected by outliers, that is to say, displays too frequent extreme values (-128 and $+127$ for signed 8 bit integers) extending too much one of the two tails of the distribution.

The procedure to detect badly sampled distributions is thus based on the assumption that the distributions have their mean and their variance so that most frequent values are well centered on the dynamic of the oscilloscope, and that the probability to observe values close to the ends of this dynamic is very low. This leads to considering distributions as bad sampled if and only if:

$$2 < q(1-\beta) - q(\beta) < 256 - 2 \tag{16}$$

where q denotes the quantile function, with $\beta = 0.0005$, this is to say that the distribution has been correctly sampled if 99.99% of points are in the sampling range of the digital oscilloscope with a margin of 2 LSB (of the oscilloscope analog to digital converter), and if it takes at least two different values (otherwise it is a constant). This procedure, applied by temporarily shifting data values from signed 8 bits to unsigned 8 bits (to be able to assess Eq. 16), led to reject (fix the Q_i of) 675/4691/0 point(s) among 1000/8000/4100 points for Pointer, Arith, and MbedTLS datasets respectively. Table 1 shows for each step of the applied methodology the corresponding number of points remaining.

Table 1. Number of points after application of each step of the proposed methodology

	Pointer	Arith	MbedTLS
Raw (Sect. 6.1)	1000	8000	4100
Remove outliers (Sect. 6.3)	325	3309	4100
PoI selection (Sect. 4)	22	1029	861
Neighbors analysis (Sect. 5)	52	1619	1482

From there, the proposed PoI selection procedure (as described in Algorithm 1) has been applied with its parameters set to:

- the number of patterns used to estimate the GMM on subset R was set to $m = 2048$, providing a good tradeoff between the computation time and distributions quality estimation. Under this value, number of elements per class becomes insufficiently low to provide a relevant estimation of underlying components estimation.
- Q_i were estimated with $w = 100$ randomly selected subsets R. High w value is necessary to avoid biases that could be caused from random initialization for the EM algorithm and take in account the diversity of data that would affect the results of statistical testing.
- the significance level of the all statistical tests (PoI test Eq. 12, F-test Eq. 5 and Spearman correlation test Eq. 15) were set to $\alpha = 10^{-3}$.
- the number of noise components was set to $\ell = 0$ for MbedTLS and Pointer and $\ell = 2$ for Arith respectively. These values were chosen following a visual inspection of some distributions in the associated patterns. The latter showed that MbedTLS is free of any outliers.
- the value of minimum correlation in the assessment of neighboring points using Spearman correlation in Sect. 5 was set arbitrarily high $\eta = 0.8$ to ensure considering neighbor points only if they convey a sufficiently common information.

Table 2 gives the processing time for the application of the proposed method on the three considered datasets. Computation is carried out using a Intel i9-10900 CPU. The processing time observed directly depends on the number of points to assess per pattern (see datasets definitions in Sect. 6.1) as well as the filtering results from the outliers processing (see above). Mixture analysis step,

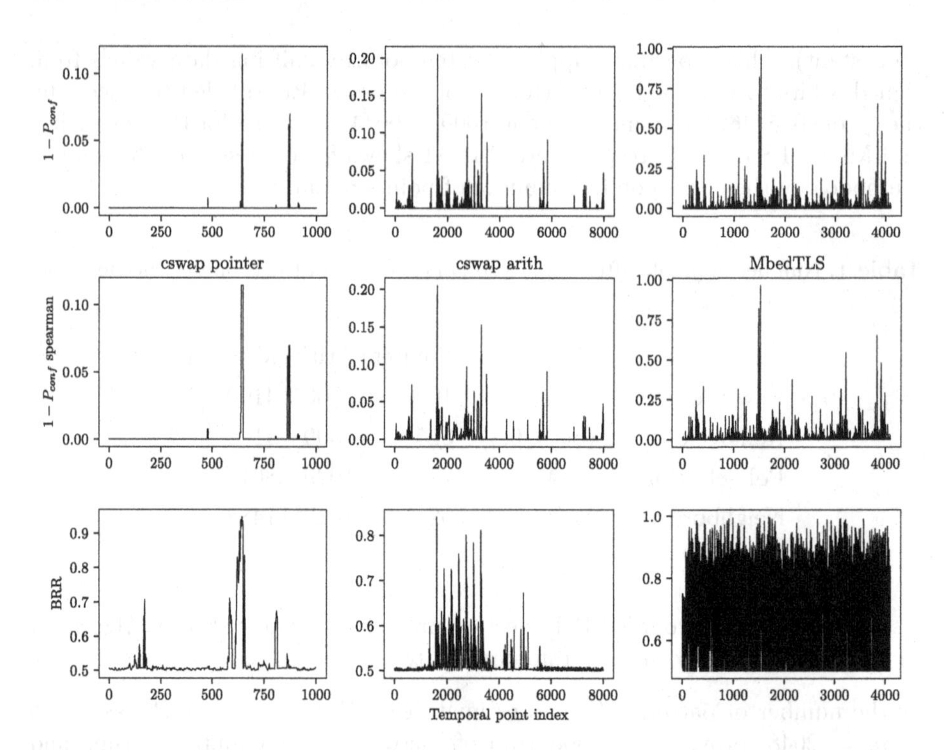

Fig. 4. Top row: $1 - P_{conf}$ traces (higher is better) for the three datasets, after application of the proposed procedure and completion of the PoI sets but without application of Spearman correlation based reinforcement. Middle row: $1 - P_{conf}$ traces for the three datasets, with application of the proposed procedure and completion of the PoI sets and application of Spearman correlation based reinforcement. Bottom row: bit recovery rates obtained by clustering individually, with the fuzzy k-means, each temporal point in patterns.

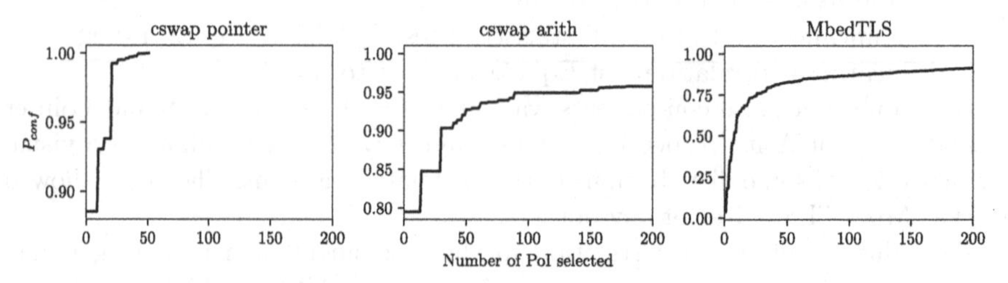

Fig. 5. Accumulative sorted values of first 200 P_{conf} associated to the sets of PoI for each dataset.

Table 2. Processing time for the application of proposed methodology

	Pointer	Arith	MbedTLS
PoI selection (Sect. 4)	∼20 m	∼2 h 30 m	∼3 h
Neighbors analysis (Sect. 5)	1 s	47 s	21 s

consisting in identifying underlying components of X_i, as well as performing statistical testing accounts for the majority of the processing time. More precisely, the application of the EM algorithm w times on R subsets accounts alone for more than 90% of the total processing time. In comparison, Spearman refinement step represents a negligible portion of total computation time, thus this step can systematically be applied to reinforce the PoI set.

Figure 4 reports, for the three datasets, the traces of $1 - P_{conf}$ obtained before and after completion (using Spearman correlation) of the sets of PoI. The application of the reinforcement procedure changes the $1 - P_{conf}$ for many points but also slightly increases the number of selected PoI from 22/1029/861 to 52/1619/1482 respectively as shown on Table 1.

Nevertheless, the most important result to observe is that peaks of the three $1 - P_{conf}$ traces coincide with the most prominent peaks of the T-values traces in Fig. 2. This sustains the soundness of the proposed PoI selection method although some points are identified as PoI while they do not carry significant information. It should be also observed that some points carrying some leakage are not detected by the proposed procedure. However, these points, according to Fig. 2, carry a low amplitude leakage (have lower T-values). This observation is sustained by the bottom row of Fig. 4 that displays the BRR obtained by applying the fuzzy k-means of each point of the patterns.

Despite this first result, it should be observed that the scale of the three $1-P_{conf}$ traces is not the same. Some points in the $1-P_{conf}$ trace associated with MbedTLS have values close to 0.8, while the highest values observed for Arith and Pointer datasets remain below 0.2. This suggests the leakage is limited in these two sets of traces. This point is also sustained by the BRR traces reported in the bottom row of Fig. 4.

Eventually, attacks were performed following the exact same procedure as in Sect. 6.2, but this time by selecting PoI from obtained set $\mathcal{I}'_{PoI}$ in increasing values of P_{conf}. Figure 6 sums up the results for the three datasets by reporting the minimal and maximal values of the BRR as well as the quartiles and mean. This representation was adopted because the adversary has to guess the scalar or the exponent bits (255 or 2048) for each newly considered PoI. This representation thus allows direct reading in the figure of the probability to get a given percentage of the secret bits each time a new PoI is processed.

At first glance, one can observe that the results, when considering the first accumulated points sorted according to P_{conf}, are satisfactory with a minimum BRR value well above 80% and on average above 90% on the first 40 PoI selected. One can also observe that, as expected, accumulating PoI with increasing P_{conf} inevitably leads to a fall of the BRR as is the case when considering the T-values (computed for the supervised approach). One could have expected this fall to occur when considering PoI with $P_{conf} \geq 0.75$ or lower. However, this is not the case and high BRR values are obtained even when considering PoI with $P_{conf} \geq 0.95$. This is probably because a sufficient number of points with high P_{conf} values are required to blind the clustering algorithm guided by the

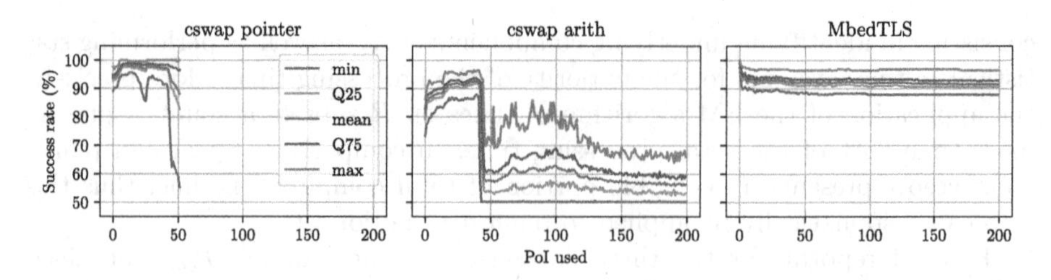

Fig. 6. Minimal, maximal, mean and quartiles of the BRR obtained for the three sets with the proposed methodology.

first points of interest with lower P_{conf} values. However, this point has not been further analyzed and remains an open question.

To assess the effectiveness of the method with respect to the state of the art, let us remind that authors of [16] reported average bit recovery rate values equal to 52% for both Pointer and Arith sets after the application of [17] methodology. Through the proposed methodology, we also obtained similar results with the solution proposed in [14] and agree with the statement of these authors according to which few PoI must be considered to get the highest BRR. Indeed, Fig. 6 shows that 20 to 40 PoI are sufficient. This raises the question of the number of PoI to keep in practice.

If we do not have a theoretical solution or explanation, one can however observe the BRR values remain high in Fig. 6 until PoI with a P_{conf} value close to the knee of the related traces (Fig. 5) are involved in the analysis. This knee appears around 20 PoI for Pointer, 40 for Arith and 50 for MbedTLS. In our opinion, there is a knee in the P_{conf} traces because non informative as well as low information points define an upper asymptotic value for P_{conf} which is necessarily lower than 1 because of the variance in the estimate of Q. Thus, we herein suggest setting the number of PoI to consider to a slightly lower value than corresponding the knee of the $1 - P_{conf}$ trace, a number that can be computed in an unsupervised way.

7 Conclusion

Starting from a commonly accepted leakage model, this paper has described a technique for selecting points of interest for Horizontal attacks. The latter has been progressively defined by successively considering how a Bernoulli process leaks in side-channel traces and then how Gaussian noise and untimely outliers change these leaking statistical distributions. This leads to consider Gaussian mixture distributions with a weighting of components meeting constraints imposed by the Bernoulli process at the origin of the leakage. The resulting technique mainly relies on two statistical tests and thus hyper-parameters are quite easy to set. In addition, analysis of neighbors of selected PoI allows to reinforce identified leakage.

The application of the methodology on three datasets, two of them being public, has demonstrated its efficiency and interest. Indeed, the obtained results are close to those of a supervised analysis while being completely unsupervised. This technique thus enables reducing the gap between supervised and unsupervised Horizontal attacks.

Future work aims at exploiting the same guiding principle with multivariate techniques and especially those based on entropy which are expected to be more efficient at exploiting multivariate leakages.

References

1. MbedTLS library. https://github.com/Mbed-TLS/mbedtls/
2. Batina, L., Hogenboom, J., van Woudenberg, J.G.J.: Getting more from PCA: first results of using principal component analysis for extensive power analysis. In: Dunkelman, O. (ed.) CT-RSA 2012. LNCS, vol. 7178, pp. 383–397. Springer, Heidelberg (2012). https://doi.org/10.1007/978-3-642-27954-6_24
3. Bauer, A., Jaulmes, E., Prouff, E., Wild, J.: Horizontal collision correlation attack on elliptic curves. In: Lange, T., Lauter, K., Lisoněk, P. (eds.) SAC 2013. LNCS, vol. 8282, pp. 553–570. Springer, Heidelberg (2014). https://doi.org/10.1007/978-3-662-43414-7_28
4. Brier, E., Clavier, C., Olivier, F.: Correlation power analysis with a leakage model. In: Joye, M., Quisquater, J.-J. (eds.) CHES 2004. LNCS, vol. 3156, pp. 16–29. Springer, Heidelberg (2004). https://doi.org/10.1007/978-3-540-28632-5_2
5. Chari, S., Rao, J.R., Rohatgi, P.: Template attacks. In: Kaliski, B.S., Koç, K., Paar, C. (eds.) CHES 2002. LNCS, vol. 2523, pp. 13–28. Springer, Heidelberg (2003). https://doi.org/10.1007/3-540-36400-5_3
6. Clavier, C., Feix, B., Gagnerot, G., Roussellet, M., Verneuil, V.: Horizontal correlation analysis on exponentiation. In: Soriano, M., Qing, S., López, J. (eds.) ICICS 2010. LNCS, vol. 6476, pp. 46–61. Springer, Heidelberg (2010). https://doi.org/10.1007/978-3-642-17650-0_5
7. Clavier, C., Feix, B., Gagnerot, G., Roussellet, M., Verneuil, V.: Square always exponentiation. In: Bernstein, D.J., Chatterjee, S. (eds.) INDOCRYPT 2011. LNCS, vol. 7107, pp. 40–57. Springer, Heidelberg (2011). https://doi.org/10.1007/978-3-642-25578-6_5
8. Coron, J.-S.: Resistance against differential power analysis for elliptic curve cryptosystems. In: Koç, Ç.K., Paar, C. (eds.) CHES 1999. LNCS, vol. 1717, pp. 292–302. Springer, Heidelberg (1999). https://doi.org/10.1007/3-540-48059-5_25
9. Diop, I., Linge, Y., Ordas, T., Liardet, P.-Y., Maurine, P.: From theory to practice: horizontal attacks on protected implementations of modular exponentiations. J. Cryptogr. Eng. **9**(1), 37 (2019)
10. Heyszl, J., Ibing, A., Mangard, S., De Santis, F., Sigl, G.: Clustering algorithms for non-profiled single-execution attacks on exponentiations. In: Francillon, A., Rohatgi, P. (eds.) CARDIS 2013. LNCS, vol. 8419, pp. 79–93. Springer, Cham (2014). https://doi.org/10.1007/978-3-319-08302-5_6
11. Johnson, D., Menezes, A., Vanstone, S.: The elliptic curve digital signature algorithm (ECDSA). Int. J. Inf. Secur. **1**(1), 36–63 (2001)
12. Kocher, P., Jaffe, J., Jun, B.: Differential power analysis. In: Wiener, M. (ed.) CRYPTO 1999. LNCS, vol. 1666, pp. 388–397. Springer, Heidelberg (1999). https://doi.org/10.1007/3-540-48405-1_25

13. Kocher, P.C.: Timing attacks on implementations of Diffie-Hellman, RSA, DSS, and other systems. In: Koblitz, N. (ed.) CRYPTO 1996. LNCS, vol. 1109, pp. 104–113. Springer, Heidelberg (1996). https://doi.org/10.1007/3-540-68697-5_9
14. Nascimento, E., Chmielewski, L.: Horizontal clustering side-channel attacks on embedded ECC implementations (extended version). Number 1204 (2017)
15. Perin, G., Chmielewski, Ł: A semi-parametric approach for side-channel attacks on protected RSA implementations. In: Homma, N., Medwed, M. (eds.) CARDIS 2015. LNCS, vol. 9514, pp. 34–53. Springer, Cham (2016). https://doi.org/10.1007/978-3-319-31271-2_3
16. Perin, G., Chmielewski, Ł., Batina, L., Picek, S.: Keep it unsupervised: horizontal attacks meet deep learning. IACR Trans. Cryptogr. Hardw. Embedded Syst., 343–372 (2021)
17. Perin, G., Imbert, L., Torres, L., Maurine, P.: Attacking randomized exponentiations using unsupervised learning. In: Prouff, E. (ed.) COSADE 2014. LNCS, vol. 8622, pp. 144–160. Springer, Cham (2014). https://doi.org/10.1007/978-3-319-10175-0_11
18. Benadjila, R., Prouff, E., Strullu, R., Cagli, E., Dumas, C.: Study of deep learning techniques for side-channel analysis and introduction to ASCAD database (2018). Report Number: 053
19. Rivest, R.L., Shamir, A., Adleman, L.: A method for obtaining digital signatures and public-key cryptosystems. Commun. ACM **21**(2), 120–126 (1978)
20. Schindler, W., Walter, C.D.: Optimal recovery of secret keys from weak side channel traces. In: Parker, M.G. (ed.) IMACC 2009. LNCS, vol. 5921, pp. 446–468. Springer, Heidelberg (2009). https://doi.org/10.1007/978-3-642-10868-6_27

Blind Side Channel Analysis Against AEAD with a Belief Propagation Approach

Modou Sarry[1(✉)], Hélène Le Bouder[1(✉)], Eïd Maalouf[1], and Gaël Thomas[2]

[1] IMT-Atlantique, OCIF, IRISA, Rennes, France
helene.le-bouder@imt-atlantique.fr

[2] DGA Maîtrise de l'Information, Bruz, France

Abstract. This paper present two new attacks on two lightweight authenticated encryption with associated data (AEAD): SPARKLE and Elephant. These attacks are blind side channel analysis (BSCA). The leakage is considered as an Hamming weight (HW) with a Gaussian noise. In both attacks, a belief propagation (BP) algorithm is used to link the different leaks. Another objective is to present BSCA as a new tool for evaluating the robustness of a symmetric cryptographic primitive subfunctions.

Keywords: authenticated encryption with associated data (AEAD) · side channel analysis (SCA) · blind side channel analysis (BSCA) · belief propagation (BP) · SPARKLE · Alzette · LFSR · Elephant

1 Introduction

Today, the Internet of things (IoT) interconnects every aspect of our lives, leading to a proliferation of embedded systems. These devices play an essential role in many sectors such as industry, intelligent transport, smart cities and healthcare, to name but a few. However, due to the sensitive nature of the data they process, it is imperative to ensure the security of these systems. Protecting confidential information and preventing cyber-attacks are major concerns when deploying these circuits. Securing these systems is therefore of crucial importance.

In this context, the National Institute of Standards and Technology (NIST) started a competition [1] for the standardization of a lightweight authenticated encryption with associated data (AEAD).

Physical attacks rely on the interaction of the computing unit with the physical environment. They are mainly divided in two families: side channel analysis (SCA) and fault injection attacks. SCA [2] are based on observations of the circuit behaviour during the computation. They exploit the fact that some physical values of a circuit depend on intermediary values of the computation. This is the so-called leakage of information of the circuit. Blind side channel analysis (BSCA) is a new family of SCA, where no inputs nor outputs are used to recover a cryptographic key, only the traces and the knowledge of the algorithm is used.

S. Bhasin and T. Roche (Eds.): CARDIS 2023, LNCS 14530, pp. 127–147, 2024.
https://doi.org/10.1007/978-3-031-54409-5_7

Motivation. Many attacks already exist on older symmetric cryptography algorithms like the advanced encryption standard (AES) [3]. Since many AEAD algorithms are relatively new, there are far fewer attacks against them [4–7]. Moreover, these new algorithms are often designed to be more resistant to physical attacks than AES. Therefore, one motivation is to test the resistance of these new algorithms against SCA.

Another main motivation is to build an attack which uses only traces, no text and no profiling. We are in the case of an attacker who can just observe a leakage but has no access to the device's input/output, hence the name of blind side channel analysis (BSCA). Once has to remark that retrieving the key without the ciphertext in BSCA may seem absurd in the context of confidentiality. However, in the context of integrity, an attacker could forge a tag for their message after obtaining the secret key.

The last motivation is to show that the BSCA context can be seen as a new tool to evaluate cryptosystems. This method is a good way to select the subfunction of an algorithm at a mathematical level. Thinking about the security of the cryptographic algorithm is an important approach.

Contribution. In this paper, we evaluate the security of two AEADs: Elephant [8] and SPARKLE [9] against BSCA. The attack on Elephant is a major improvement of a first BSCA [10,11]. The attack on SPARKLE is a new attack. The results are based on a belief propagation (BP) algorithm [12].

Organization. Targeted algorithms SPARKLE and Elephant are described in Sect. 2. State of the art of BSCA attacks are summarised in Sect. 3. Our attack methodology is described in Sect. 4. The application of this attack against the Elephant algorithm and the corresponding results are presented in Sect. 5. Similarly, the application of this attack against the SPARKLE algorithm is discussed in Sect. 6. Finally, the conclusion is drawn in Sect. 7.

2 Authenticated Encryption with Associated Data (AEAD)

2.1 NIST Competition

Authenticated encryption with associated data (AEAD), illustrated in Fig. 1, should ensure confidentiality and integrity. It takes as input different parameters: a plaintext denoted M, the associated data denoted A, a secret key K, and an initialisation vector N also called a nonce. The nonce is public but must be different for each new plaintext. The algorithms ensure confidentiality of the plaintext M and integrity of both the plaintext M and the associated data A. It returns a ciphertext C and a tag T.

In this context, the NIST started the competition [1] for lightweight cryptography candidates for AEAD. In August 2019, NIST received 57 submissions

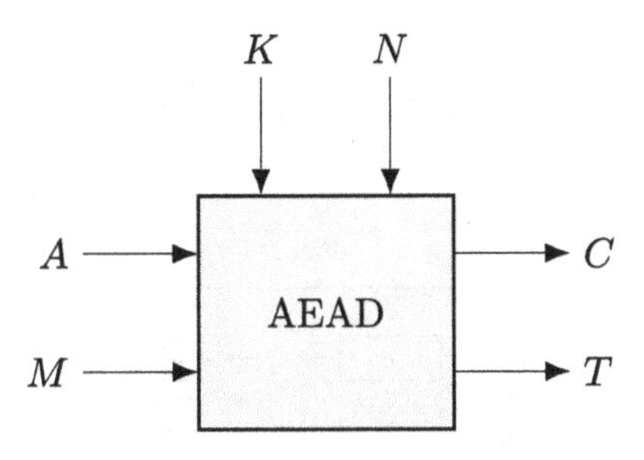

Fig. 1. Overview of AEAD.

to be considered for standardization. In March 2021, only 10 finalists remained of which Elephant [8] and SPARKLE [9] which are targeted in this paper. Finally, in February 2023, NIST decided to standardize ASCON [13].

2.2 Elephant

Elephant [8,14] was a finalist to the NIST lightweight cryptography competition. It is a nonce-based AEAD. Its construction is based on an Encrypt-then-MAC that combines counter (CTR) mode encryption with a variant of the protected counter sum [15,16]. Elephant uses a cryptographic permutation masked with linear feedback shift registers (LFSRs) in an Even-Mansour-like fashion [17] in place of a block cipher.

Let P be an n-bit cryptographic permutation, and φ an n-bit LFSR. The function $\mathsf{mask} : \{0,1\}^{128} \times \mathbb{N} \times \{0,1,2\} \rightarrow \{0,1\}^n$ used to mask the permutation P is defined as follows:

$$\mathsf{mask}_K^{j,i} = (\varphi \oplus \mathsf{id})^i \circ \varphi^j \circ \mathsf{P}(K||0^{n-128}); \tag{1}$$

where id is the identity function. One important thing to note is that the values of $\mathsf{mask}_K^{j,i}$ depend only on the secret key K and not on any other input of Elephant.

Encryption enc under Elephant gets as input a 128-bit key K, a 96-bit nonce N, associated data $A \in \{0,1\}^*$, and a plaintext $M \in \{0,1\}^*$. It outputs a ciphertext C as large as M, and a τ-bit tag T. The algorithm is depicted on Fig. 2.

Elephant comes in three flavours which differ on the n-bit cryptographic permutation P and the n-bit LFSR φ used, as well as the tag size τ.

In this paper, we focus on the smallest and main instance Dumbo. It uses the 160-bit permutation Spongent-π[160] [18], the 160-bit LFSR φ_{Dumbo}, and has tag size $\tau = 64$ bits.

Let $\lll$, $\ll$, and $\gg$ denote the left rotation, left shift, and right shift operators respectively. The update equation of LFSR φ_{Dumbo} is given at the byte level by Eq. (2) and illustrated on Fig. 3.

$$\varphi_{\mathsf{Dumbo}} : (x_0, \cdots, x_{19}) \mapsto (x_1, \cdots, x_{19}, x_0 \lll 3 \oplus x_3 \ll 7 \oplus x_{13} \gg 7) \tag{2}$$

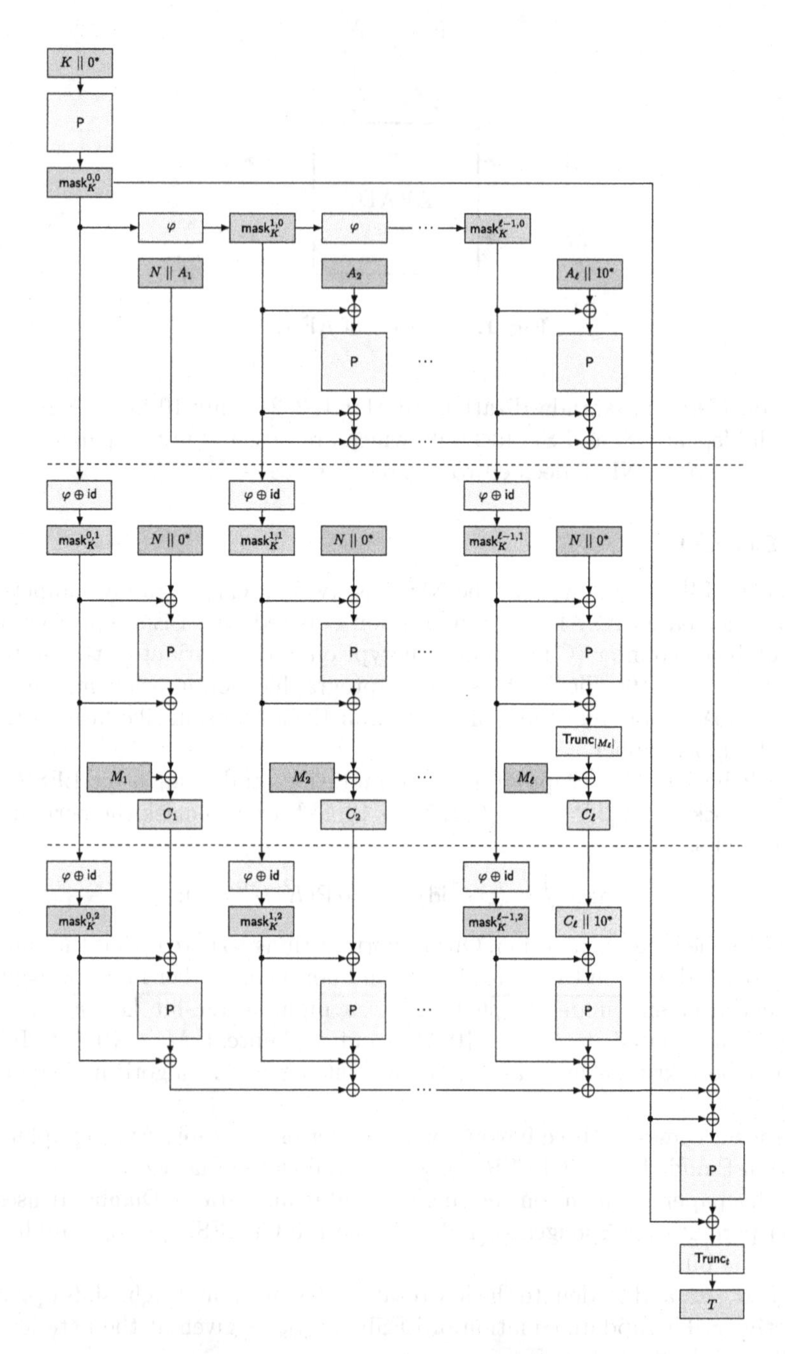

Fig. 2. Elephant associated data authentication (left), plaintext encryption (middle), and ciphertext authentication (right). This figure comes from [10,11] according to the description of Elephant [8].

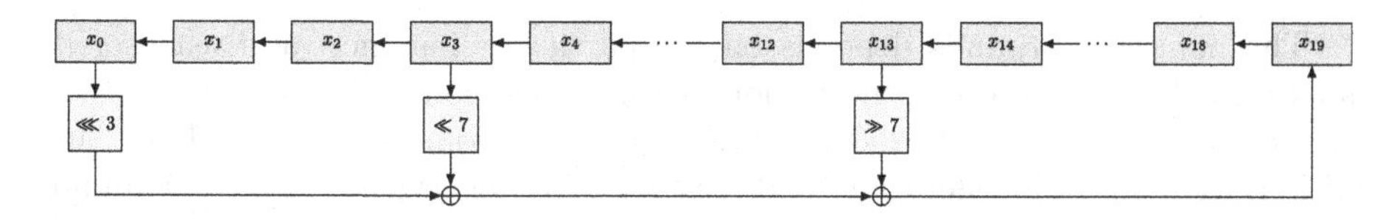

Fig. 3. The 160-bit LFSR φ_{Dumbo}. This figure comes from [10,11] according to the description of Elephant [8].

2.3 SPARKLE

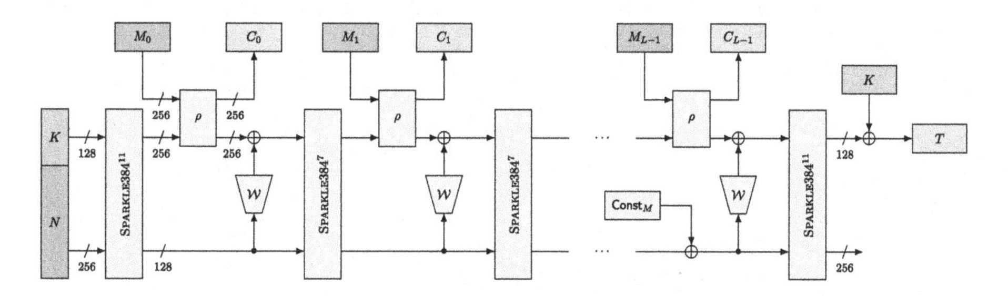

Fig. 4. Encryption using SCHWAEMM256-128, where SPARKLE384^R is the SPARKLE384 permutation with R rounds, ρ is a linear application, $\mathcal{W}$ the linear whitening layer, and the constant Const_M indicating whether the plaintext M was padded or not.

SCHWAEMM is the AEAD of the SPARKLE [9] submission to the NIST lightweight cryptography competition. It is a permutation-based AEAD that builds upon the Beetle [19] variant of the Duplex [20] construction by adding an extra linear whitening layer $\mathcal{W}$, from the inner state into the outer state. The underlying cryptographic permutation is named SPARKLE which gives the whole submission its name.

Several instances of both algorithms with various performance and security trade-offs are proposed.

The main instance is SCHWAEMM256-128. The rest of this paper then focuses on this particular instance. It is illustrated in Fig. 4.

SCHWAEMM256-128 takes a 256-bit nonce N, a 128-bit key K, and produces a 128-bit authentication tag T. The corresponding SPARKLE permutation is the 384-bit-wide variant, denoted SPARKLE384.

Zoom on SPARKLE. The SPARKLE permuations are based on a substitution permuation network (SPN) structure. The linear layer is a Feistel-type transformation operating on 64-bit branches. It is based on a linear function $\mathcal{M}$ whose description is not necessary for the understanding of this paper. The non-linear layer is a 64-bit S-box named Alzette $\mathcal{A}_i$, applied in parallel to every branch of the Feistel, where i denotes the index of the branch. More details on Alzette are given in the Sect. 2.3.

The number of rounds depends on the particular instance, and whether it is used in the initialisation, encryption or tag generation phase of SCHWAEMM. In the case of SCHWAEMM256-128, the number of rounds is $R = 11$ for the initialisation and tag generation, and only $R = 7$ for encryption. The first round of SPARKLE384 of the initialisation phase of SCHWAEMM256-128 is depicted on Fig. 5. One important thing to note is that the state is initialised with the concatenation of the nonce N and the secret key K.

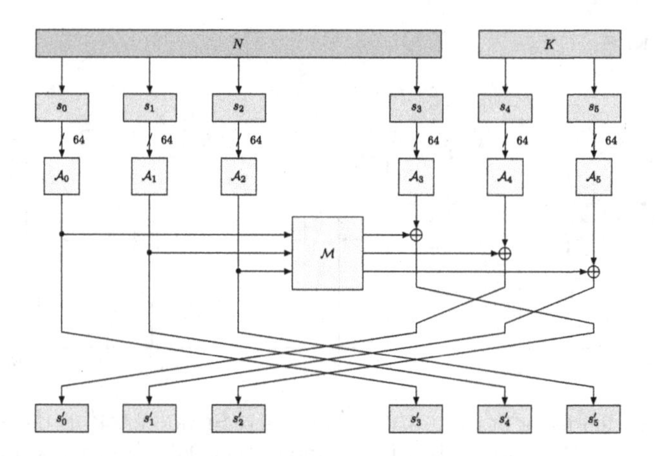

Fig. 5. The first round of the SPARKLE384 permutation during the initialisation of SCHWAEMM256-128.

Alzette. The Alzette S-box $\mathcal{A}_i$ is a 64-bit permutation parametrised by a 32-bit known value α_i that depends only on the index i of branch inside the SPARKLE permutation. It is built as a 4-round addition-rotation-xor (ARX) Feistel cipher with all round keys equal to α_i. As the name ARX suggests, it uses a combination of 32-bit modular additions, rotations, and xors. The whole process is depicted on Fig. 6.

3 BSCA Context

While it has been proven that an algorithm is mathematically secure, its implementation can still expose vulnerabilities known as physical attacks. Physical attacks, such as SCA, constitute a specific subcategory of these vulnerabilities. They exploit the fact that certain physical characteristics of a device depend on intermediate values of the calculations performed, resulting in an information leakage within the circuit. This information leakage could be utilized to retrieve secrets, such as a secret key.

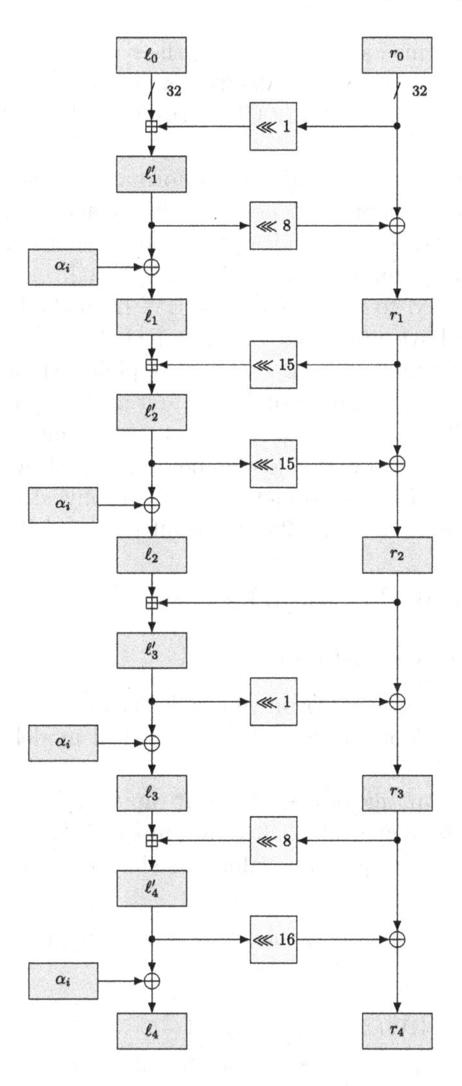

Fig. 6. The Alzette S-box $\mathcal{A}_i$ used in SPARKLE.

Many SCA belong to the family of correlation power analysis (CPA) [21–23]. To succeed these attacks need data like plaintext or ciphertext. Then these attacks use a mathematical model for the leakage. A confrontation is performed between measurements and predictions built with the model and known data. More precisely, a statistic tool called distinguisher gives score to the different targets.

In the case of sponge and duplex building, to the best of our knowledge, there are very few CPA attacks [24]. In a classical block cipher, the secret key is directly used in each round. In a duplex or sponge case, the secret key is used only once or twice, at the start and at the end of the whole algorithm. The first difficulty is caused by this fact. Moreover, in the specific use case SPARKLE, there is no small S-box. On the contrary, Alzette is a very big S-box. That is one other reason, we have not chosen to study CPA attacks on SPARKLE, but blind SCA.

BSCA is a type of SCA where some information that is usually known is unknown. The main concept is to perform the attack solely based on leakage measurements. The attacker model is different from that in a classic CPA. They have access to less information. Often BSCA is based on a strong assumption: the attacker is supposed to retrieve a noisy HW from the leakage.

Linge *et al.* [25] have presented the first BSCA. The goal is to attack the block cipher AES without using data such as plaintext or ciphertext. At the same time, Le Bouder *et al.* [26] published a first attack. Then, these works have been improved by Clavier *et al.* [27]. Moreover, their contribution introduces for the first time, the name of blind side channel analysis (BSCA). Now it is a new family of SCA, and different symmetric cryptography algorithms are targeted by these attacks such as AES [25–28], Elephant [10], and PRINCE [29].

4 Description of the Attacks

4.1 Simulated Leakage Model

The power consumption or electromagnetic leakages are very correlated to the Hamming weight (HW) of the data. It is a classical model used in the domain of SCA.

One important advantage of HW is that it rapidly reduces guesses. For example, let x be a byte, so x can take 256 values in $[\![0, 255]\!]$. With the HW of x, the attacker reduces the list of possible values, as shown in Table 1.

Table 1. Number of possible values for a byte x according its HW. This table comes from [26]

HW(x)	0	1	2	3	4	5	6	7	8
$\#x$	1	8	28	56	70	56	28	8	1

Another model for the leakage is a HW with an additive Gaussian noise. For a given discrete random variable byte x, and its $\mathrm{HW}(x) \in [\![0, 8]\!]$; $\tilde{\mathrm{HW}}(x)$ denotes the continuous random variable representing the simulated Hamming weight so called **noisy**; defined in $\mathbb{R}$ as:

$$\tilde{\mathrm{HW}}(x) = \mathrm{HW}(x) + \sigma_{x,t}; \tag{3}$$

with $\sigma_{x,t}$ an event of the Gaussian random variable $\mathcal{N}\left(0, \sigma^2\right)$ at a time t. The probability density function $\mathbf{F}$ associated to $\mathcal{N}\left(0, \sigma^2\right)$ is given by:

$$\mathbf{F}_\sigma(x) = \frac{1}{\sigma \cdot \sqrt{2\pi}} \cdot \exp\left(-\frac{1}{2} \cdot \left(\frac{x}{\sigma}\right)^2\right). \tag{4}$$

In this paper, BSCA are studied, so the attacker uses only simulated leakage. The simulation of the power consumption or electromagnetic leakage is generated with Gaussian noise added to the HW obtained from the intermediate computations at the byte level. To simplify the simulation, it is assumed that the noise level remains the same throughout the system. The simulations, which were performed using the Python programming language (v3.9), represent a typical unprotected implementation on an 8-bit processor.

4.2 Belief Propagation (BP)

In our attack approach, relations between the different values for which a nosy HW is simulated, and exploited. For that belief propagation (BP) is used. BP was first used by Gallager [30,31] for decoding low-density parity-check (LDPC) codes [32]. It was then rediscovered by Tanner [33] and formalized by Pearl [34]. The first time that BP was used in SCA on symmetric encryption, in the attack of Veyrat-Charvillon *et al.* [35], then it is studied in [26,36].

Tanner Graph. A BP algorithm relies on a bipartite graph called a factor graph (or Tanner graph). To each node in the factor graph is associated some information. The nodes of a factor graph as are of two kinds:

- variable nodes V representing the variables handled by the algorithm under attack;
- factor nodes, representing the equations E between these variables.

An edge links a variable node V with a factor node E, when the equation represented by the factor node E involves the variable node V. An example of a factor graph is illustrated in Fig. 7.

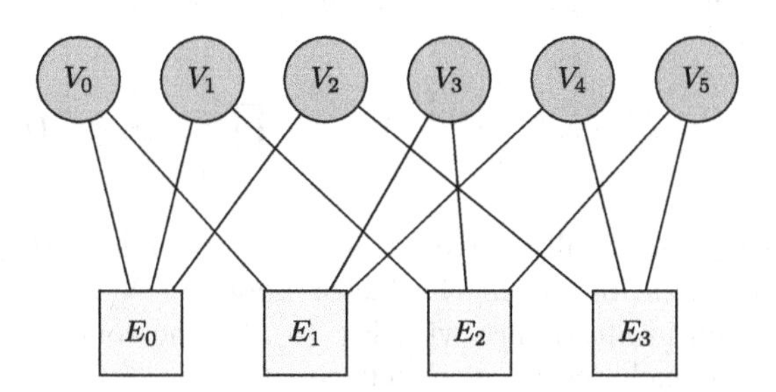

Fig. 7. Example of a Tanner factor graph part. Circles are variable nodes and squares are factor nodes.

N denotes the set of neighbours for a node. Thus the set $\mathsf{N}(E)$ is made of the variable nodes V involved in equation E, and the set $\mathsf{N}(V)$ is made of factor nodes E that depend on V.

BP Algorithm. The belief propagation (BP) algorithm inputs are

- the Tanner graph,
- prior probabilities $\mathbb{P}_A(V = v)$ on the different variable nodes V.

The BP algorithm returns a better belief, from the initial values $\mathbb{P}_A(V = v)$. More precisely, BP computes probabilities a posteriori $\mathbb{P}_P(V = v)$ on the different variable nodes according prior probabilities.

The values $\mathbb{P}_P(V = v)$ are computed according to the input prior probability $\mathbb{P}_A(V = v)$ and to the probabilities $\mathbb{P}(V = v|E)$ conditional on factor nodes $E \in \mathsf{N}(V)$ to be satisfied using the following equation:

$$\mathbb{P}_P(V = v) \propto \mathbb{P}_A(V = v) \times \prod_{\mathcal{E} \in \mathsf{N}(V)} \mathbb{P}(V = v|E) \tag{5}$$

In practice, nodes in the factor graph exchange information messages with their neighbours. More precisely, since the graph is bipartite, two types of messages are exchanged:

- factor to variable messages between a factor node E and a variable node V, denoted $\mu_{E\to V}$:

$$\mu_{E\to V}(v) = \sum_{v_1, v_2, \cdots, v_{\#\mathsf{N}(E)-1}} E(v, v_1, \cdots, v_{\#\mathsf{N}(E)-1}) \times \prod_{V_i \in \mathsf{N}(E)\setminus\{V\}} \mu_{V_i \to E}(v_i) \tag{6}$$

  This equation comes from the law of total probability. Roughly speaking, the quantity $\mu_{E\to V}(v)$ represents the belief from the perspective of factor node E that variable node V take the value v. The information is gathered from the other variables nodes in $\mathsf{N}(E)$.
- variable to factor messages between a variable node V and a factor node E, denoted $\mu_{E\to V}$.

$$\mu_{V\to E}(v) \propto \mathbb{P}_A(V = v) \times \prod_{E' \in \mathsf{N}(V)\setminus\{E\}} \mu_{E'\to V}(v) \tag{7}$$

  This equation is essentially Bayes' rule. Here variable node V updates its belief using informations from the factor nodes in $\mathsf{N}(V)$. This equation is meant to be used in alternance with Eq. (6). So the contribution from node E is deliberately excluded to avoid self-persuasion.

Both equations make independence assumptions to make computations feasible in practice at the expanse of the risk to converge to an incorrect value.

To complete the description of the BP algorithm, an initialization step is done before applying the above equations. The variable to factor messages $\mu_{V\to E}(v)$ are initialized with the prior probabilities $\mathbb{P}_A(V = v)$. In summary, after an initialization phase, BP works by alternatively applying Eqs. (6) then (7) for every edge (V, E) in the Tanner graph. At the end of the execution, the returned value $\mathbb{P}_P(V = v)$ is computed using Eq. (5) where the probabilities $\mathbb{P}(V = v|E)$ are replaced by their approximations $\mu_{E \to V}(v)$. The number of iterations depends of the rapidity of convergence, which in turn depends on the Tanner graph.

5 Attack on Elephant

5.1 Attack Path

The attack path is based on a similar approach as presented in the theoretical work by Meraneh *et al.* [10,11]. However, their attack focuses on obtaining HW information and is limited to scenarios without noise. It is a theoretical and mathematical assumption that is very useful for testing ideas, but it never happens in the real world. In this paper, we improve this attack to consider noisy HW more close than real measurements and propose the use of BP as a solution to address this challenge.

Linear feedback shift registers (LFSRs) find applications in various lightweight cryptography candidates, where their initial state is typically determined by both a key and a nonce. Since the nonce must be altered for each encryption request, attacks against such schemes are confined to the decryption algorithm. However, in the case of Elephant, the LFSR solely relies on the secret key. As a result, our attack can be employed in an encryption scenario as well.

The objective of the described attack is to recover the secret initial state of the LFSR. Three crucial points should be noted as following.

- Retrieving the initial state of the LFSR, which is equal to $\mathsf{mask}_K^{0,0}$, is equivalent to retrieving the secret key. Indeed, the initial state is the result of the known permutation P applied to the key.
- As the retroaction polynomial is publicly known, it is possible to shift the LFSR backwards: an attacker who recovers enough consecutive bytes of the secret stream is able to reconstruct the initial state.
- The smaller the LFSR is, the more the attack is able to succeed. As a consequence, the Dumbo instance (see Fig. 3) is the most vulnerable one: the following of this section is focused on Dumbo.

Since the LFSR generates a single new byte at each iteration, let the content of the Dumbo LFSR be denoted as follows:

$$(x_j, \cdots, x_{j+19}) = \mathsf{mask}_K^{j,0} \tag{8}$$

and let x_{j+20} be the byte generated at iteration j. Similarly, let $(y_j, \cdots, y_{j+19}) = \mathsf{mask}_K^{j,1}$ and $(z_j, \cdots, z_{j+19}) = \mathsf{mask}_K^{j,2}$. By definition of mask, the following hold for all $j \geq 0$:

$$y_j = x_j \oplus x_{j+1} \tag{9}$$

$$z_j = x_j \oplus x_{j+2}. \tag{10}$$

The attacker can thus exploit two attack vectors: on the one hand, Eqs. (2) coming from iterating the LFSR, and on the other hand, Eqs. (9) and (10) coming from the different masks used for domain separation.

In this use case, the attacker obtain a leakage, simulated by noisy HW on the different bytes of the Dumbo LFSRs.

Once has to remark that a classic CPA is impossible on the $\mathsf{mask}_K^{0,0}$, because the attacker knows zero data in this function.

5.2 Tanner Graph

In the case of Elephant, the Tanner graph is defined as follows. Variable nodes are the bytes x_j, y_j and z_j of the LFSRs. There are four types of factor nodes:

- $E_{y_0}, E_{y_1}, \ldots, E_{y_{20}}$ represent the Eq. (9).
- $E_{z_0}, E_{z_1}, \ldots, E_{z_{19}}$ represent the Eq. (10).
- We have divided the feedback Eq. (2) into two sub-equations. This helps reduce the amount of computation in BP. Indeed, the number of terms in Eq. (6) is exponential in the number of neighbours in $\mathsf{N}(E)$. The factor nodes corresponding to Eq. (2) are given in Table 2.

The Elephant Tanner graph is illustrated in Fig. 8.

Table 2. Intermediate and final feedback equation

intermediate feedback equation	final feedback equation
$E_{tx_{20}} : tx_{20} = (x_0 \lll 3) \oplus (x_3 \ll 7)$	$E_{ux_{20}} : x_{20} = tx_{20} \oplus (x_{13} \gg 7)$
$E_{tx_{21}} : tx_{21} = (x_1 \lll 3) \oplus (x_4 \ll 7)$	$E_{ux_{21}} : x_{21} = tx_{21} \oplus (x_{14} \gg 7)$
$E_{ty_{20}} : ty_{20} = (y_0 \lll 3) \oplus (y_3 \ll 7)$	$E_{uy_{20}} : y_{20} = ty_{20} \oplus (y_{13} \gg 7)$

5.3 Results

Table 3 presents the average and median rank of the correct key byte for various noise levels σ. The statistics are done for all 20 bytes and repeated with 1000 different randomly generated keys.

Results indicate that when the noise level is low, BP helps pushing up the ranks of the correct key bytes. An attacker can then leverage the posterior probabilities returned by BP using smart key enumeration techniques to try and find the whole key. Note that since the LFSR computation only depend on the key, it is possible to reduce the noise level by averaging several traces.

Another important thing to note is that, finding the correct value of *any* $\mathsf{mask}_K^{j,i}$, not just the first one, counts as a successful attack. Depending on the actual value of $\mathsf{mask}_K^{j,i}$, some may be easier to attack than others, as was already noticed by Meraneh *et al.* [10] in the noise-free case.

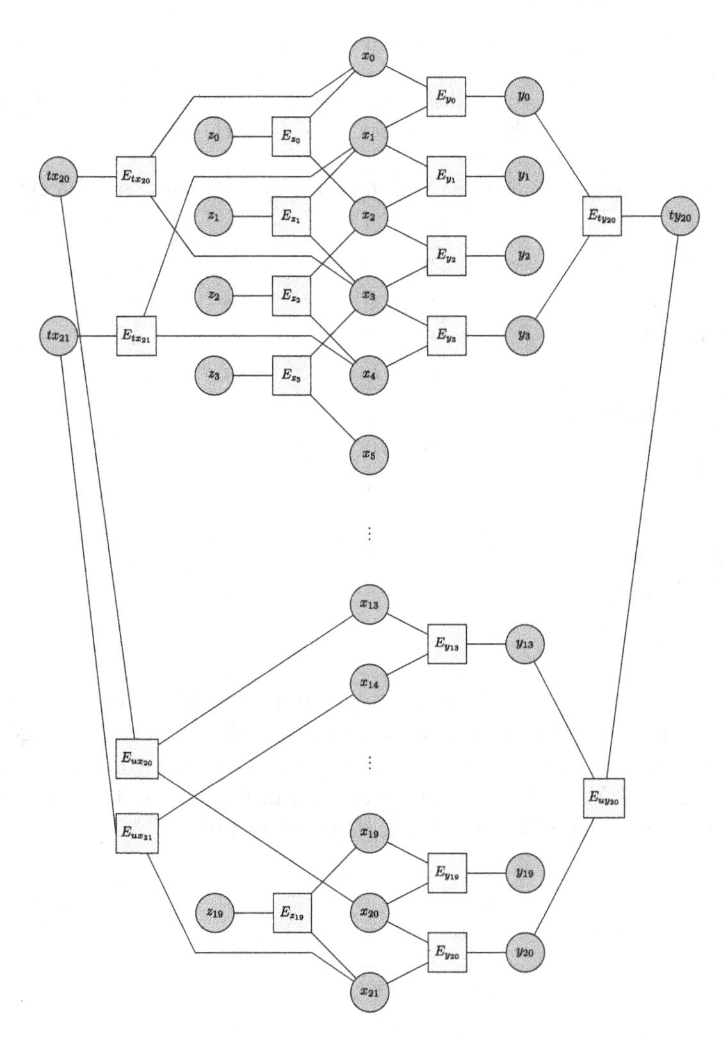

Fig. 8. Tanner factor graph on LFSR Dumbo of Elephant.

6 Attack on SPARKLE

6.1 Attack Path

During the initialisation phase of SCHWAEMM256-128, the key K is loaded into the last 128 bits of the SPARKLE384 state, see Fig. 4. Let split $K = K_1||K_2$ into two 64-bits halves. Then, in the first round, K_1 is the input of Alzette $\mathcal{A}_4$, and K_2 of Alzette $\mathcal{A}_5$.

The rest of this section only describes the attack for a single Alzette since they work exactly the same but for the constant α.

Table 3. Rank of all key bytes on 1000 different keys of 20 bytes for different noise levels σ. More precisely, BP algorithm returns probabilities a posteriorifor each possible value of each byte. These probabilities can be sorted from most likely to least likely. This table calculates the rank of the correct value for each byte.

σ	Mean	Standard Deviation	Min	Quartile Q1	Median	Quartile Q3	Max
0.1	3.54	5.97	0	0	3	3	27
0.15	3.54	5.97	0	0	3	3	27
0.2	3.67	6.11	0	0	3	3	31
0.25	4.95	9.31	0	0	3	3	97
0.3	6.00	10.76	0	0	3	3	97
0.35	7.46	13.10	0	0	3	8	97
0.4	9.59	15.72	0	0	3	8	97
0.5	15.97	23.03	0	3	8	31	153
0.6	23.65	31.41	0	3	8	31	157
0.7	33.73	40.11	0	3	27	36	213
1	70.68	61.42	0	31	36	92	246

As stated in Subsect. 2.3, Alzette is a 4-round ARX Feistel cipher, illustrated in see Fig. 6. In this use case, the attacker can obtain simulated leakage on the different internal values ℓ_i, ℓ'_i, and r_i of Alzette. Here again, we assume a leakage model as a noisy HW of the bytes, as described in Eqs. (3) and (4). The leakage observed in the first round of Alzette is depicted in Fig. 9.

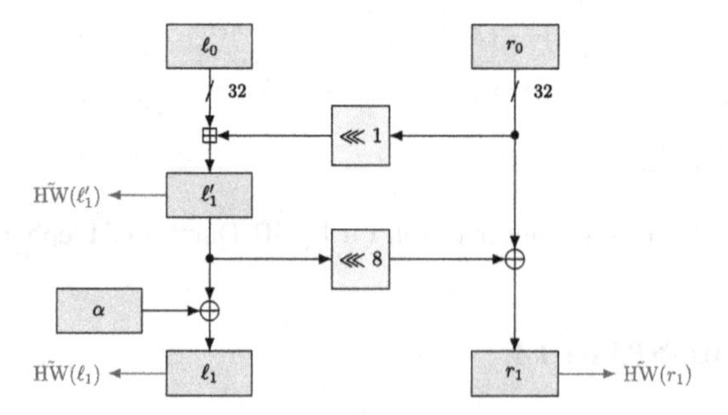

Fig. 9. Measurement leakage in the first round of Alzette described in Sect. 2.3 and Fig. 6.

The different internal values related to each other thanks to the following Alzette equations:

$$\begin{array}{lll} \ell_1' = \ell_0 \boxplus (r_0 \lll 1)\ , & \ell_1 = \ell_1' \oplus \alpha, & r_1 = (\ell_1' \lll 8) \oplus r_0 \\ \ell_2' = \ell_1 \boxplus (r_1 \lll 15), & \ell_2 = \ell_2' \oplus \alpha, & r_2 = (\ell_2' \lll 15) \oplus r_1 \\ \ell_3' = \ell_2 \boxplus r_2 \qquad\quad\ , & \ell_3 = \ell_3' \oplus \alpha, & r_3 = (\ell_3' \lll 1) \oplus r_2 \\ \ell_4' = \ell_3 \boxplus (r_3 \lll 8)\ , & \ell_4 = \ell_4' \oplus \alpha, & r_4 = (\ell_4' \lll 16) \oplus r_3 \end{array} \tag{11}$$

To the best of our knowledge, this attack is the first BSCA against SPARKLE.

6.2 Tanner Graph

In the case of SPARKLE, the variable node in the Tanner graph are defined at the bit level. They are the bits of the different ℓ_i, ℓ_i', and r_i. for $0 \leq i \leq 4$. Since modular additions are involved, a 31-bit carry variable c_i is also added at each round.

As for the equations, they are of two kinds:

- the bitwise translation of Eqs. (11) which describe Alzette,
- the HW equations (sum of bits gives the HW).

In this case, the HW are represented in the graph because they induce new relations between the different bits. The Fig. 10 shows a part of the Tanner graph of the first round of Alzette.

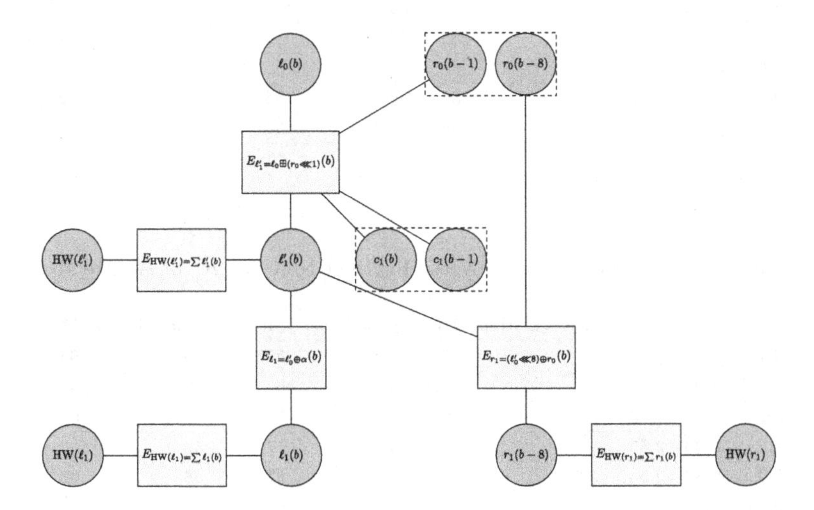

Fig. 10. Tanner graph of the first round of Alzette. Only a single bit with index $b \in [\![0, 31]\!]$ of each 32-bit register is shown. Relative indices are given mod 32. The carries in the modular operation are denoted $c_1(b)$ and $c_1(b-1)$. Bits belonging to the same 32-bit register are enclosed in dashed lines.

6.3 Results

We ran our attack on both Alzettes in a parallel manner, the only difference being the α_4 constant used by $\mathcal{A}_4$ and α_5 used by $\mathcal{A}_5$.

For the different values of the noise σ and with a single simulated measurement, the number of correct decisions on the input bits, or the number of recovered bits was calculated. The experiment was repeated on 1000 uniformly drawn random keys.

Table 4 presents the number of recovered bits of the 64-bit K_1 input of Alzette $\mathcal{A}_4$. Table 5 does the same for K_2 and $\mathcal{A}_5$. Finally, Table 6 presents the number of bits recovered for the whole 128-bit master key K of SPARKLE.

Table 4. Number of bits recovered for the 64-bit key K_1 of Alzette $\mathcal{A}_4$.

σ	Mean	Standard Deviantion	Min	Quartile Q1	Median	Quartile Q3	Max
0.1	57.08	3.02	36	56	57	59	63
0.15	57.16	2.66	38	56	57	59	63
0.2	57.20	2.57	40	56	57	59	64
0.25	57.15	2.76	38	56	57	59	64
0.3	57.07	2.74	40	56	57	59	64
0.35	56.77	2.94	36	55	57	58	64
0.4	56.63	2.81	39	55	57	58	64
0.45	56.12	3.01	35	55	56	58	64
0.5	55.81	2.81	39	54	56	57	64
0.6	54.83	2.94	36	53	55	56	64
0.7	54.19	2.68	36	53	54	56	62
1	52.86	3.47	35	52	54	55	59

Table 5. Number of bits recovered for the 64-bit key K_2 of Alzette $\mathcal{A}_5$.

σ	Mean	Standard Deviantion	Min	Quartile Q1	Median	Quartile Q3	Max
0.1	56.98	2.78	36	56	57	59	63
0.15	57.05	2.40	39	56	57	59	63
0.2	57.05	2.43	39	56	57	59	63
0.25	56.96	2.55	39	55	57	59	63
0.3	56.82	2.53	40	55	57	58	63
0.35	56.59	2.64	39	55	57	58	64
0.4	56.29	2.67	40	55	56	58	64
0.45	55.92	2.75	37	54	56	58	64
0.5	55.53	2.66	38	54	55	57	64
0.6	54.64	2.70	36	53	55	56	63
0.7	53.86	2.79	36	53	54	55	61
1	52.78	3.29	33	52	54	54	59

Table 6. Number of bits recovered for the 128-bit master key K.

σ	Mean	Standard Deviation	Min	Quartile Q1	Median	Quartile Q3	Max
0.1	114.06	4.07	89	112	114	117	123
0.15	114.22	3.59	92	112	114	116	123
0.2	114.26	3.51	97	112	114	117	124
0.25	114.11	3.76	92	112	114	116	124
0.3	113.89	3.67	97	112	114	116	125
0.35	113.36	3.94	90	111	113	116	124
0.4	112.92	3.82	95	111	113	115	124
0.45	112.05	4.09	89	110	112	115	123
0.5	111.35	3.85	92	109	111	114	123
0.6	109.48	3.97	89	108	110	112	121
0.7	108.06	3.77	87	106	108	110	120
1	105.65	4.65	85	105	107	108	115

First, note that as in the case of the attack on Elephant, it is possible to reduce the noise level by averaging several traces, since the variables handled only depend on the key.

Second, it is important to consider that the correct key is often not the first attempt, as there may be a few unknown bits for the attacker. Therefore, allowing the attacker to perform an exhaustive search can be relevant to uncover the remaining bits. That is why we grant the attacker a certain computational power.

Let n be the number of bits the attacker can flip on 64 bits. This means that they can try up to $\mathcal{P}$ keys where $\mathcal{P}$ is given by:

$$\mathcal{P} = \binom{64}{0} + \binom{64}{1} + \binom{64}{2} + \cdots + \binom{64}{n}.$$

Since they must do this for both half-keys, the total number of keys they can guess is $\mathcal{P}^2$. Table 7 indicates the number of keys found out of 1000 with an attacker possessing a computing power of $\mathcal{P}^2$ as a function of the number of bits n they can flip in a reasonnable amount of time.

It is essential to point out that this exhaustive search does not exploit the a posteriori probabilities returned by BP. Therefore, the use of smarter key enumeration techniques, exploiting these probabilities, will certainly lead to improved results without the need for a specific computing power to perform an exhaustive search. Future projects include the use of key enumeration.

Table 7. Number of keys, out of 1000, found for both Alzettes by an attacker with computational power $\mathcal{P}^2$, that can exhaustively flip up to n bits in each 64-bit subkey.

	$n=4, \mathcal{P}^2=2^{39}$		$n=5, \mathcal{P}^2=2^{46}$		$n=6, \mathcal{P}^2=2^{53}$		$n=7, \mathcal{P}^2=2^{59}$		$n=8, \mathcal{P}^2=2^{65}$	
σ	$\mathcal{A}_4$	$\mathcal{A}_5$	$\mathcal{A}_4$	$\mathcal{A}_5$	$\mathcal{A}_4$	$\mathcal{A}_5$	$\mathcal{A}_4$	$\mathcal{A}_5$	$\mathcal{A}_4$	$\mathcal{A}_5$
0.1	175	169	274	283	494	422	610	586	831	787
0.15	174	168	273	281	492	420	611	585	831	786
0.2	175	172	284	285	485	421	611	586	818	776
0.25	171	170	284	280	470	417	612	566	801	740
0.3	177	133	284	246	447	389	582	554	751	716
0.35	155	136	248	229	368	365	540	512	703	664
0.4	137	105	243	187	368	306	523	468	652	618
0.45	110	84	180	167	285	265	431	407	584	559
0.5	88	53	154	112	246	216	380	342	527	495
0.6	34	23	86	51	132	116	225	213	360	326
0.7	9	7	25	18	65	43	150	100	266	211
1	0	0	2	1	15	6	47	35	119	88

7 Conclusion

In this paper, two attacks on two lightweight AEAD algorithms have been successfully conducted and for each attack we used a single simulated measurement.

Firstly, an improvement on the attack against Elephant [10] incorporating a noisy Hamming weight model is presented. This enhanced approach allows for intelligent backtracking of the key, increasing the chances of finding the correct value.

Next, the first BSCA on SPARKLE has been described. By exploiting the attacker computational power, we have shown that the attack can indeed recover a significant portion of the secret keys when the noise level is moderate. Using backtracking on the key in an intelligent way that could improve results without doing an exhaustive search for the few remaining bits not predicted by our attack.

The power of the BP algorithm was also highlighted in our research, prompting us to develop a generic tool for it in the future. In future work, we can speed up the computation time of BP by using Walsh-Hadamard transform [37].

Our future research has to focus on transitioning these attacks into practical implementations and to develop a key backtracking algorithm that is intelligent with posterior probabilities. Additionally, we plan to explore the security of ASCON, the NIST competition winner, as a potential target for our future investigations.

Acknowledgments. This research is part of the APCIL project found by the Brittany region. The authors would like to thank Laurent Toutain.

References

1. NIST. Lightweight Cryptography Standardization Process (2018)
2. Ouladj, M., Guilley, S.: Side-channel analysis of embedded systems. Springer, Cham (2021). https://doi.org/10.1007/978-3-030-77222-2
3. NIST. Specification for the Advanced Encryption Standard. FIPS PUB 197 (2001)
4. Berti, F., et al.: A finer-grain analysis of the leakage (non) resilience of OCB. IACR T CHES (2022)
5. Sim, S.M., Jap, D., Bhasin, S.: Differential analysis aided power attack on (non-) linear feedback shift registers. IACR TCHES (2021)
6. Adomnicai, A., Masson, L., Fournier, J.J.A.: Practical algebraic side-channel attacks against ACORN. In: Lee, K. (ed.) ICISC 2018. LNCS, vol. 11396, pp. 325–340. Springer, Cham (2019). https://doi.org/10.1007/978-3-030-12146-4_20
7. Banciu, V., Oswald, E., Whitnall, C.: Exploring the resilience of some lightweight ciphers against profiled single trace attacks. In: Mangard, S., Poschmann, A.Y. (eds.) COSADE 2014. LNCS, vol. 9064, pp. 51–63. Springer, Cham (2015). https://doi.org/10.1007/978-3-319-21476-4_4
8. Beyne, T., Chen, Y.L., Dobraunig, C., Mennink, B.: Elephant v2. NIST lightweight competition (2021)
9. Beierle, C., et al.: Schwaemm and esch: lightweight authenticated encryption and hashing using the sparkle permutation family. NIST Round **2** (2019)
10. Meraneh, M.H., Clavier, C., Le Bouder, H., Maillard, J., Thomas, G.: Blind side channel on the elephant LFSR (2022)
11. Maillard, J., Meraneh, A.H., Sarry, M., Clavier, C., Bouder, H.L., Thomas, G.: Blind side channel analysis on the Elephant LFSR extended version. In: In: Van Sinderen, M., Wijnhoven, F., Hammoudi, S., Samarati, P., Vimercati, S.D.C.d. (eds.) E-Business and Telecommunications. ICSBT SECRYPT 2022. Communications in Computer and Information Science, vol. 1849, pp. 20–42. Springer, Cham (2023). https://doi.org/10.1007/978-3-031-45137-9_2
12. Barber, D.: Bayesian Reasoning and Machine Learning. Cambridge University Press, Cambridge (2011)
13. Dobraunig, C., Eichlseder, M., Mendel, F., Schläffer, M.: Ascon. Submission to the CAESAR Competition (2014)
14. Beyne, T., Chen, Y.L., Dobraunig, C., Mennink, B.: Dumbo, jumbo, and delirium: parallel authenticated encryption for the lightweight circus. IACR Trans. Symmetric Cryptology. **2020**, 5–30 (2020)
15. Bernstein, D.J.: How to stretch random functions: Secur. Protected Counter Sums. J. Cryptol. (1999)
16. Luykx, A., Preneel, B., Tischhauser, E., Yasuda, K.: A MAC mode for lightweight block ciphers. In: Peyrin, T. (ed.) FSE 2016. LNCS, vol. 9783, pp. 43–59. Springer, Heidelberg (2016). https://doi.org/10.1007/978-3-662-52993-5_3
17. Granger, R., Jovanovic, P., Mennink, B., Neves, S.: Improved masking for tweakable blockciphers with applications to authenticated encryption. In: Fischlin, M., Coron, J.-S. (eds.) EUROCRYPT 2016. LNCS, vol. 9665, pp. 263–293. Springer, Heidelberg (2016). https://doi.org/10.1007/978-3-662-49890-3_11
18. Bogdanov, A., Knežević, M., Leander, G., Toz, D., Varıcı, K., Verbauwhede, I.: SPONGENT: a lightweight hash function. In: Preneel, B., Takagi, T. (eds.) CHES 2011. LNCS, vol. 6917, pp. 312–325. Springer, Heidelberg (2011). https://doi.org/10.1007/978-3-642-23951-9_21

19. Chakraborti, A., Datta, N., Nandi, M., Yasuda, K.: Beetle family of lightweight and secure authenticated encryption ciphers. IACR TCHES. (2018)
20. Bertoni, G., Daemen, J., Peeters, M., Van Assche, G.: Duplexing the sponge: single-pass authenticated encryption and other applications. In: Miri, A., Vaudenay, S. (eds.) SAC 2011. LNCS, vol. 7118, pp. 320–337. Springer, Heidelberg (2012). https://doi.org/10.1007/978-3-642-28496-0_19
21. Kocher, P., Jaffe, J., Jun, B.: Differential power analysis. In: Wiener, M. (ed.) CRYPTO 1999. LNCS, vol. 1666, pp. 388–397. Springer, Heidelberg (1999). https://doi.org/10.1007/3-540-48405-1_25
22. Brier, E., Clavier, C., Olivier, F.: Correlation power analysis with a leakage model. In: Joye, M., Quisquater, J.-J. (eds.) CHES 2004. LNCS, vol. 3156, pp. 16–29. Springer, Heidelberg (2004). https://doi.org/10.1007/978-3-540-28632-5_2
23. Gierlichs, B., Batina, L., Tuyls, P., Preneel, B.: Mutual information analysis. In: Oswald, E., Rohatgi, P. (eds.) CHES 2008. LNCS, vol. 5154, pp. 426–442. Springer, Heidelberg (2008). https://doi.org/10.1007/978-3-540-85053-3_27
24. Samwel, N., Daemen, J.: DPA on hardware implementations of Ascon and Keyak. In: Computing Frontiers Conference. ACM (2017)
25. Linge, Y., Dumas, C., Lambert-Lacroix, S.: Using the joint distributions of a cryptographic function in side channel analysis. In: Prouff, E. (ed.) COSADE 2014. LNCS, vol. 8622, pp. 199–213. Springer, Cham (2014). https://doi.org/10.1007/978-3-319-10175-0_14
26. Le Bouder, H., Lashermes, R., Linge, Y., Thomas, G., Zie, J.-Y.: A multi-round side channel attack on AES using belief propagation. In: Cuppens, F., Wang, L., Cuppens-Boulahia, N., Tawbi, N., Garcia-Alfaro, J. (eds.) FPS 2016. LNCS, vol. 10128, pp. 199–213. Springer, Cham (2017). https://doi.org/10.1007/978-3-319-51966-1_13
27. Clavier, C., Reynaud, L.: Improved blind side-channel analysis by exploitation of joint distributions of leakages. In: Fischer, W., Homma, N. (eds.) CHES 2017. LNCS, vol. 10529, pp. 24–44. Springer, Cham (2017). https://doi.org/10.1007/978-3-319-66787-4_2
28. Clavier, C., Reynaud, L., Wurcker, A.: Quadrivariate improved blind side-channel analysis on Boolean masked AES. In: Fan, J., Gierlichs, B. (eds.) COSADE 2018. LNCS, vol. 10815, pp. 153–167. Springer, Cham (2018). https://doi.org/10.1007/978-3-319-89641-0_9
29. Yli-Mäyry, V., et al.: Diffusional side-channel leakage from unrolled lightweight block ciphers: a case study of power analysis on PRINCE. IEEE Trans. Inf. Forensics Secur. **16**, 1351–1364 (2020)
30. Gallager, R.G.:. Low-density parity-check codes. IRE Trans. Inf. Theory **8**, 21–28 (1962)
31. Gallager, R.G.: Low Density Parity check codes. PhD thesis, MIT, Cambridge, MA (1963)
32. Chung, S.-Y., Forney Jr, G.D., Richardson, T.J., Urbanke, R.L.: On the design of low-density parity-check codes within 0.0045 dB of the Shannon limit. IEEE Commun. Lett. **5**, 58–60 (2001)
33. Tanner, R.M.: A recursive approach to low complexity codes. IEEE Trans. Inf. Theory **27**, 533–547 (1981)
34. Pearl, J.: Reverend bayes on inference engines: a distributed hierarchical approach. In: National Conference on Artificial Intelligence. AAAI Press (1982)
35. Veyrat-Charvillon, N., Gérard, B., Standaert, F.-X.: Soft analytical side-channel attacks. In: Sarkar, P., Iwata, T. (eds.) ASIACRYPT 2014. LNCS, vol. 8873, pp.

282–296. Springer, Heidelberg (2014). https://doi.org/10.1007/978-3-662-45611-8_15
36. Grosso, V., Standaert, F.-X.: ASCA, SASCA and DPA with enumeration: which one beats the other and when? In: Iwata, T., Cheon, J.H. (eds.) ASIACRYPT 2015. LNCS, vol. 9453, pp. 291–312. Springer, Heidelberg (2015). https://doi.org/10.1007/978-3-662-48800-3_12
37. Ouyang, W., Cham, W.K.: Fast algorithm for Walsh Hadamard transform on sliding windows. Trans. Pattern Anal. Mach. Intell. **32**, 165–171 (2009)

Leveraging Coprocessors as Noise Engines in Off-the-Shelf Microcontrollers

Balazs Udvarhelyi[1,2](✉) and François-Xavier Standaert[1]

[1] UCLouvain, ICTEAM, Crypto Group, Louvain-la-Neuve, Belgium
{balazs.udvarhelyi,francois-xavier.standaert}@uclouvain.be
[2] ST Microelectronics, Diegem, Belgium
balazs.udvarhelyi@st.com

Abstract. Securing low-cost microcontrollers against side-channel attacks is an important challenge. One core issue for this purpose is that such devices may exhibit leakages with very limited noise. As a result, standard countermeasures like shuffling or masking, which emulate or amplify noise, have limited effectiveness. In this paper, we investigate the possibility to run hardware coprocessors in parallel to a masked software implementation, in order to generate algorithmic noise. We detail the conditions for such a noise generation to be effective and show experimental evidence that it leads to security improvements compared to masked software implementations running without activated coprocessors. While masking remains expensive, the gains we show in number of traces to recover the key are systematic: an approximate factor two in our experiments, that is raised to the number of masking shares.

1 Introduction

Side-channel attacks are an important threat against which various countermeasures exist. Among generic solutions that can be applied independent of the algorithms, a standard idea is to emulate and amplify the physical noise inherently present in the measurements. Masking [12,22] and shuffling [24,38] are typical examples of such countermeasures. In particular, it is now established that assuming shares' leakages that are sufficiently independent, masking can amplify the noise (variance) exponentially in the number of shares [3,16,17,30,31].

In general, such countermeasures can of course only work if there is indeed some noise inherently present in the leakages. This requirement turns out to be quite easily achieved in hardware implementations contexts (e.g., FPGAs, ASICs), since designers then have a good level of control on the design. As a result, the primary constraint in those cases is to ensure the independence of the leakages, which is reflected by the security order (i.e., the smallest statistical moment of the leakage distribution that depends on the target secret). This is still not trivial and requires dealing with physical defaults like glitches or transitions, but it is now quite well understood and design principles exist to deal with such defaults. See for example [8,20,28] for glitches and [1,10,15] for transitions.

S. Bhasin and T. Roche (Eds.): CARDIS 2023, LNCS 14530, pp. 148–165, 2024.
https://doi.org/10.1007/978-3-031-54409-5_8

Unfortunately, the situation quite significantly differs in the context of software implementations running on Commercially available Off-The-Shelf (COTS) Micro-Controller Units (MCUs). On the one hand, such (small) devices usually exhibit limited physical noise. On the other hand, their serial nature lets the adversary mounting so-called horizontal attacks that combine the leakages of multiple operations in order to further reduce the noise [2]. As a result, a recent work showed that state-of-the-art masked implementations of the AES can be broken nearly independently of the security order (i.e., that exploiting the low noise is the best strategy independent of physical defaults like transitions that may show up in software implementations) [7].

In this paper, we are therefore interested in possibilities to improve the noise level of COTS devices. Our general idea relies on the presence of peripherals alongside the main CPU of recent MCUs, which could be run in parallel and provide additional algorithmic noise. Such an algorithmic noise could in turn make masking more effective and possibly reduce the number of shares needed to reach a given security level. Natural candidates for this purpose are cryptographic coprocessors as they have the advantage of producing seemingly random computations, can be supplied via the same power supply and may rely on large (parallel) architectures generally leading to large(r) levels of noise.

In order to evaluate the effectiveness of such a possibility, we run state-of-the-art attacks against a masked bitslice implementation of the AES with different number of shares. We show that running dummy coprocessor operations in parallel to the main (masked) computations indeed improves the noise level by a factor $f \approx 2$ in our experiments, leading to an increase of the best attacks' complexity by a factor $\approx 2^n$ for n shares. As a side result, we also show that the MCU we consider, which uses a more advanced technology than the one of [7], leads to slightly less informative leakages even without running coprocessors in parallel. These combined observations lead to improved security for masked software implementations. Reaching high security levels remains expensive but overheads are reduced. For example, we were not able to attack a 4-share implementation with additional noise in less than one million traces.

We note that our work is focused on power measurements (or more generally, global leakages). This excludes advanced adversaries taking advantage of high-resolution localized measurements like [35,36], which generally work best in an invasive attack setting (i.e., after depackaging). We leave the investigation of such advanced attacks as an interesting open problem.

2 Background

In this section, we describe the necessary tools that are used in the paper. Namely, we first describe out notations, then the Signal-to-Noise Ratio (SNR), the Regression-based Linear Discriminant Analysis (RLDA) leakage model, Soft Analytical Side-Channel Attacks (SASCA) and the Perceived Information (PI) metric. Finally, we describe the masking countermeasure.

2.1 Notations

We use capital letters X for random variables, bold letters $\boldsymbol{x}$ for vectors, capital bold letters $\boldsymbol{X}$ for matrices and calligraphic letters $\mathcal{X}$ for sets. We use subscript to indicate shares, if relevant. We additionally use the following conventions: n_s is the number of samples in a trace, n_p and n_a are respectively the number of profiling and attack traces, b is the number of bits of the profiled variable, and p is the dimensionality of the subspace used by RLDA.

2.2 Signal-to-Noise Ratio

The SNR is a common univariate metric in side-channel analysis. It was first defined by Mangard in [26]. For an intermediate variable X, it models the signal as the variance of the mean leakage of each class $x \in \mathcal{X}$, and the noise as the mean of the variance of the leakage of each class:

$$\hat{\text{SNR}} = \frac{\hat{\mathsf{Var}}_x\left(\hat{\mathsf{E}}_i\left(l_i^x\right)\right)}{\hat{\mathsf{E}}_x\left(\hat{\mathsf{Var}}_i\left(l_i^x\right)\right)}, \tag{1}$$

where l_i^y corresponds to the ith leakage sample of variable y. $\hat{\mathsf{Var}}_i$ and $\hat{\mathsf{E}}_i$ (resp., $\hat{\mathsf{Var}}_x$ and $\hat{\mathsf{E}}_x$) are the sample variance and the sample mean over the leakages (resp. the classes). We note here that the modeled noise is a combination of physical and algorithmic noise. Mangard's SNR is a good estimator for the complexity of univariate attacks like Correlation Power Analysis or (univariate) template attacks as their complexity can be directly linked to the SNR [18,26,27]. The SNR can also be used as tool to detect Points-of-Interest (POIs) as the time samples in a measured leakage trace where the SNR is high represent samples for which first-order information can be extracted.

2.3 Regression-Based Linear Discriminant Analysis

RLDA has been introduced in [14]. Its core idea was to replace the mean of each class in the equations of Linear Discriminant Analysis (LDA) [33] by a value obtained through linear regression [32]. We use the efficient implementation of Cassiers et al. [11], which combines the efficient profiling of large states enabled by linear regression and the ability to profile models for long traces enabled by the dimensionality reduction embedded into LDA.

Internally, RLDA first fits a linear regression model with n_b basis functions β_i corresponding to the b bits of the target variable and the intercept:

$$\beta_i(x) = \begin{cases} 1 & \text{if } i = 0, \\ 1 & \text{if } \lfloor x/2^{i-1} \rfloor \mod 2 = 1, \\ -1 & \text{otherwise}, \end{cases} \tag{2}$$

with $n_b = b + 1 = \lceil \log_2(|\mathcal{X}|) \rceil + 1$. The regression model to fit for each leakage sample s is:

$$\mathsf{m}_s(x) = \sum_{i=0}^{n_b-1} a_{i,s}\beta_i(x),$$

where $a_{i,s}$ corresponds to the ith coefficient of leakage sample s. We define the mean vector of x as $\mathsf{m}(x) = \boldsymbol{A}\beta(x)$ with $\boldsymbol{A} \in \mathbb{R}^{n_s \times n_b}$ the matrix of all coefficients. Then, this model is used to calculate the inter and intra-class scatter matrices:

$$\boldsymbol{S_B} = \sum_{x \in \mathcal{X}} |\mathcal{L}(x)| \, (\mathbf{m}(x) - \hat{\boldsymbol{\mu}})(\mathbf{m}(x) - \hat{\boldsymbol{\mu}})^T, \tag{3}$$

$$\boldsymbol{S_W} = \sum_{x \in \mathcal{X}} \sum_{\boldsymbol{l} \in \mathcal{L}(x)} (\boldsymbol{l} - \mathbf{m}(x))(\boldsymbol{l} - \mathbf{m}(x))^T, \tag{4}$$

where $\mathcal{L}(x)$ defines the subset of the trace set $\mathcal{L}$ where $x = X$. By solving the following problem :

$$\boldsymbol{W} = \underset{\boldsymbol{W} \in \mathbb{R}^{n_s \times p}}{\arg\max} \frac{|\boldsymbol{W}^T \boldsymbol{S_B} \boldsymbol{W}|}{|\boldsymbol{W}^T \boldsymbol{S_W} \boldsymbol{W}|},$$

we find the projection matrix maximizing the SNR in the projected subspace, where p is a parameter corresponding to the number of dimensions in the subspace [19]. Next, the covariance matrix of the Gaussian model in the subspace is estimated from the intra-class scatter $\hat{\Sigma}_{\boldsymbol{W}} = |\mathcal{L}|^{-1} \boldsymbol{W}^T \boldsymbol{S_W} \boldsymbol{W}$, and a second, normalizing projection is applied such that the covariance matrix becomes identity. To do so, by computing the eigendecomposition of the symmetric positive-definite covariance matrix:

$$\hat{\Sigma}_{\boldsymbol{W}} = \boldsymbol{V} \boldsymbol{\Lambda} \boldsymbol{V}^T,$$

where $\boldsymbol{V}$ and $\boldsymbol{\Lambda}$ are respectively the matrix of eigenvectors and eigenvalues, we define the normalizing projection $\boldsymbol{W}^{\mathsf{norm}} = \boldsymbol{V}\boldsymbol{\Lambda}^{-1/2}$. Eventually, the RLDA model is obtained by:

$$\hat{\mathsf{f}}[\boldsymbol{l}|X = x] = \frac{1}{\sqrt{(2\pi)^p}} \exp\left(-\frac{1}{2} \left\|\boldsymbol{W}^{\mathsf{RLDA}}\boldsymbol{l} - \boldsymbol{A}^{\mathsf{RLDA}}\beta(x)\right\|^2\right), \tag{5}$$

with $\boldsymbol{W}^{\mathsf{RLDA}} = \boldsymbol{W}\boldsymbol{W}^{\mathsf{norm}}$ the combined projection matrix and $\boldsymbol{A}^{\mathsf{RLDA}} = \boldsymbol{W}^{\mathsf{RLDA}}\boldsymbol{A}$ the projected coefficients. The likelihood of X conditioned on the leakages are obtained using Bayes' law:

$$\hat{\mathsf{f}}[X = x|\boldsymbol{l}] = \frac{\hat{\mathsf{f}}[\boldsymbol{l}|X = x]\mathsf{Pr}[X = x]}{\sum_{x' \in \mathcal{X}} \hat{\mathsf{f}}[\boldsymbol{l}|X = x']\mathsf{Pr}[X = x']}.$$

In the following sections, we also use Gaussian templates with LDA-based dimensionality reduction [13,33], of which the calculation is essentially the same except that the means in Eq. 3 are calculated exhaustively.

2.4 Perceived Information

Evaluating the success rate of a multivariate attack cannot be done with the SNR (which is a univariate metric). The Mutual Information (MI) between the leakage and target variable is a good candidate for this purpose. However, it is notoriously hard to estimate. We therefore use the PI as a surrogate, which provides an easy to estimate lower bound [5]. It can be computed by sampling the model $\hat{\mathsf{f}}$ as follows:

$$\hat{\mathsf{PI}}(X, \boldsymbol{L}) = \mathrm{H}(X) + \sum_{x \in \mathcal{X}} \mathsf{Pr}[x] \sum_{\boldsymbol{l}' \in \mathcal{L}'(x)} \frac{1}{|\mathcal{L}'(x)|} \log_2 \hat{\mathsf{f}}[x|\boldsymbol{l}'], \tag{6}$$

where $\mathrm{H}(X)$ is the entropy of X and $\mathcal{L}'$ is the set of traces used to estimate the PI, which must differ from the set of traces $\mathcal{L}$ used to fit the model.

2.5 Boolean Masking

Masking is a common countermeasure against side-channel attacks. It relies on an encoding of any sensitive variable x into a tuple of n shares:

$$x = x_1 \oplus x_2 \oplus \ldots \oplus x_n,$$

where the $n-1$ first shares are selected uniformly at random, so that an adversary needs knowledge of all the shares in order to recover the sensitive information. This security guarantee is easily expressed in the abstract probing model [25], which states security if an adversary who can probe up to $n-1$ wires in the circuit does not learn anything about x.

From an implementation viewpoint, linear operations are performed share-by-share and have a linear complexity. Multiplications of two encodings are more complex: they have quadratic complexity in the number of shares and require additional randomness to remain d-probing secure. Popular algorithms are the ISW multiplication [25] and the PINI gadgets introduced in [9].

Probing security reduces to noisy leakage security [16], which is closer to real-world measurements. The noisy leakage model assumes that all the shares leak noisy observations to the adversary. In this case, the data complexity N of the attack is inversely proportional to the product of the MI between each share and the leakage [3], so that:

$$N \geq \frac{c}{\prod_{i=1}^{n} \mathrm{MI}(X_i, L)}, \tag{7}$$

increases exponentially with the number of shares given that the MI per share is small enough. As mentioned in introduction, this guarantee only holds as long as the leakage function ensures some independence between the shares' leakages (i.e., as long as it does not recombine the shares). When replacing the MI by the PI in this equation, we no longer have a lower bound but an estimate that measures the particular model used to estimate the PI.

Attacking a masked implementation requires the adversary to model each share independently to obtain probability densities on the shares $\hat{\mathsf{f}}[x_i|\boldsymbol{l}]$, and then to recombine them to obtain probability densities on the target variable:

$$\hat{\mathsf{f}}[x|\boldsymbol{l}] \propto \sum_{\{x_0,\ldots,x_{n-1}|\sum x_i = x\}\in\mathcal{X}^n} \prod_{i=1}^{n} \hat{\mathsf{f}}[x_i|\boldsymbol{l}] \tag{8}$$

2.6 Soft Analytical Side-Channel Attacks

The previous sections showed how we can profile the leakage of an intermediate variable X. However, during the computation of a masked encryption algorithm, several intermediate states exist which can all leak useful information. Directly profiling many variables in the same template is rapidly impractical as the number of classes grows exponentially. SASCA has been introduced as an efficient way to profile several variables independently and exploit them jointly [37]. For this purpose, it models the relations between the variables with a factor graph and leverages the Belief Propagation (BP) algorithm.

Concretely, the factor graph is a bipartite graph containing the variables (circles) on one side and the relations (squares) on the other. The edges represent the relations between the variables. An example is given in Fig. 1.

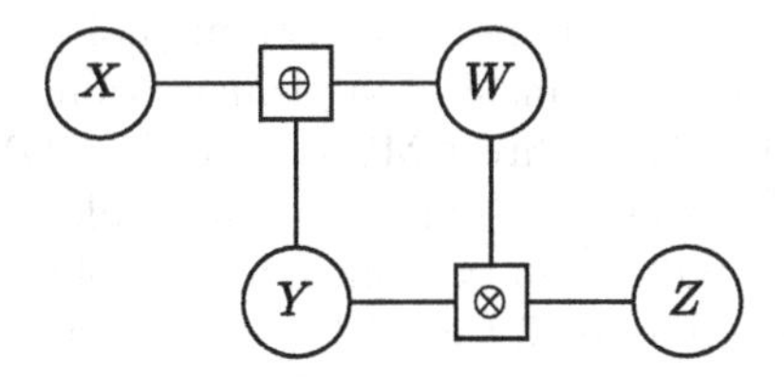

Fig. 1. Exemplary factor graph for $X \oplus Y = W$ and $Y \otimes Z = W$.

Internally, the BP algorithm iterates over 3 steps. At iteration t, we have:

1. The belief of a variable X is computed as the product of its likelihood and the beliefs coming from its neighbors ∂X.

$$P_X^{t+1}(x) = \hat{\mathsf{p}}(X = x) \prod_{R\in\partial X} m_{R\rightarrow X}^t(x),$$

2. From variable X to a relation R, the belief is the product of the beliefs on the variable divided by the belief from the destination relation.

$$m_{X\rightarrow R}^{t+1}(x) = P_X^t / m_{R\rightarrow X}^t(x).$$

3. From relation R to variable X, the belief is the sum over all compatible values for the relation of the products of the beliefs from all neighbors of R except X, where we denote the compatibility function with ψ_R, which has a value of 1 if the values are compatible with the relation and 0 otherwise:

$$m_{R \to X}^{t+1}(x) = \sum_{x_i \in \mathcal{X}_i \text{ s.t. } \{X_1,\ldots,X_k\} = \partial R \setminus X} \psi_R(x, x_1, \ldots, x_k) \prod_{i=1}^{k} m_{X \to R}^{t}(x_i),$$

This algorithm is proven to converge on a tree-like structure with a number of iterations corresponding to twice the diameter of the graph. With graphs containing cycles like Fig. 1, the algorithm becomes a heuristic.

3 Noise Generation

In this section, we detail the implementation of our coprocessor-based noise engine. In order to be characterized as noise, the coprocessor should process hard-to-predict data, with ideally random states at each cycle. This way, an adversary is not able to predict the data and its leakage, which would make the noise deterministic and easy to filter. It should ideally be slow (i.e., take many cycles of computation) but quick to configure such that interrupts of the main task are sparse and short. The coprocessor should also rely on a large (parallel) architecture in order to generate more algorithmic noise and to better hide the leakage of the CPU that runs a masked implementation.

In practice, we configured a Direct Memory Access (DMA) peripheral which loads the buffer to process into to coprocessor and unloads it without the CPU being interrupted. The buffer can contain multiple blocks of data to process. When the buffer is fully processed, the CPU gets interrupted, the DMA and the coprocessor are reset and the cycle can restart. The buffer size therefore gives a trade-off parameter between the memory used and the overhead in execution time due to the interrupts. We note that the suitable coprocessors of the MCU we used did not allow circular DMA transfers.[1] Hence we needed to reset after processing the buffer. A representation of this process is shown in Fig. 2.

The coprocessor we choose for our experiments was an AES-128 implementation with a 128-bit architecture, performing the encryption of one block of data in 16 cycles (i.e., one cycle per round). In order for its states to remain unpredictable, we put this coprocessor in a leakage-resilient mode of operation for which it is expected that a large parallel implementation provides sufficient security. Concretely, we used a construction similar to the one in [6] where it is shown that the 128-bit coprocessor maintains 64 bits of security as long as we re-key it every 100 encryptions. We used the True Random Number Generator (TRNG) of the MCU to generate the random buffer and initializing the coprocessor with a random key. The TRNG generated a 32-bit random word every

[1] Which are DMA transfers where the DMA automatically resets its pointer to the start address at the end of the buffer, without input from the CPU.

40 cycles. After each completed buffer, we generate a new random key with the TRNG and encrypt the buffer with the new key. Overall, this strategy keeps performance overheads due to resets quite small (a few percents).

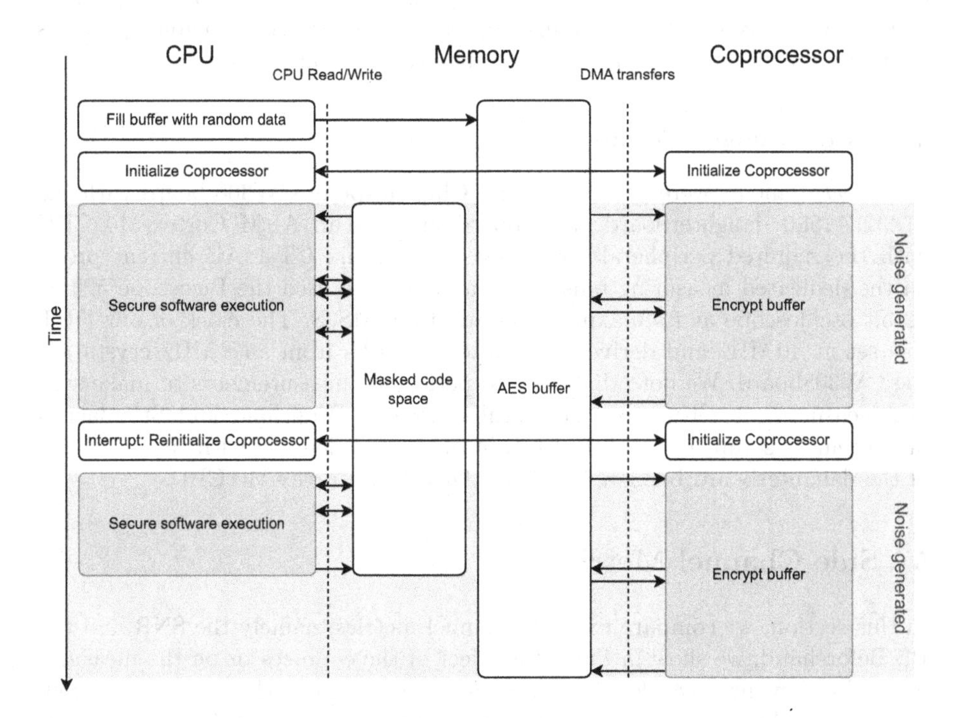

Fig. 2. Execution scheme of the noise engine.

We note that reducing the buffer size would increase the computation time of our masked implementation, but improve the security of the leakage-resilient mode of operation running in parallel. It could also improve the security of the masked implementation due to the desynchronization of the traces that it would imply (since interrupts could then act as somewhat random delays).

4 Evaluation Setup

We now describe the implementation we analyze and our measurement setup.

4.1 Description of the Target

The implementation under test is the bitsliced AES implementation of Goudarzi and Rivain [23] with state-of-the-art PINI gadgets from [9]. We precisely used the assembly code provided in [4]. Bitslicing allows efficient masked software implementations. It works by splitting each word and placing each bit into a different register, so that bitwise operations can be applied at the register level. Given the size of the CPU, we are not limited to one bit per register and can easily parallelize all the 16 S-boxes of the AES in 32-bit registers. In the case of masked implementations, each register is shared into n registers such that in one register, only bits of one share are present. This avoids recombining shares in the barrel shifter, also known as the shareslicing issue [21].

4.2 Measurement Setup

Our measurements were performed on a Chipwhisperer CW308 board with the `STM32F415RG` daughterboard. This board includes an ARM Cortex-M4 CPU with the required peripherals. We used a Tektronix CT-1 AC current probe on the dedicated measuring pins.[2] For sampling, we used the Picoscope 5244D 12-bit oscilloscope at its maximum speed of 500 MS/S. The clock of our DUT was set at 40 MHz and derived with internal PLLs from an 8 MHz crystal on the CW308board. We note that we also performed measurements to make sure our measurement setup produces results close to Bronchain and Standaert's measurement setup [7]. For this, we kept our setup identical with the exception of the daughterboard being a `STM32F051R4` with a Cortex-M0 CPU.

5 Side-Channel Metrics

In this section, we compare two side-channel metrics, namely the SNR and the PI. Beforehand, we show in Fig. 3 the effect of the coprocessor on the measurements by showing mean leakage traces with and without the noise generation. The mean leakage traces were computed over 100k non-averaged traces and represent the execution of the first S-box layer of the AES. We clearly see the impact of the co-processor: without it, groups of operations are distinguishable; when activated, they are hidden within the generated noise.

5.1 SNR Comparison

First, we compare the SNR with and without the noise engine for each state in the computation of the first S-box Layer. The target states have a 16-bit width, corresponding to the 16 S-Boxes computed in parallel. In each word, 16 bits correspond to the 16 S-boxes and the 16 remaining ones are set to zero. Thus, each bit of the bus is modeled and no algorithmic noise comes from the bus. Every state corresponds to a share of an intermediate variable and is processed independently. We used 500k traces with random inputs to compute the SNR.

In Fig. 4, we show the (123×2) SNR curves calculated for each intermediate state of the AES bitslice S-box for $n = 2$ shares, with (top) and without (bottom) noise generation. We observe two types of SNR peaks in the upper figure: the tighter ones are the XOR (and NXOR) gadgets, the more spread out, and slightly

[2] This effectively shorts the onboard shunt resistor.

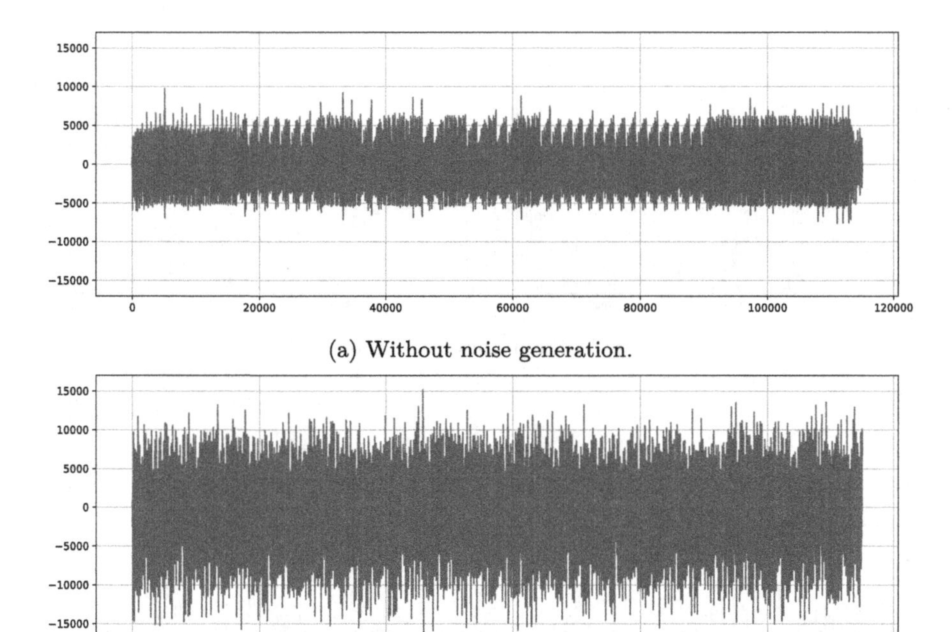

(a) Without noise generation.

(b) With noise generation.

Fig. 3. Comparison of mean leakage traces.

higher ones correspond to the AND gadgets. The impact of the noise generation appears on the bottom figure, where all time samples exhibit a reduced SNR. There is no significant alteration of the shape of the SNR curves. We observed a similar reduction for implementations with more shares.

5.2 PI Comparison

In order to evaluate the reduction of the PI due to the coprocessor, we first detail the steps followed to generate our leakage models. First, for each intermediate state, we find our Points-Of-Interest (POIs) which we define as the time samples that are higher than the noise floor of the SNR. It leads to around 2000 samples for each state share. We then profile these intermediate state shares with 500k traces using the RLDA leakage model on 16-bits, with 10 dimensions in the subspace. Finally, 5000 fresh traces are used to evaluate the PI.

To justify these parameters, Fig. 5 shows the extracted information for an exemplary state share as a function of the number of profiling traces used, with and without the noise generation. It can be seen that all models converge with $\approx$ 100k traces. Furthermore, increasing the number of dimensions (e.g., beyond $p = 4$) has limited impact, especially in the noisy case (bottom figure).

We next show the PI per share of the input of the S-box layer in Table 1, computed from a 2-share implementation. We observe that the 8 words of the

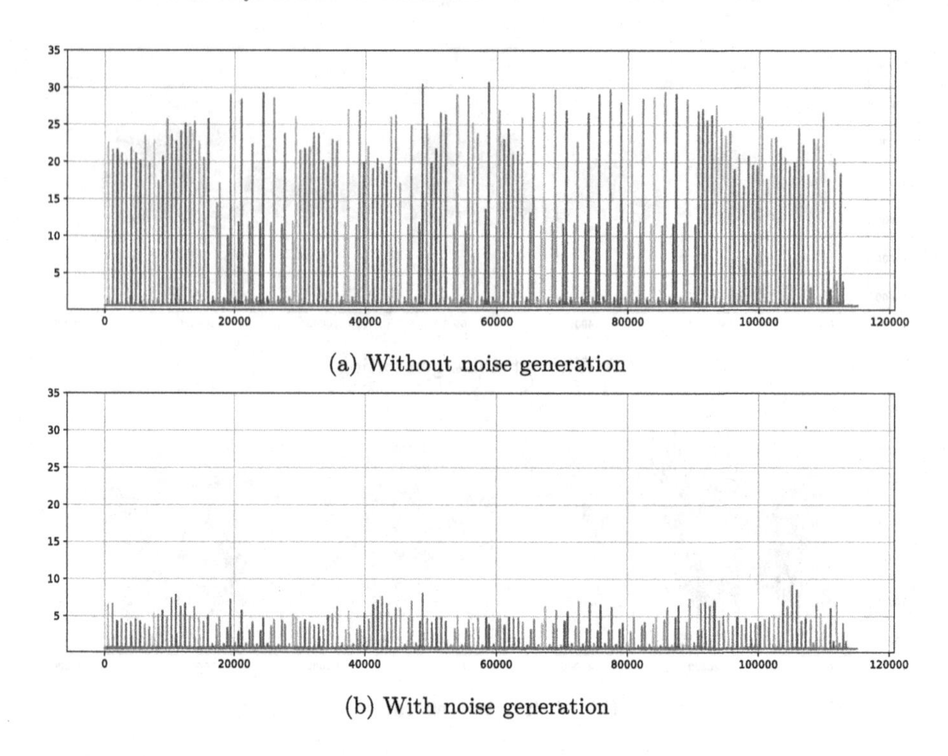

(a) Without noise generation

(b) With noise generation

Fig. 4. SNR for each share of the first S-box layer masked with two shares.

AES state have slightly different levels of information, but the noise produced by the coprocessor consistently reduces it. This is independent of the share that is being modeled. We also note that the PI per share of the linear operations is stable with the number of shares (since these operations are applied share-wise). Since our following attacks will primarily exploit this information, we do not detail the evaluation of the PI per share for the non-linear operations that takes place for larger numbers of shares (due to multiple shares manipulations). On average, we observed a reduction by an approximate factor 2 for the PI per share when activating the noise generation. Based on this value, we can extrapolate the security gains that the noise generation brings thanks to Eq. 7: it should increase the data complexity by an approximate factor 2^n with n shares.[3]

For completeness, we also computed the results obtained with two 8-bit LDA-based models, in order to compare them with the 16-bit RLDA-based model. As shown in Table 2, this leads to significant reduction of the perceived information. This last table is interesting since it uses a similar model as [7] and shows significantly lower PI values that this previous work obtained with a Cortex-M0

[3] This is assuming that the independence condition holds to a sufficient degree. Since the gadgets we use were tested against such defaults in [4] and the possible reduction of the security order due to transitions is orthogonal to the noise issue we discuss, we did not reproduce this part of the experiments in the paper.

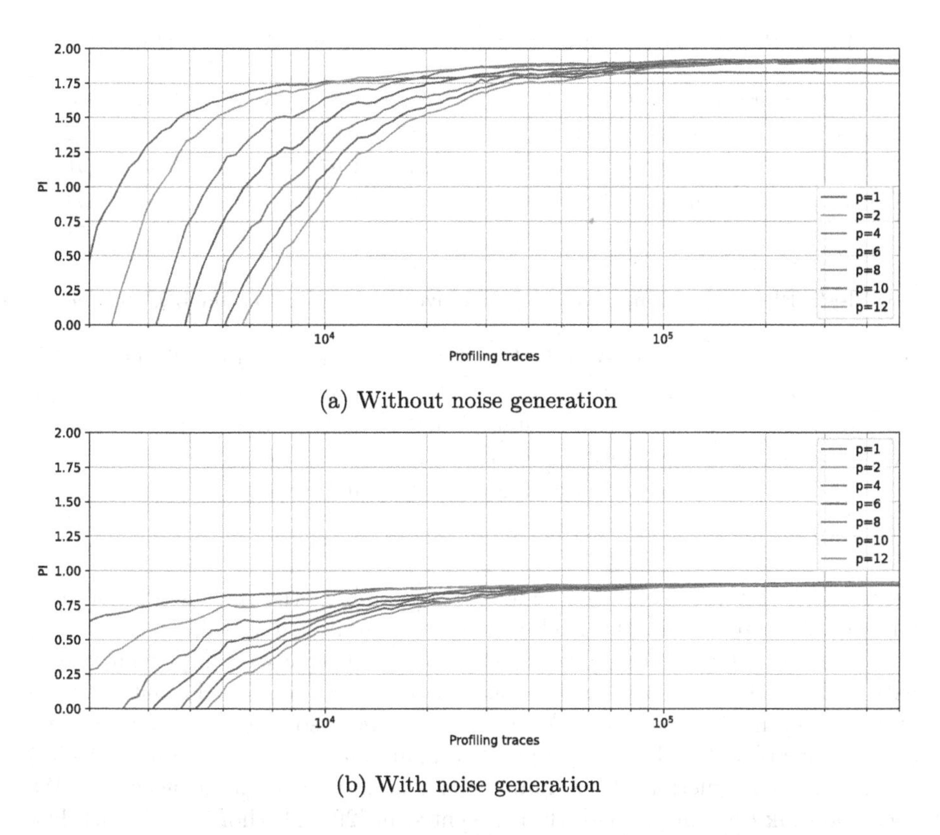

(a) Without noise generation

(b) With noise generation

Fig. 5. PI per share obtained for an exemplary variable.

STM32F0 MCUs. We repeated the measurements of [7] with our measurement setup (just plugging/unplugging the devices) and obtained similar results. So we posit that the changes are due to the higher complexity of the Cortex-M4 and a change of technology node, from 180 nm for the F0 line to 90 nm for the more recent (and lower power) F4 line we evaluate in this paper.[4]

6 Attack Results

In this section, we finally present concrete attacks against implementations with different masking orders, and discuss the effectiveness of the proposed noise generation. First, a baseline template attack is shown against the key addition. Next, we compare our results to the attack presented in [7]: a SASCA exploiting the leakages of all the intermediate states of the S-box layer.

This baseline attack is a textbook template attack on each of the 8 words of the key addition. We profile each share of each word as explained in Subsect. 5.2,

[4] https://blog.st.com/stm32g0-mainstream-90-nm-mcu/
https://www.st.com/en/microcontrollers-microprocessors/stm32f405-415.html.

Table 1. PI per share: input of the S-box layer, 2 shares, 16-bit RLDA model.

	Share #	Word 0	Word 1	Word 2	Word 3	Word 4	Word 5	Word 6	Word 7
No noise	0	1.90	1.53	1.21	1.89	0.81	2.00	2.47	3.12
	1	1.97	1.64	1.29	1.97	0.67	2.12	2.60	3.91
With noise	0	0.95	0.71	0.42	0.72	0.37	0.92	1.06	1.69
	1	1.06	0.67	0.46	0.76	0.30	1.10	1.12	1.78

Table 2. PI per share: input of the S-box layer, 2 shares, two 8-bit LDA models.

	Share #	Word 0	Word 1	Word 2	Word 3	Word 4	Word 5	Word 6	Word 7
No noise	0	0.89	0.75	0.76	0.98	0.57	0.87	1.25	1.56
	1	0.98	0.96	0.86	1.05	0.50	1.18	1.56	2.39
With noise	0	0.62	0.49	0.35	0.52	0.30	0.59	0.75	1.01
	1	0.70	0.49	0.40	0.55	0.26	0.79	0.77	1.21

using the 16-bit RLDA models. Then, we recombine the likelihoods obtained on our shares to obtain the likelihood of the unmasked variable as shown in Eq. 8. Figure 6 shows the results of attacks against 3 masked implementations, with 2, 3 and 4 shares. For each number of shares, we measured a dataset with and without noise generation. As a reference, we also ran the attack against an implementation without masking. The figure shows the mean rank (in bits) of the key as a function of the number of traces used in the attacks [34]. We used the rank estimation algorithm presented in [29]. The horizontal black line represents a rank of 32 bits, set as a limit for the success of the attack (below that rank, enumeration is almost instantaneous).

First, looking at the dashed lines, the impact of masking is clearly visible. For each additional share, roughly 10 times more traces are needed. Next, the impact of the noise generation can be seen in the rightwards shift when moving from the dashed curves to the plain curves. As expected from our previous extrapolation, the security gain brought by the additional algorithmic noise is amplified as the number of shares increases. This is reflected by the gap between curves of the same color, that grows roughly with 2^n. For the 4-share design, we were not able to reduce the key rank below 2^{32} with one million attack traces.

For completeness, we repeated the SASCA of [7], where the adversary computes probabilities on the intermediate variables using Eq. 8 and combines them thanks to BP. We performed this attack using the same datasets as the baseline attack. All variables were profiled using the same RLDA models and parameters. The factor graph of the AES being cyclic, we studied the number of iterations and observed the best results after 5 or 6 iterations, depending on the cases. Iterating the BP algorithm further lead to worse ranks on the key.

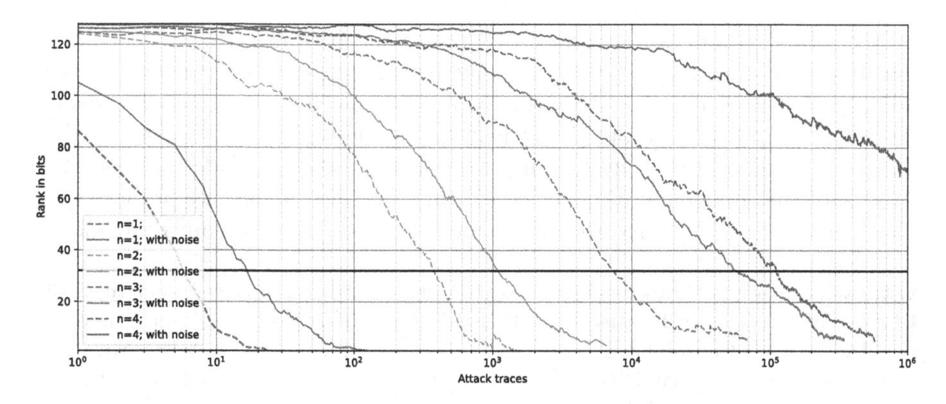

Fig. 6. Key rank for ≠ number of shares, with and without noise generation.

The results of the SASCA and baseline attacks are compared in Table 3, which gives the number of traces required to reach the rank of 2^{32} (mostly because the improvements are hardly distinguishable from Fig. 6). The improvement of the SASCA over the baseline attack is limited and decreases with the masking order. We posit that this limited effectiveness is due to the lower information obtained on the target intermediate variables combined with the information loss when propagating beliefs through XOR operations.

Table 3. Data complexity to obtain a 2^{32} rank with SASCA (vs. template attack).

Noise generation	$n = 2$		$n = 4$		$n = 4$	
	Off	On	Off	On	Off	On
TA	373	1101	7500	36500	110500	NA
SASCA	295	1024	7000	36000	109000	NA
Gain in %	21	7	7	1.3	1.3	NA.

7 Conclusions

The versatility of low-end MCUs makes them appealing solutions for deployment in various applications. Yet, their simplicity also makes them good targets for physical attacks. Previous works showed that their low-noise makes them vulnerable to horizontal side-channel analysis. In this work, we show that the move to more advanced manufacturing technologies combined with the generation of algorithmic noise thanks to coprocessors running in parallel to masked software implementations can improve this situation with limited performance overheads.

Activating co-processors admittedly has a cost in terms of power consumption. However, we expect that this higher power consumption can be compensated by the lower number of shares required to reach a given security level – the concrete confirmation of this tradeoff being an interesting scope for further research. It would also be interesting to further improve the side-channel security of software implementations by combining the noise engines we propose with other heuristic means to reduce the leakage (e.g., via time randomizations).

Acknowledgments. François-Xavier Standaert is a senior research associate of the Belgian Fund for Scientific Research (FNRS-F.R.S.). This work has been funded in parts by the ERC Consolidator grant 724725 (acronym SWORD), the ERC Advanced Grant 101096871 (acronym BRIDGE) and by the EU-Walloon FEDER Project USER-Media (convention 501907-379156). Views and opinions expressed are those of the authors only and do not necessarily reflect those of the European Union or the European Research Council. Neither the European Union nor the granting authority can be held responsible for them.

References

1. Balasch, J., Gierlichs, B., Grosso, V., Reparaz, O., Standaert, F.: On the cost of lazy engineering for masked software implementations. In: Joye, M., Moradi, A. (eds.) CARDIS. Lecture Notes in Computer Science, vol. 8968, pp. 64–81. Springer, Heidelberg (2014). https://doi.org/10.1007/978-3-319-16763-3_5
2. Battistello, A., Coron, J., Prouff, E., Zeitoun, R.: Horizontal side-channel attacks and countermeasures on the ISW masking scheme. In: Gierlichs, B., Poschmann, A. (eds.) CHES. Lecture Notes in Computer Science, vol. 9813, pp. 23–39. Springer, Heidelberg (2016). https://doi.org/10.1007/978-3-662-53140-2_2
3. Béguinot, J., et al.: Removing the field size loss from duc et al'.s conjectured bound for masked encodings. In: Kavun, E.B., Pehl, M. (eds.) COSADE. Lecture Notes in Computer Science, vol. 13979, pp. 86–104. Springer, Heidelberg (2023). https://doi.org/10.1007/978-3-031-29497-6_5
4. Bronchain, O., Cassiers, G.: Bitslicing arithmetic/boolean masking conversions for fun and profit with application to lattice-based kems. IACR Trans. Cryptogr. Hardw. Embed. Syst. **2022**(4), 553–588 (2022)
5. Bronchain, O., Hendrickx, J.M., Massart, C., Olshevsky, A., Standaert, F.: Leakage certification revisited: Bounding model errors in side-channel security evaluations. In: Boldyreva, A., Micciancio, D. (eds.) CRYPTO (1). Lecture Notes in Computer Science, vol. 11692, pp. 713–737. Springer, Heidelberg (2019). https://doi.org/10.1007/978-3-030-26948-7_25
6. Bronchain, O., Momin, C., Peters, T., Standaert, F.: Improved leakage-resistant authenticated encryption based on hardware AES coprocessors. IACR Trans. Cryptogr. Hardw. Embed. Syst. **2021**(3), 641–676 (2021)
7. Bronchain, O., Standaert, F.: Breaking masked implementations with many shares on 32-bit software platforms or when the security order does not matter. IACR Trans. Cryptogr. Hardw. Embed. Syst. **2021**(3), 202–234 (2021). https://doi.org/10.46586/tches.v2021.i3.202-234
8. Cassiers, G., Grégoire, B., Levi, I., Standaert, F.: Hardware private circuits: from trivial composition to full verification. IEEE Trans. Comput. **70**(10), 1677–1690 (2021)

9. Cassiers, G., Standaert, F.: Trivially and efficiently composing masked gadgets with probe isolating non-interference. IEEE Trans. Inf. Forensics Secur. **15**, 2542–2555 (2020)
10. Cassiers, G., Standaert, F.: Provably secure hardware masking in the transition- and glitch-robust probing model: better safe than sorry. IACR Trans. Cryptogr. Hardw. Embed. Syst. **2021**(2), 136–158 (2021)
11. Cassiers, G., Devillez, H., Standaert, F.X., Udvarhelyi, B.: Efficient regression-based linear discriminant analysis for side-channel security evaluations: towards analytical attacks against 32-bit implementations. IACR Trans. Cryptogr. Hardw. Embed. Syst. **2023**(3), 270–293 (2023)
12. Chari, S., Jutla, C.S., Rao, J.R., Rohatgi, P.: Towards sound approaches to counteract power-analysis attacks. In: Wiener, M. (ed.) CRYPTO. Lecture Notes in Computer Science, vol. 1666, pp. 398–412. Springer, Heidelberg (1999). https://doi.org/10.1007/3-540-48405-1_26
13. Chari, S., Rao, J.R., Rohatgi, P.: Template attacks. In: CHES. Lecture Notes in Computer Science, vol. 2523, pp. 13–28. Springer, Heidelberg (2002). https://doi.org/10.1007/3-540-36400-5_3
14. Choudary, M.O., Kuhn, M.G.: Efficient stochastic methods: Profiled attacks beyond 8 bits. In: Joye, M., Moradi, A. (eds.) CARDIS. Lecture Notes in Computer Science, vol. 8968, pp. 85–103. Springer, Heidelberg (2014). https://doi.org/10.1007/978-3-319-16763-3_6
15. Coron, J., et al.: Conversion of security proofs from one leakage model to another: a new issue. In: Schindler, W., Huss, S.A. (eds.) COSADE. Lecture Notes in Computer Science, vol. 7275, pp. 69–81. Springer, Heidelberg (2012). https://doi.org/10.1007/978-3-642-29912-4_6
16. Duc, A., Dziembowski, S., Faust, S.: Unifying leakage models: From probing attacks to noisy leakage. In: Nguyen, P.Q., Oswald, E. (eds.) EUROCRYPT. Lecture Notes in Computer Science, vol. 8441, pp. 423–440. Springer, Heidelberg (2014). https://doi.org/10.1007/978-3-642-55220-5_24
17. Duc, A., Faust, S., Standaert, F.: Making masking security proofs concrete - or how to evaluate the security of any leaking device. In: Oswald, E., Fischlin, M. (eds.) EUROCRYPT (1). Lecture Notes in Computer Science, vol. 9056, pp. 401–429. Springer, Heidelberg (2015). https://doi.org/10.1007/978-3-662-46800-5_16
18. Duc, A., Faust, S., Standaert, F.: Making masking security proofs concrete (or how to evaluate the security of any leaking device), extended version. J. Cryptol. **32**(4), 1263–1297 (2019)
19. Duda, R.O., Hart, P.E., Stork, D.G.: Pattern Classification, 2nd Edn. Wiley, Hoboken (2001)
20. Faust, S., Grosso, V., Pozo, S.M.D., Paglialonga, C., Standaert, F.: Composable masking schemes in the presence of physical defaults & the robust probing model. IACR Trans. Cryptogr. Hardw. Embed. Syst. **2018**(3), 89–120 (2018)
21. Gao, S., Marshall, B., Page, D., Oswald, E.: Share-slicing: friend or foe? IACR Trans. Cryptogr. Hardw. Embed. Syst. **2020**(1), 152–174 (2020)
22. Goubin, L., Patarin, J.: DES and differential power analysis (the "duplication" method). In: Koc, C.K., Paar, C. (eds.) CHES. Lecture Notes in Computer Science, vol. 1717, pp. 158–172. Springer (1999). https://doi.org/10.1007/3-540-48059-5_15
23. Goudarzi, D., Rivain, M.: How fast can higher-order masking be in software? In: Coron, J.S., Nielsen, J. (eds.) EUROCRYPT (1). Lecture Notes in Computer Science, vol. 10210, pp. 567–597. Springer, Heidelberg (2017). https://doi.org/10.1007/978-3-319-56620-7_20

24. Herbst, C., Oswald, E., Mangard, S.: An AES smart card implementation resistant to power analysis attacks. In: Zhou, J., Yung, M., Bao, F. (eds.) ACNS. Lecture Notes in Computer Science, vol. 3989, pp. 239–252. Springer, Heidelberg (2006). https://doi.org/10.1007/11767480_16
25. Ishai, Y., Sahai, A., Wagner, D.A.: Private circuits: securing hardware against probing attacks. In: Boneh, D. (ed.) CRYPTO. Lecture Notes in Computer Science, vol. 2729, pp. 463–481. Springer, Heidelberg (2003). https://doi.org/10.1007/978-3-540-45146-4_27
26. Mangard, S.: Hardware countermeasures against DPA ? A statistical analysis of their effectiveness. In: Okamoto, T. (ed.) CT-RSA. Lecture Notes in Computer Science, vol. 2964, pp. 222–235. Springer, Heidelberg (2004). DOI: https://doi.org/10.1007/978-3-540-24660-2_18
27. Mangard, S., Oswald, E., Standaert, F.: One for all - all for one: unifying standard differential power analysis attacks. IET Inf. Secur. **5**(2), 100–110 (2011)
28. Nikova, S., Rijmen, V., Schläffer, M.: Secure hardware implementation of nonlinear functions in the presence of glitches. J. Cryptol. **24**(2), 292–321 (2011)
29. Poussier, R., Standaert, F., Grosso, V.: Simple key enumeration (and rank estimation) using histograms: an integrated approach. In: Gierlichs, B., Poschmann, A. (eds.) CHES. Lecture Notes in Computer Science, vol. 9813, pp. 61–81. Springer, Heidelberg (2016). https://doi.org/10.1007/978-3-662-53140-2_4
30. Prest, T., Goudarzi, D., Martinelli, A., Passelègue, A.: Unifying leakage models on a rényi day. In: Boldyreva, A., Micciancio, D. (eds.) CRYPTO (1). Lecture Notes in Computer Science, vol. 11692, pp. 683–712. Springer, Heidelberg (2019). https://doi.org/10.1007/978-3-030-26948-7_24
31. Prouff, E., Rivain, M.: Masking against side-channel attacks: a formal security proof. In: Johansson, T., Nguyen, P.Q. (eds.) EUROCRYPT. Lecture Notes in Computer Science, vol. 7881, pp. 142–159. Springer, Heidelberg (2013). https://doi.org/10.1007/978-3-642-38348-9_9
32. Schindler, W., Lemke, K., Paar, C.: A stochastic model for differential side channel cryptanalysis. In: Johansson, T., Nguyen, P.Q. (eds.) CHES. Lecture Notes in Computer Science, vol. 3659, pp. 30–46. Springer, Heidelberg (2005). https://doi.org/10.1007/978-3-642-38348-9_9
33. Standaert, F., Archambeau, C.: Using subspace-based template attacks to compare and combine power and electromagnetic information leakages. In: Oswald, E., Rohatgi, P. (eds.) CHES. Lecture Notes in Computer Science, vol. 5154, pp. 411–425. Springer, Heidelberg (2008). https://doi.org/10.1007/978-3-540-85053-3_26
34. Standaert, F., Malkin, T., Yung, M.: A unified framework for the analysis of side-channel key recovery attacks. In: Joux, A. (ed.) EUROCRYPT. Lecture Notes in Computer Science, vol. 5479, pp. 443–461. Springer, Heidelberg (2009). https://doi.org/10.1007/978-3-642-01001-9_26
35. Unterstein, F., Heyszl, J., Santis, F.D., Specht, R.: Dissecting leakage resilient prfs with multivariate localized EM attacks - a practical security evaluation on FPGA. In: Guilley, S. (ed.) COSADE. Lecture Notes in Computer Science, vol. 10348, pp. 34–49. Springer, Heidelberg (2017). https://doi.org/10.1007/978-3-319-64647-3_3
36. Unterstein, F., Heyszl, J., Santis, F.D., Specht, R., Sigl, G.: High-resolution EM attacks against leakage-resilient prfs explained - and an improved construction. In: Smart, N. (ed.) CT-RSA. Lecture Notes in Computer Science, vol. 10808, pp. 413–434. Springer, Heidelberg (2018). https://doi.org/10.1007/978-3-319-76953-0_22

37. Veyrat-Charvillon, N., Gérard, B., Standaert, F.: Soft analytical side-channel attacks. In: Sarkar, P., Iwata, T. (eds.) ASIACRYPT (1). Lecture Notes in Computer Science, vol. 8873, pp. 282–296. Springer, Heidelberg (2014). https://doi.org/10.1007/978-3-662-45611-8_15
38. Veyrat-Charvillon, N., Medwed, M., Kerckhof, S., Standaert, F.: Shuffling against side-channel attacks: a comprehensive study with cautionary note. In: Wang, X., Sako, K. (eds.) ASIACRYPT. Lecture Notes in Computer Science, vol. 7658, pp. 740–757. Springer (2012). https://doi.org/10.1007/978-3-642-34961-4_44

Smartcards and Efficient Implementations

The Adoption Rate of JavaCard Features by Certified Products and Open-Source Projects

Lukas Zaoral[1], Antonin Dufka[2], and Petr Svenda[2](✉)

[1] Red Hat, Raleigh, USA
lzaoral@redhat.com
[2] Masaryk University, Brno, Czech Republic
{xdufka1,svenda}@fi.muni.cz

Abstract. JavaCard is the most prevalent platform for cryptographic smartcards nowadays. Despite having more than 20 billion smartcards shipped with it and thirteen revisions since the JavaCard API specification was first published more than two decades ago, uptake of newly added features, cryptographic algorithms or their parameterizations, and systematic analysis of overall activity is missing. We fill this gap by mapping the activity of the JavaCard ecosystem from publicly available sources with a focus on 1) security certification documents available under Common Criteria and FIPS140 schemes and 2) activity and resources required by JavaCard applets released in an open-source domain (Paper supplementary materials, full results of analysis and open tools are available at https://crocs.fi.muni.cz/papers/cardis2023).

The analysis performed on all certificates issued between the years 1997–2023 and on more than 200 public JavaCard applets shows that new features from JavaCard specification are adopted slowly, typically taking six or more years. Open-source applets utilize new features even later, likely due to the unavailability of recent performant smartcards in smaller quantities. Additionally, almost 70% of constants defined in JavaCard API specification are completely unused in open-source applets. The applet portability improves with recent cards, and transient memory requirements (scarce resource on smartcards) are typically small.

While twenty or more products have been consistently certified every year since 2009, the open-source ecosystem became more active around 2013 but seemed to decline in the past two years. As a result, the whole smartcard ecosystem might be negatively impacted by limited exposure to new ideas and usage scenarios, serving only well-established domains and potentially harming its long-term competitiveness.

Keywords: Smartcard · JavaCard · Security certification · Open-source

1 Introduction

A recent paper [13] mapped the difference between features (cryptographic algorithms, key lengths) described in JavaCard API specification and features

S. Bhasin and T. Roche (Eds.): CARDIS 2023, LNCS 14530, pp. 169–189, 2024.
https://doi.org/10.1007/978-3-031-54409-5_9

actually supported by the real-world physical smartcards and documented a large disparity. As implementation of cryptographic algorithms frequently requires dedicated hardware co-processors to achieve acceptable performance and secure execution, the decision about the level of support is left to a smartcard vendor, with API specification intentionally listing the majority of the features as *optional* to allow for such flexibility. But while JavaCard applets are supposed to be binary portable between different smartcards, the uneven level of feature support directly influences the actual portability and adoption of specific algorithms by applets developed for the smartcards.

The analysis performed in paper [13] has two main limitations – although more than 100 smartcards are analyzed, only the ones available to community-maintained JCAlgTest database are considered (typically missing the recent or uncommon smartcards), and actual use of algorithms in applet code running on these cards is not evaluated. As a result, the impact of uneven support of JavaCard API features on the whole JavaCard ecosystem and the rate of new features adoption are not studied enough. We aim to fill this gap by answering the following research questions:

Q1: *What is the level and delay in adoption for cryptographic algorithms introduced by specific versions of JavaCard specification by a) certified smartcards and b) implementation of open-source applets?*

Q2: *How does the JavaCard open-source ecosystem evolve with respect to development activity, requested JavaCard features, and memory utilization?*

To answer these questions, we first performed an automatic analysis of documents accompanying security certification performed under the Common Criteria and FIPS140 schemes for JavaCard API-related keywords and analyzed the evolution in time for different API versions and cryptographic algorithms included. That still left one angle unanswered – what algorithms are typically required by applications (applets) running on these smartcards?

We addressed this question by analysis of more than 200 open-source JavaCard projects. The resulting insight covers the evolution of cryptographic algorithms in time, the expected time span before the algorithms from the given version of JavaCard specification start to be available in practice, and the overall activity of the JavaCard open-source community. The results can be used to inform the creators of the JavaCard specification regarding the features popularity and the possibility of features deprecation, the vendors of smartcards with respect to the support for new features and comparison with competition, and the developers with respect to the expected time of practical availability of features included in a new version of the specification.

Paper Contribution:

- Analysis of JavaCard API features adoption rate by products certified under Common Criteria and FIPS140 schemes.
- Analysis of the evolution of the JavaCard open-source ecosystem of more than 200 projects with respect to resources required (packages, crypto. algs., memory) and applets portability between multiple physical smartcards.

The rest of this section introduces the JavaCard specification and surveys the related work. Section 2 analyzes mentions of JavaCard technology by certification artifacts issued under Common Criteria and FIPS140 certification schemes. Section 3 surveys the JavaCard open-source ecosystem and analyzes its activity in time and the portability of applets between different physical cards. Section 4 analyzes their memory and cryptographic resource requirements. Discussion about observed trends and limitations is provided in Sect. 5 with conclusions following in Sect. 6.

1.1 Related Work

The closest work to ours is a recent publication by Svenda et al. [13]. In contrast to this paper, only support of JavaCard smartcards by direct testing is performed in [13], certification reports are not analyzed and usage of the JavaCard features in actual source code is not analyzed. A work by Hajny et al. [6] presented performance measurement of selected basic cryptographic operations on the three major smartcard platforms (JavaCard, .NET, and MultOS).

Apart from works focusing on smartcards, there have been a number of works analyzing various software ecosystems. A systematic review of software ecosystem literature was given in works by Barbosa and Alves [1] and Manikas and Hansen [8]. An analysis focusing on mobile software ecosystems was performed by Fontão et al. [3]. Ecosystem of open-source software for drones was surveyed by Glossner et al. [5]. Furthermore, there have been a number of analyses performed by mainstream developer platforms like StackOverflow[1] and GitHub[2] who produce yearly reports of their ecosystems and developers.

2 JavaCard in Common Criteria and FIPS140

The necessary prerequisite for an algorithm to be used in a production applet's code is its support by the used smartcard platform. As smartcards tend to be used in environments imposing strict compliance requirements, the platform often needs to be certified, typically under FIPS140 or Common Criteria schemes.

We therefore performed an analysis of certification documents for any mentions of JavaCard API versions and algorithms to obtain further insight into the current state of the certified devices ecosystem. These results support open-source applet analysis presented later in Sect. 4. While existing work [13] analyzed supported algorithms by direct testing of physical cards, only cards available to authors were tested. The analysis based on certification documents is therefore complementary to [13], as we analyze not only certificates for the JavaCard platform and cryptographic libraries but also certificates for applets built atop a base platform. Only certified items are analyzed here while the authors of [13] test also smartcards with non-certified platforms (although frequently based on certified underlying hardware).

[1] https://survey.stackoverflow.co/2023/.

[2] https://octoverse.github.com/.

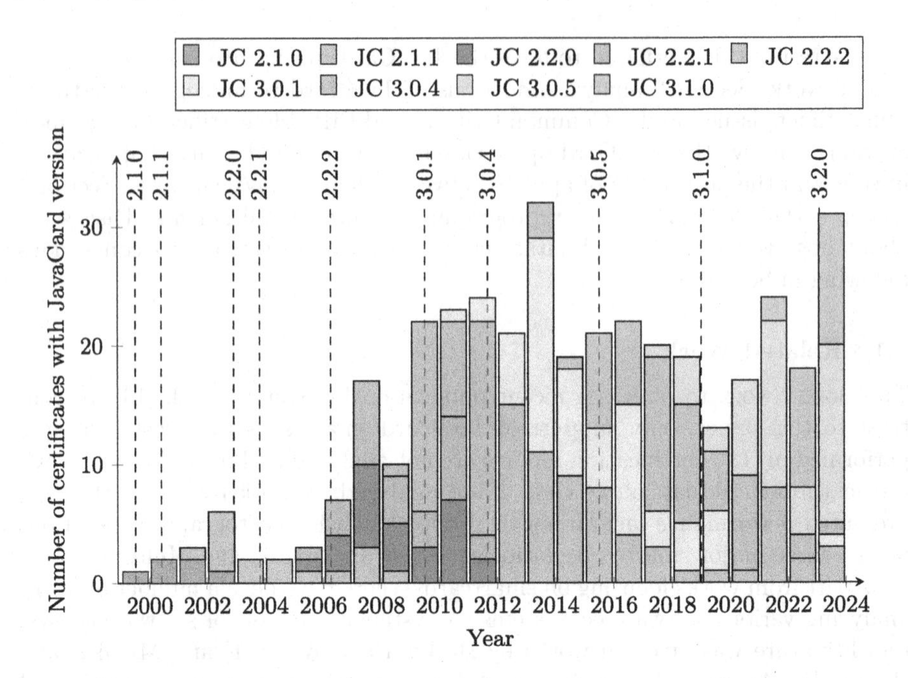

Fig. 1. The number of certification documents mentioning specific JavaCard API version per year (the year 2023 only till the end of October). In case multiple versions were detected in a document, only the latest one was included in the chart.

To perform the analysis, we downloaded all public certification documents, conveniently collected by the *sec-certs* project[3] [7]. The dataset already contained JavaCard API version numbers extracted from the documents using regular expressions. However, the second part of the analysis required more precise data processing, which involved semi-manual filtering of raw data. We matched the regular expression `"ALG_[0-9A-Z_]+"` against text content of all certification documents and filtered only those that matched the pattern and corresponded a JavaCard target (list of all matched certificates can be found in Table 5 in the Appendix). In case a match was found, we compared the matching string to known JavaCard API names and corrected values that were clearly incorrectly read or input.

2.1 Analysis of Certification Documents

First, we analyzed the dependency of mentions of JavaCard API versions in certification documents on a certificate issuance year. The distribution is shown in Fig. 1, where we display only the latest version detected per certificate document (406 documents in total). We see that JavaCard API version 2.2.2 was in use for the longest period, starting in 2008, 2 years after its introduction, and being

[3] https://seccerts.org/.

commonly mentioned till 2017. Even in the year 2022, certificates mentioning this API version were issued. Still, since the year 2010, API with major version number 3 started gradually appearing, and by 2013, they were mentioned in the majority of JavaCard certification documents. For the analysis of new algorithm adoption, we focused on these newer versions, namely 3.0.1, 3.0.4, 3.0.5, and 3.1.0.

Among the 10197 certificate documents in the dataset, we have identified 66 that describe a JavaCard target and include references to strings prefixed with `ALG_`, typical naming of JavaCard algorithm constants. Out of the 66 documents, those issued before the year 2009 were certified only under FIPS140 scheme (6 certificates). After that, all 60 subsequent documents were issued under the Common Criteria scheme. In the documents, we identified 139 unique strings that match the pattern; however, only 104 of them correspond to algorithms included in the JavaCard API[4].

Figure 2 displays mentions of JavaCard algorithms introduced in JavaCard API 3.0.1 and later, grouped by the certificate issuance year. In general, algorithms start being mentioned in certification documents only two or more years after the specification version is published. The only exception to this is the certificate of an Oberthur smartcard[5], which mentions algorithm `ALG_SHA_224` even before its specification was released.

JavaCard API version 3.0.5 was released in 2015, and the first algorithm added to the specification appeared in a certification document in 2017. Later, an algorithm (`ALG_EC_SVDP_DH_PLAIN_XY`), allowing for efficient ECDH computation outputting full point, started appearing in 2018. This algorithm was the most frequently mentioned in the documents in the year 2022, together with its variant (`ALG_EC_SVDP_DH_PLAIN`) outputting only X-coordinate that was introduced in API version 3.0.1. Version 3.0.4 introduced two new algorithms, which first appeared in certification documents in 2016, five years after its publication. Apart from the Oberthur certificate, mentions of algorithms introduced in JavaCard API version 3.0.1, released in May 2009, started appearing in 2011, but their mentions became more frequent only after 2015.

We also inspected the remaining strings that matched the JavaCard algorithm pattern but are not part of the JavaCard API. Many of these algorithms are likely incorrectly input in the source documents; however, some of them clearly denote algorithms unsupported by JavaCard API. The algorithm names are presented in Table 1 with several interesting items like `ALG_EC_SVDP_DH_GK` whose function is not known due to missing public documentation, referring to Oberthur's proprietary API extensions[6]. Another interesting mentions are `ALG_ED25519PH_SHA_512` in seven certificates by NXP[7], which most likely refers to Ed25519 signature algorithm with prehashed input and SHA512 function. The

[4] JavaCard API contains in total 151 constants with such name.
[5] https://seccerts.org/fips/6d094db49a6e2242/.
[6] https://seccerts.org/fips/6d094db49a6e2242/, /a5bef651c8e3fd6c/.
[7] https://seccerts.org/cc/03aded94fb04c62e/, /45098872448f5816/, /03aded94fb04c62e/, /b0e6f667d52402df/.

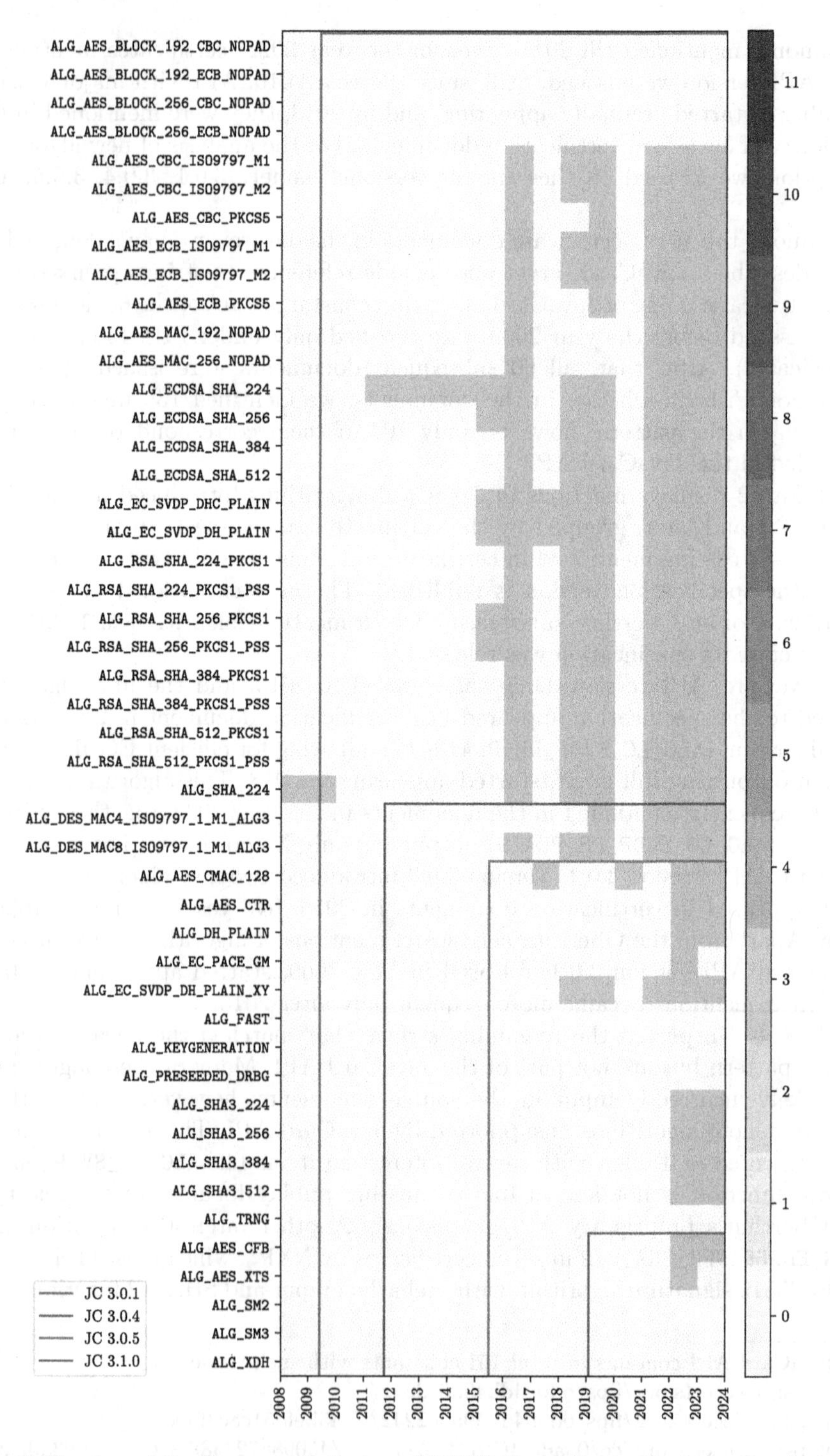

Fig. 2. Number of the Common Criteria and FIPS140 certificates that mention JavaCard API 3.0.1+ algorithms, by year.

same seven certificates also include string `ALG_MONT_DH_25519`, likely referring to Diffie-Hellman key agreement on Curve25519 in the Montgomery form.

Table 1. Algorithms detected in CC and FIPS certificates by search of prefix `ALG_` that are not included in the official JavaCard API specification.

ALG_AES_BLOCK_128_CBC_NOPAD_STANDARD	ALG_DES_CMAC	ALG_EC_SVDP_DHC_GK
ALG_AES_CBC_ISO9797_M2_STANDARD	ALG_DES_CMAC8	ALG_EC_SVDP_DHC_PACE
ALG_AES_CBC_ISO9797_STANDARD	ALG_DES_ECB_PKCS7	ALG_EC_SVDP_DH_GK
ALG_AES_CBC_PKCS7	ALG_DES_MAC128_ISO9797_1_M2_ALG3	ALG_ED25519PH_SHA_512
ALG_AES_CMAC128	ALG_DES_MAC_8_NOPAD	ALG_MONT_DH_25519
ALG_AES_CMAC16	ALG_ECDSA_RAW	ALG_RSA_SHA256_PKCS1
ALG_AES_CMAC16_STANDARD	ALG_ECDSA_SHA224	ALG_RSA_SHA256_PKCS1_PSS
ALG_AES_CMAC8	ALG_ECDSA_SHA256	ALG_RSA_SHA_1_RFC2409
ALG_AES_ECB_PKCS7	ALG_ECDSA_SHA256_LDS	ALG_RSA_SHA_256_ISO9796
ALG_AES_MAC_128_ISO9797_1_M2_ALG3	ALG_ECDSA_SHA384	ALG_SHA2_CHAIN
ALG_AES_OFB	ALG_ECDSA_SHA384_LDS	ALG_SHA_CHAIN
ALG_DES_CBC_PKCS7	ALG_ECDSA_SHA_LDS	
ALG_DES_CMAC	ALG_EC_SVDP_DHC_GK	

3 JavaCard Open-Source Ecosystem

To analyze the overall focus, activity, and evolution of the JavaCard open-source ecosystem, we built a database of all available public repositories relevant to the JavaCard platform, performed a static and dynamic analysis of contained JavaCard applets, and tested deployability on several physical smartcards.

3.1 Database of Open-Source Projects

Since 2017, we periodically searched for JavaCard open-source repositories on GitHub, SourceForge, and GitLab for the occurrence of *'javacard.framework'* keyword and collected hits into a public curated list hosted on GitHub[8]. We include almost all discovered repositories without assessing their maturity – only obviously trivial, testing or unfinished applets are excluded. As some projects are occasionally moved or removed, we create backups by forking each repository when included. While a large majority (>95%) of records were inserted by us, some contributions were also made by the community.

We have manually examined all repositories from the curated list to create a refined list with an explicit enumeration of JavaCard applets that we have found. The refined list contains 139 repositories containing at least one JavaCard applet or library. The repositories are mainly found on GitHub, with a very small number on SourceForge as well, and no repository has been detected on GitLab so far. Many repositories are monorepos with multiple projects. Sources

[8] https://github.com/crocs-muni/javacard-curated-list.

that could not be parsed even with import resolution disabled were omitted from further analysis. The list contains 206 JavaCard projects with 223 possible applet entry points, of which 36 cannot be parsed precisely due to missing dependencies or programming mistakes preventing code compilation with standard *javac* compiler. For analysis of the usage of specification features, we excluded the applet corresponding to JCAlgTest project[9], which is designed to test the target smartcard and intentionally references (almost) all cryptographic algorithms and parameterization constants found in the JavaCard specification [10].

3.2 Activity of JavaCard Open-Source Ecosystem in Time

To analyze the activity of the JavaCard open-source ecosystem in time, we analyzed each project's state with respect to its git commits closest to (but prior) the next date interval. For a yearly activity, we analyzed the commit closest to the 1st of January of the next year, whereas for monthly activity, we used the commit closest to the 1st day of the next month.

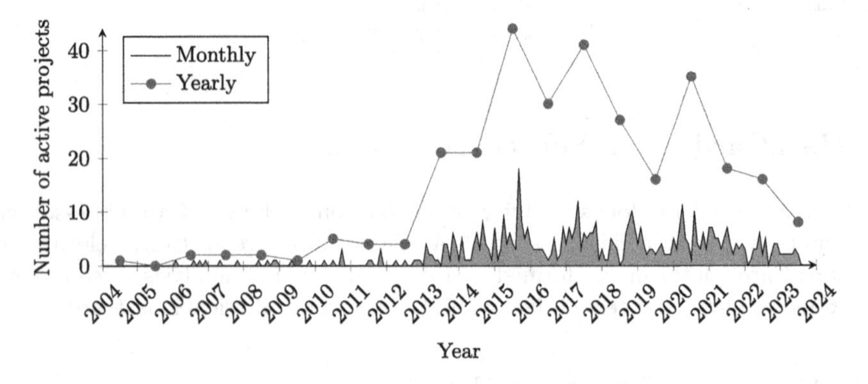

Fig. 3. Number of open-source projects with at least one commit per month (black line) or per year (red line) respectively. The year 2023 is only till end of October. (Color figure online)

The number of projects with at least one commit in a given period (year or month) is shown in Fig. 3. While the oldest commits are from the year 2004, the number of active projects was very low (around three projects) until the year 2013, when the number of active projects increased significantly to more than 20 projects, reaching more than 40 active projects in years 2015 and 2017. Figure 4 visualizes included repositories based on a number of forks and stars with the most popular repositories annotated.

While the development activity fluctuates significantly between years (the year 2019 had only around 50% of active projects with respect to 2017 and 2020), there seems to be a significant activity decrease following the year 2020, with only 8 active projects till the end of October of the year 2023. Would the trend continue, we may be witnessing a decline of open-source activity in the

[9] https://github.com/crocs-muni/jcalgtest.

JavaCard ecosystem, despite a still high number of newly certified JavaCard products (see Sect. 2).

An intriguing possibility might be a correlation with the trends observed in the JavaCard-related certification artifacts but with a lag of about four to six years. The lag can be explained by the delayed availability of capable certified smartcards to open-source developers for purchase in smaller quantities. If this correlation is genuine, we may observe increased activity in the open-source ecosystem in the following years again, as activity in the certification ecosystem dipped in the year 2019 and increased again since then.

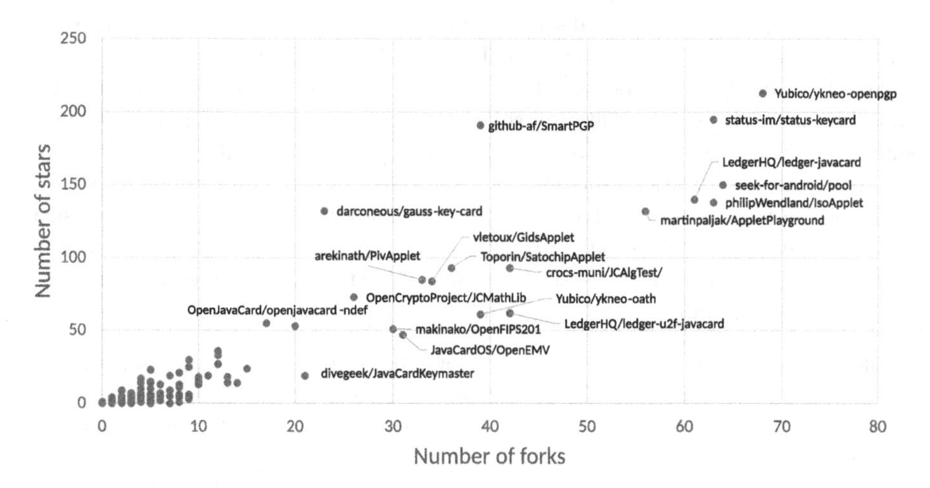

Fig. 4. The scatter plot of JavaCard repositories popularity using number of forks and stars on the GitHub platform. Projects with at least 20 forks are annotated by name.

3.3 Open-Source Applets Compatibility and Portability

To test the possibility of deployment of open-source applets to physical smartcards, we attempted the following steps for all open-source applets from the open-source database described in Sect. 3.1:

1) **Compile** applet using standard *javac* compiler, resulting in **.class* file(s). The step fails if the source code is not compatible with Java language (typically 1.8) or references proprietary packages.
2) **Convert** compiled **.class* files using Oracle JavaCard *converter* for specific version of JC SDK, resulting in a **.cap* file. The step fails if the bytecode violates JavaCard language specification or references proprietary packages.
3) **Upload** contents of the *cap* file to smartcard using the GlobalPlatform interface via *gppro* tool [11]. The step fails if *cap* file references unsupported packages or is rejected by the on-card verifier for other reasons.
4) **Install** applet instance from previously uploaded package (content of the *cap* file). The step fails if an exception is emitted in the applet's constructor, typically caused by instantiation of an unsupported algorithm

(e.g., ALG_SHA3_512), unsupported parameters (e.g., of LENGTH_RSA_4096 or invalid curve domain parameters), or excessive memory allocation (typically RAM memory via JCSystem.getTransientByteArray()). Occasionally, the exception may be caused by other programming mistakes or platform limitations (e.g., the limited size of the card's transaction buffer).

5) **Select** the installed applet instance, making it active for subsequent commands from a host controller. The step typically fails when additional custom code like delayed allocation or reset of the temporary state is executed.

The next step is performed only if the previous one succeeds. Steps 1) and 2) do not require any physical smartcard. Steps 3), 4), and 5) were performed on three physical smartcards and on the virtual card via jCardSim [2].

We first analyzed the possibility of performing applet conversion resulting in *cap* file, but with load, install, and select steps performed only for the jCardSim simulator. Such test detects applets that are compatible with (some) version of JavaCard specification and can be converted into a valid *cap* file. Yet no limitations of actual support of the required algorithms by the target physical card are imposed. Additionally, applets with programming errors that could be fixed easily (e.g., missing typecast from int to short, etc.) are listed in a separate column. The results are shown in Table 2.

Table 2. The number of applets with successful finalization of particular development step on virtual card simulated by jCardSim. Steps 1) and 2) are independent of any smartcard; steps 4) and 5) are tested on a simulated card provided by jCardSim.

Step	Count	(with fixes)
1) Compilation	174	187
2) Conversion (any JC SDK version)	145	176
3) Upload (does not apply for jCardSim)	-	-
4) Installation	121	150
5) Applet selection	111	144

Table 3. The portability of applets to different smartcards. The jCardSim simulator is listed as a theoretical upper bound of convertable and executable applets without resource limitations imposed by a particular physical platform. The numbers presented in this table also correspond to the applets with success/failure of measurement of entry point class constructor memory usage on physical smartcards. All *skips* correspond to applets that require a newer JavaCard API than supported by a given card.

Card	API	Success	Failure			Skip
			Upload	Install	Select	
jCardSim simulator	3.0.5	144	0	0	0	0
NXP JCOP4 J3R180 (2020)	3.0.5	124	7	13	0	0
Feitian JavaCOS A22	3.0.4	98	3	41	0	2
G+D StarSign Crypto USB Token S	3.0.4	64	3	73	1	2

Secondly, we tested the *deployability* and *portability* of applets among three different smartcards, all relatively commonly available to open-source developers. The results are shown in Table 3 with the number of applets deployable to cards simulated via jCardSim serving as a baseline. The largest number of applets (124) are deployable to NXP JCOP4 J3R180 (CC certificate[10] issued on 01. 03. 2020), which is currently the most performant card available at online shops (2023 YTD) in small quantities. The decreasing number of deployable applets to other smartcards illustrates the limited portability of open-source applets among smartcards of different vendors.

4 Resources Required by JavaCard Applets

Having the large list of open-source JavaCard projects available via git repositories allows us to analyze the packages, classes, methods, and constants using static analysis of their source code. The analysis relies heavily on the functionality provided by the Spoon library [12] designed for analysis and transformation of Java source code. Spoon parses the input source code into a complete abstract syntax tree (AST). Therefore, the source code can be analyzed and instrumented more reliably than in the case of the regex-based approaches searching for target keywords. The results are provided in Sect. 4.1.

The Spoon library also allows for automatic instrumentation of the applet's code usable for the dynamic analysis of memory requirements by the target applet with results presented in Sect. 4.2.

4.1 JavaCard Packages and Cryptographic Algorithms Required

The combination of git history and algorithms extracted from the source code at the given date (e.g., the latest commit in a given year) allows us to analyze the rate of adoption for *every* constant defined in the JavaCard specification.

Figure 5 shows the popularity of both symmetric and asymmetric cryptography signature schemes. The ECDSA signature algorithm, together with RSA with PKCS#1 padding, is the most popular, followed by the 3DES-based and HMAC-based MAC schemes. Relatively few applets utilize other options. Related Fig. 6 shows the adoption of different variants of hash functions in combination with ECDSA over time, and Fig. 10 compares time of introduction and use of two frequently used signature algorithms – while `ALG_ECDSA_SHA_256` is still added into new code, `ALG_RSA_SHA_PKCS1` no longer is.

The popular key lengths are shown in Fig. 7 and they are mostly as expected – AES-128/256b and 3DES-112/168b are used frequently for symmetric cryptography, EC-FP-256b and RSA-2048b for asymmetric one. The only surprise is still a high usage of RSA-1024b. Related Fig. 8 demonstrates the trend for DES algorithm with one (`LENGTH_DES`), two (`LENGTH_DES3_2KEY`) and three (`LENGTH_DES3_3KEY`) keys in time, showing adoption of longer keys later.

[10] https://seccerts.org/cc/2a45531c2dbd1ab8/.

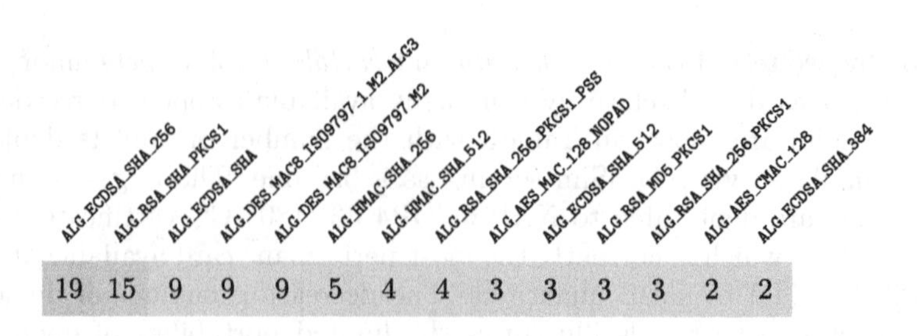

Fig. 5. `javacard.security.Signature` `ALG` constants with at least two usages.

The usage of 3rd-party, sometimes proprietary, APIs is shown in Fig. 9. GlobalPlatform/OpenPlatform API is used mostly for the handling of secure channel messages or accessing global card state. If proprietary API is used (NXP, Giesecke+Devrient), the applet can be statically analyzed for the usage of JavaCard features but cannot be dynamically analyzed due to the missing proprietary SDK. The results for other relevant classes are available in supplementary materials[11].

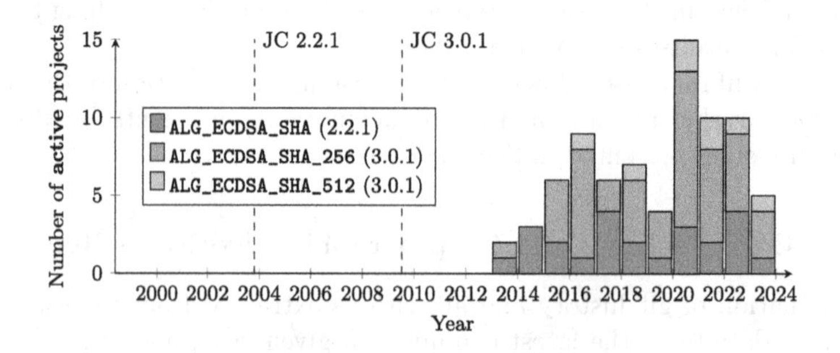

Fig. 6. Histogram of the adoption of `ALG_ECDSA_SHA` constants.

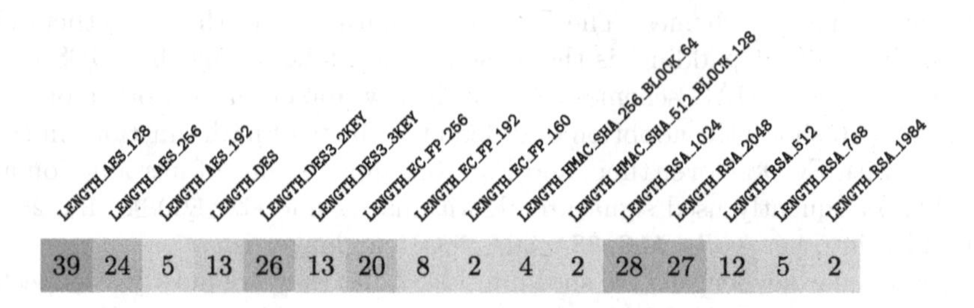

Fig. 7. `javacard.security.KeyBuilder` `LENGTH` constants with at least two usages sorted by the corresponding algorithms.

[11] https://crocs.fi.muni.cz/papers/cardis2023.

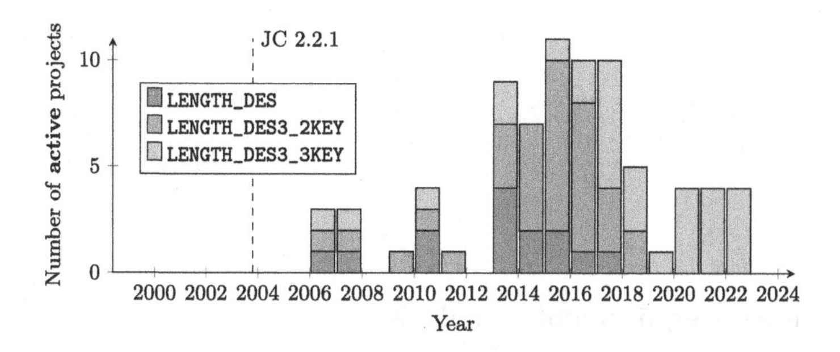

Fig. 8. Histogram of the adoption of LENGTH_DES constants from JavaCard 2.2.1.

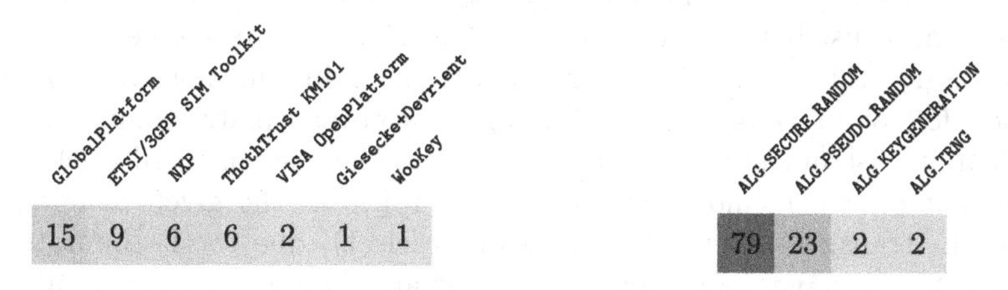

Fig. 9. Usage of third-party APIs and javacard.security.RandomData constants.

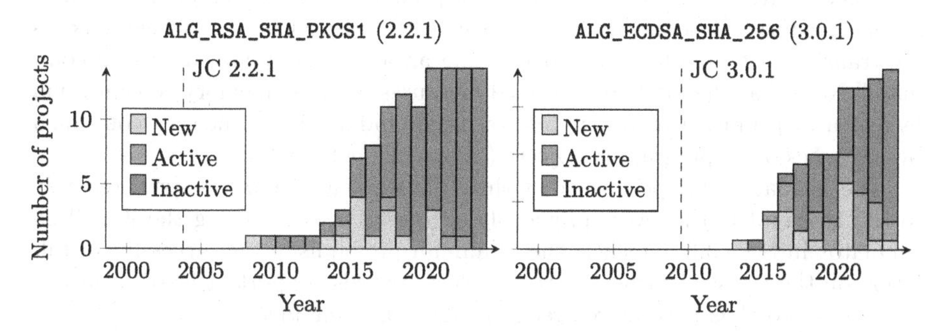

Fig. 10. Comparison of the adoption of ALG_RSA_SHA_PKCS1 and ALG_ECDSA_SHA_256 constants. The ECDSA signature was adopted only in 2013 and is still being added to the source code of active projects, while the RSA signature was included already since 2008, but overall less frequently and not after the year 2020.

We also analyzed the adoption of different constants from JavaCard API up to 3.2 with 583 values in total. A surprisingly high number of 404 constants are completely unused (69.3%), and 488 constants are used in less than six projects (83.7%). Only 42 constants are used in more than 25 (7.2%) projects. Table 4 provides complete statistics.

Table 4. The usage of constants from JavaCard API by the set of all applets.

# applets using constant	no use	1	2-5	6-10	11-25	26-50	51-75	76-100	100+
# API constants (583)	404	36	48	20	33	21	7	4	10

4.2 Memory Requirements Analysis

The memory requirements by a given applet can be inferred from the difference in reported available memory right after the applet's constructor method is entered and just before it is finished. Automatic code instrumentation using Spoon library [12] was used to insert the corresponding memory measurements methods `JCSystem.getAvailableMemory()` into every analyzed applet.

When an instrumented code line is reached, the instrumented applet stores the amount of free memory for all memory types at the given point in time (persistent, transient reset, and transient deselect), due to the limitations of the `JCSystem.getAvailableMemory()` JavaCard API, the gathered values are capped at 32 kB (3.0.3 and older) or 2 GB (3.0.4 and newer). Hence, usage of smartcards with JavaCard 3.0.4 or newer is preferable for analysis of applet's persistent memory. After the applet instance creation is finished, all measurements are transferred to the host controller using an additionally added custom command. Note that the method described measures not only memory consumption by allocated primitive arrays but also transient and persistent memory consumed by instances of cryptographic objects (keys, engines) created.

The evaluation was performed on three physical cards and was restricted only to the 144 applets that were previously successfully tested using the jCardSim simulator in Sect. 3.3, which is the minimum prerequisite for successful installation on the physical cards. All three selected cards support at least JavaCard API 3.0.4 and thus have more precise memory measurements.

The representative results for the NXP JCOP4 J3R180 card with the most applets installable (124 in total) are shown in Fig. 11 for the transient and persistent memory types, respectively. The results obtained for other cards were closely matching these results with only small differences. The differences are caused by slightly different memory requirements of cryptographic objects on different cards. The large majority of applets require less than 1 kB of transient memory, with 59 applets requiring less than 70 bytes. Only a handful of applets required more than 2 kB of transient memory, notably the project eVerification-2[12] (2.6 kB).

Note that we cannot detect applets requiring *more* memory than available on the target smartcard using this methodology as the installation will fail, and the measured memory values are not retrieved. We have manually analyzed the 20 applets, which failed to install and assessed the reason for failure, and found

[12] https://github.com/CRISES-URV/eVerification-2.

out that nine applets likely failed due to transient memory requirements larger than what is available on the JCOP4 card.

Similarly, a large majority of applets required less than 10 kB of persistent memory, with only ten applets requiring more, notably applets from the PinSentry-OTP[13] project (67 kB) and the smart_card_TLS[14] project (34 kB). Twenty-seven applets required only 100 B or less of persistent memory.

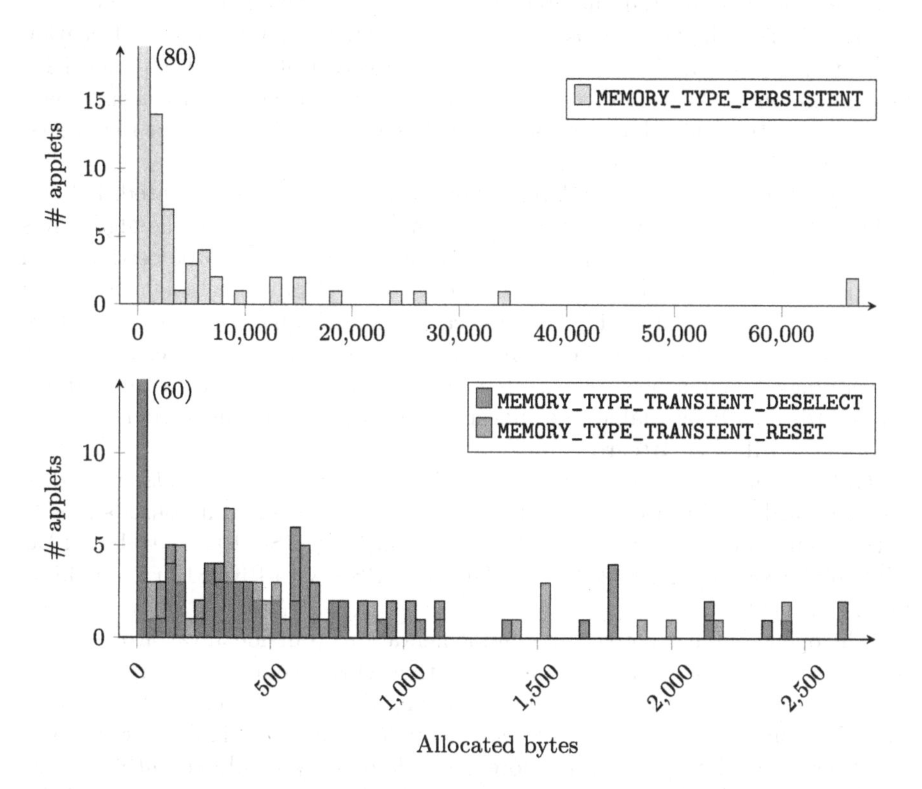

Fig. 11. Histograms of memory allocation in entry point class constructors per applet on NXP JCOP4 J3R180 with maximum available transient memory around 3.8 kB. Note that the first largest bin is clipped in both graphs.

5 Discussion and Limitations

JavaCard's open-source ecosystem is relatively long-running, with the first commits made already in the year 2004, although only infrequently. It became sharply more active from the year 2013, reaching more than 30 active projects (code changes done to actual JavaCard code) in the years 2015, 2017, and 2020 (maximum of 44 in 2015), although with significant variability between the years.

[13] https://github.com/Celliwig/PinSentry-OTP.

[14] https://github.com/gilb/smart_card_TLS.

However, there seems to be a significant decline after the year 2020, with only 18 and 16 projects active in years 2021 and 2022, respectively, and only 8 till November 2023. We do not have a good explanation for this trend, especially given the increased supply of smartcards available in small quantities, which are often used by open-source developers. It will be interesting to observe if the trend confirms in coming years. A decrease in the development activity of the JavaCard open-source ecosystem may be only temporary if the open-source ecosystem is generally copying the trends in the certification ecosystem. The certification activity increased in 2009, peaked in 2013 (32 certificates), and experienced a local minimum in 2019 (12 certificates issued), increasing again since then. Open-source ecosystem gained activity similarly, but with a delay of about four years, so we may see growth in activity again soon.

New algorithms from JavaCard specification are slow to be adopted. It takes at least two and up to five years for smartcards with support for the algorithms introduced in the new version of the JavaCard specification to be certified. An additional delay of an algorithm used in open-source implementations is caused by the unavailability of freshly certified cards in small quantities to open-source developers. Even if it is reasonable to assume that commercial closed-source applets may adopt a new algorithm sooner as pre-certification testing sample smartcards may be available, the widespread use of an implementation based on a not-yet-certified smartcard is unlikely.

Only a small subset of around 30% of algorithms and constants defined in the JavaCard specification is used by open-source applets and it consists mostly of the common ones – SHA1 and SHA256 for hashing, RSA 1024/2048b and EC FP 256b for signing, AES128/256b, 3DES and RSA with PKCS1 or no padding for the encryption.

The open-source cryptographic implementations (outside JavaCard domain) tend to adopt new algorithms sooner than their closed-source counterparts serving more established and, thus, slower-moving domains. An example might be the adoption of Curve25519 and related algorithms like Ed25519 or X25519, which are long-time popular in more open domains yet only recently become considered for adoption in mainstream domains like government-recognized signatures or signatures of PDF documents in Adobe Reader. Similarly, JavaCard open-source projects would likely utilize Curve25519 and related algorithms but were not able to due to missing direct support for these features (optional support for `X25519` and `Ed25519` was introduced only in JavaCard API 3.1). As Sect. 2 demonstrates, only a handful of products for API version 3.1 are now certified and likely still not available to open-source developers. As implementation of these algorithms requires access to cryptographic co-processors, compensating around missing features is not easy or even possible. For example, open-source implementation of Ed25519[15] become available only in 2022 and is based on Curve25519 implementation[16] utilizing host-side computation and JCMathLib library [9]; having non-trivial RAM requirements and not providing

[15] https://github.com/dufkan/JCEd25519.
[16] https://github.com/david-oswald/jc_curve25519.

a high level of security against side-channel attacks due to higher leakage of JCVM in comparison to fully native implementation. We can conclude that the JavaCard open-source ecosystem is significantly held back by the slow introduction of new features in the specification and further by their availability in actual physical smartcards. As a result, the whole smartcard ecosystem is likely negatively impacted by limited exposure to new ideas and usage scenarios, serving only well-established domains and harming its long-term competitiveness. The gradual replacement of functionality once delivered by smartcards by software-only implementations is not caused only by easier deployment but also by the inability of smartcards to deliver the functionality required.

Algorithms tend to stay in the source code once included as typical example in Fig. 10 demonstrates and is observed for almost all constants used. As a result, the legacy algorithms typically continue to be supported by smartcards, although they may be discouraged from use (e.g., single DES or MD5) to maintain backward compatibility.

The open-source applets usually do not depend on the proprietary extension packages as their use is typically limited under a non-disclosure agreement (NDA) which would prevent public availability of the applet's source code. The exceptions are GlobalPlatform, SIM toolkit, and ThothTrust KM101 packages where NDA is not required. Despite NDA limitations, NXP and G+D proprietary packages are referenced from publicly available code, although without export files necessary for the cap file conversion.

Memory requirements tend to be low due to the simplicity of applet tasks and to fit into restrictions of widely available smartcards. Open-source applets tend to utilize less than 1 kB of transient memory (RAM) and less than 10 kB of persistent memory (EEPROM/flash). No measured applet uses more than 2.7 kB of RAM and 70 kB of EEPROM. The exceptions are applets trying to implement advanced algorithms like Ed25519 or protocols like FROST [4], otherwise unavailable on smartcards.

5.1 Known Limitations

The analysis performed in this paper has its limitations, mainly stemming from the limited access to non-public closed-source applets and detailed proprietary documents about certified products.

The final date of the specification release is not the date when the specification becomes available for the first time to the smartcard manufacturers, as manufacturers are typically involved in the draft preparation of the new release. The actual span for adoption from the first availability of a given feature by a specific card is, therefore, likely longer than reported in this paper. Similarly, the date of the certification is not the date when the card becomes available to potential customers. Engineering preview samples or production-ready but not-yet-certified cards might be available earlier. This decreases the time span between the availability of the given feature by a specific card.

An algorithm might be available sooner in the proprietary packages provided by a vendor than standardized in the public JavaCard API. Closed-source applets may adopt features earlier than open-source ones or keep using the proprietary variants after the feature is available in the public API.

Only open-source applets are analyzed. The closed-source proprietary code may exhibit different usage patterns and are likely to use proprietary extension packages more frequently. However, the usage will still be limited by the algorithms available on the underlying platform as partially mapped by the JCAlgTest project [13] and certification analysis from this paper in Sect. 2.

Memory consumption for closed-source applets may be smaller than for some more complex open-source ones, especially when features unavailable in the public API are required. The prime example is the utilization of the JCMathLib library [9] with relatively large RAM requirements to expose lower-level `ECPoint` operations, which are otherwise easily available in proprietary packages.

We tested the deployability of applets on physical smartcards only up to the `select()` method. The applet functionality may still systematically fail later, even if provided with correct input data, for example, if the required engine is not allocated in the applet's constructor, but only when the functionality triggering command is received. However, such a situation shall be relatively uncommon as a good programming practice in the JavaCard world is to allocate all required resources in the applet's constructor. The full examination would require the usage of the host-side controlling application, which is frequently not available.

Some algorithms from certification documents may be missed due to PDF parsing errors, and not all supported algorithms may be explicitly listed in the smartcard certificate. As the total number of JavaCard-related certificates is not overwhelming (66 in total), we checked all the matches for possible PDF-to-text errors while we cannot compensate for the omitted ones.

6 Conclusions

The paper primarily provides data-based insight into the adoption rate of features introduced by the JavaCard specification versions using security certification reports and a large database of open-source applets.

The analysis of JavaCard-related constants from certification artifacts shows increased certification activity since the year 2006, with around twenty certified products every year for the whole last decade and taking 1–2 years before the first certified product implements at least part of the newly released specification. However, the speed of adoption of its features is relatively slow, typically six or more years before a feature is widely supported by certified products. Around 70% of the features mentioned are never used by open-source applets.

The static and dynamic analysis of JavaCard open-source projects shows increased activity from the year 2013 but possibly declining during the last two years. Around 20 projects achieved wider popularity and development activity. The rate of adoption of new specification features is slower, likely correlated with the unavailability of recent performant smartcards in smaller quantities, which also causes limited applet portability between different available physical cards. The specification features tend to stay in the existing source code, making feature deprecation more complicated, especially when a feature intended to replace the deprecated one is not supported by the majority of cards. The open-source applets typically require only a small amount of transient and persistent memory, with some notable exceptions. The applets with large transient memory requirements are typically the ones that need to complicatedly compensate for the unavailability of some low-level features like `ECPoint` operations in the public API. JavaCard open-source ecosystem is likely held back by the slow introduction of new features into the specification and further delayed by their inaccessibility of physical smartcards with desired algorithmic support.

Acknowledgments. We would like to thank reviewers for their valuable comments. The authors were supported by Ai-SecTools (VJ02010010) project and by the European Union under Grant Agreement No. 101087529 (CHESS).

Appendix

JavaCard projects and certificates included in analysis

The following types of JavaCard open-source projects were included in the analysis[17]: Electronic passports and citizen ID (8x), Authentication and access control (29x), Payments and loyalty (20x), Key and password managers (15x), Digital signing, OpenPGP and mail security (8x), e-Health (1x), NDEF tags (6x), Cryptocurrency wallets (7x), Emulation of proprietary cards (8x), Mobile telephony (SIM) (5x), Library JavaCard code (47x), Learning (school projects, etc.) (5x) and Other (24x).

[17] https://github.com/crocs-muni/javacard-curated-list.

Table 5. Common Criteria and FIPS140 certificates of smartcards with JavaCard platform used for algorithm analysis (mentioning at least one JavaCard constant).

5G PK 5.2.2 Advanced SIM (D00233151F016B)	https://seccerts.org/cc/cb872473562ed2cf/
Athena IDPass ICAO BAC avec AA sur composant SB23YR48/80B avec librairie cryptographique NesLib v3.0	https://seccerts.org/cc/d49988efd778ca9d/
Athena IDPass ICAO EAC avec AA sur composant SB23YR48/80B avec librairie cryptographique NesLib v3.0	https://seccerts.org/cc/337ece90615ed69d/
Athena IDProtect Duo v10 (in BAC Configuration)	https://seccerts.org/cc/598e2a9978d3c79c/
Athena IDProtect Duo v10 (in EAC Configuration)	https://seccerts.org/cc/4285d9b580f8a2a6/
Athena IDProtect Duo v5 avec application IASECC en configuration ICAO BAC sur composant AT90SC28880RCFV	https://seccerts.org/cc/16b24ff1bf3c079b/
Athena IDProtect Duo v5 avec application IASECC en configuration ICAO EAC sur composant AT90SC28880RCFV	https://seccerts.org/cc/e23f9c02819688f6/
Athena IDProtect Duo v5 avec applicationIASECC en configuration ICAO BAC surcomposant AT90SC28880RCFV	https://seccerts.org/cc/797ea41cc2e95f08/
Athena IDProtect Duo v5 avec applicationIASECC en configuration ICAO EAC surcomposant AT90SC28880RCFV	https://seccerts.org/cc/a49706f436a6873d/
Athena IDProtect/OS755 (release 0355, level 0602, correctif P6) avec application IAS-ECC (version 03, build 02, correctif FA) sur composants SB23YR48/80B	https://seccerts.org/cc/b168db2651cbc1d6/
Athena IDProtect/OS755 (release 0355, level0802, correctif P8) avec application IAS-ECC(version 03, build 02, correctif FA) surcomposants SB23YR48/80B	https://seccerts.org/cc/be9ca77bec616fe5/
Athena IDProtect/OS755 (release 4016, level0101) avec application IAS-ECC (version 03,build 02, correctif FA) sur composantsSB23YR48/80B	https://seccerts.org/cc/3cb89a7fc7672364/
Athena IDProtect/OS755 Key version 9.1.2 on AT90SC25672RCT-USB Microcontroller embedding IDSign applet	https://seccerts.org/cc/82dce1546c69369d/
Athena IDProtect/OS755 avec application IAS-ECC sur composants SB23YR48/80B	https://seccerts.org/cc/79f48541164e16ff/
Athena IDProtect/OS755 avec application ICAO BAC sur composants SB23YR48/80B	https://seccerts.org/cc/ba521df2c728c5a2/
Athena IDProtect/OS755 avec application ICAO EAC sur composants SB23YR48/80B	https://seccerts.org/cc/c8fc487eec95c21e/
Athena OS755/IDProtect v6 avec application IAS-ECC sur composant AT90SC28872RCU	https://seccerts.org/cc/fca98ecd003e1b82/
Carte UpTeq NFC3.2.2_Generic v1.0 sur composant ST33G1M2-F	https://seccerts.org/cc/60f0dd83c8f32b8c/
ID-One Cosmo 32 v5	https://seccerts.org/fips/fcf1e4bf9cc9c108/
ID-One Cosmo 64 v5	https://seccerts.org/fips/068814875fcdfb8f/
ID-One Cosmo 64 v5	https://seccerts.org/fips/4a30a219a9d7663f/
ID-One Cosmo 64 v5	https://seccerts.org/fips/6e2d4d3a3d4c4bb2/
IDeal Citiz v2.1 Open platform	https://seccerts.org/cc/f67f3737a3b47208/
IDeal Citiz v2.1 STC Open Platform	https://seccerts.org/cc/05c7e778528a1596/
IDeal Citiz v2.1.1 Open platform on M7892 B11	https://seccerts.org/cc/8782c7c292ef2759/
IDeal Citiz v2.1.1 Open platform on M7893 B11	https://seccerts.org/cc/29b0321ec6b75ebd/
IDeal Citiz v2.15-i on Infineon M7892 B11 Java Card Open Platform	https://seccerts.org/cc/c301c38902477230/
IDeal Citiz v2.16-i on M7892 B11 - Java Card Open Platform	https://seccerts.org/cc/4dc023ea2e3c4115/
IDeal Citiz v2.17-i on Infineon M7892 B11 Java Card Open Platform	https://seccerts.org/cc/08408cbc394e7421/
IDeal Citiz v2.17-i on Infineon M7893 B11 Java Card Open Platform	https://seccerts.org/cc/40fa0a4cb5193977/
Infineon SECORA^{TM} ID S v1.1 (SLJ52GxxyyyzS)	https://seccerts.org/cc/8c92d4b773f61ca8/
Infineon SECORA^{TM} ID X v1.1 (SLJ52GxAyyyzX)	https://seccerts.org/cc/cd4d1f03b2d3b1cc/
MultiApp v4.0.1 with Filter Set 1.0 Java Card Open Platform on M7892 G12 chip	https://seccerts.org/cc/dcb259767b98f381/
NXP JAVA OS1 ChipDoc v1.0 SSCD (J3K080/J2K080)	https://seccerts.org/cc/30120e4f3aa2f30a/
NXP JCOP 3 EMV P60	https://seccerts.org/cc/61832cb4291c343f/
NXP JCOP 3 P60	https://seccerts.org/cc/3feda0b8b5637540/
NXP JCOP 3 SECID P60 (OSA)	https://seccerts.org/cc/d699ed2b1adc6be4/
NXP JCOP 3 SECID P60 (OSA) PL2/5	https://seccerts.org/cc/3e08a27e9d9c9b1e/
NXP JCOP 4 P71	https://seccerts.org/cc/3d8083b1e6c7b336/
NXP JCOP 4 SE050M	https://seccerts.org/cc/5a71cce42550635e/
NXP JCOP 4.0 on P73N2M0	https://seccerts.org/cc/173704f0d2b8a02f/
NXP JCOP 4.5 P71	https://seccerts.org/cc/f0d94de8beabbd84/
NXP JCOP 4.7 SE051	https://seccerts.org/cc/6cd7a7a1cffaa67e/
NXP JCOP 4.x on P73N2M0B0.2C2/2C6 Secure Element	https://seccerts.org/cc/5f12d355f8acb7cd/
NXP JCOP 5.1 on SN100.C48 Secure Element	https://seccerts.org/cc/542a1d5017a48c14/
NXP JCOP 5.1 on SN100.C48 Secure Element	https://seccerts.org/cc/c8982f2e8de39b22/
NXP JCOP 5.2 on SN100.C58 Secure Element	https://seccerts.org/cc/4526aa14337a2b0c/
NXP JCOP 5.2 on SN100.C58 Secure Element	https://seccerts.org/cc/62b62c0f0c22fd23/
NXP JCOP 6.2 on SN220 Secure Element, R1.01.1, R1.02.1, R1.02.1-1, R2.01.1	https://seccerts.org/cc/3b0d05e2fe06c803/
NXP JCOP 7.0 on SN300 Secure Element, JCOP 7.0 R1.62.0.1	https://seccerts.org/cc/b0e6f667d52402df/
NXP JCOP 7.0 with eUICC extension on SN300 Secure Element, JCOP 7.0 R1.64.0.2	https://seccerts.org/cc/45098872448f5816/
NXP JCOP 7.x on SN300 Secure Element, version JCOP 7.0 R1.62.0.1 and JCOP 7.1 R1.04.0.1	https://seccerts.org/cc/8849e07fd8bccbfb/
NXP JCOP 7.x with eUICC extension on SN300 B1.1 Secure Element, version JCOP 7.0 R1.64.0.2, JCOP 7.0 R2.04.0.2, JCOP 7.1 R1.04.0.2	https://seccerts.org/cc/81d7bbc4dfaa3aee/
NXP JCOP 8.x with eUICC extension on SN300 B2 Secure Element, version JCOP 8.0 R1.35.0.2	https://seccerts.org/cc/51116db927a18a5d/
NXP JCOP on SN100.C25 Secure Element	https://seccerts.org/cc/03aded94fb04c62e/
NXP JCOP3 P60	https://seccerts.org/cc/30e27bba1b1aece7/
NXP JCOP6.x on SN200.C04 Secure Element	https://seccerts.org/cc/ae175aa839dbf692/
Oberthur ID-One Cosmo 128 v5.5 D	https://seccerts.org/fips/6d094db49a6e2242/
Oberthur ID-One Cosmo 128 v5.5 for DoD CAC	https://seccerts.org/fips/a5bef651c8e3fd6c/
Plateforme JavaCard MultiApp V4.0.1 - PACE en configuration ouverte masque sur le composant M7892 G12	https://seccerts.org/cc/40fc6ad0aed92913/
SafeNet eToken (Smartcard or USB token) version 9.1.2 Athena IDProtect/OS755 Java Card on INSIDE Secure AT90SC25672RCTUSB Microcontroller embedding IDSign applet	https://seccerts.org/cc/87a80325171d8add/
SafeNet eToken - Athena IDProtect/OS755 Java Card on Atmel AT90SC25672RCT-USB Microcontroller embedding IDSign applet	https://seccerts.org/cc/c3d0eacd0639efb5/
Shenzhen Goodix GEOP01 on GSEA01 Security Chip version 1.0	https://seccerts.org/cc/adc792b4d5407e10/
Thales NFC422 v1.0 JCS	https://seccerts.org/cc/e38c6956b53dfd36/
Thales TESS v3.0 Platform	https://seccerts.org/cc/9be0c86102117436/
XSmart OpenPlatform V1.1 on S3CT9KW/S3CT9KC/S3CT9K9	https://seccerts.org/cc/cb203cf5d91b1ae3/

The JavaCard version reference analysis included all certification documents from the dataset containing a match of the JavaCard API version using regular expressions (406 documents in total).

References

1. Barbosa, O., Alves, C.: A systematic mapping study on software ecosystems. In: Proceedings of International Workshop on Software Ecosystems (2011)
2. Licel Corporation. jCardSim — Java Card Runtime Environment Simulator (2022). https://jcardsim.org/. Accessed 29 Sept 2023
3. de Lima Fontao, A., dos Santos, R.P., Dias-Neto, A.C.: Mobile software ecosystem (mseco): a systematic mapping study. In: 2015 IEEE 39th Annual Computer Software and Applications Conference, vol. 2, pp. 653–658. IEEE (2015)
4. Dufka, A., Švenda, P.: Enabling efficient threshold signature computation via java card API. In: Proceedings of the 18th International Conference on Availability, Reliability and Security, ARES 2023, New York, NY, USA, 2023. Association for Computing Machinery (2023)
5. Glossner, J., Murphy, S., Iancu, D.: An overview of the drone open-source ecosystem. arXiv preprint arXiv:2110.02260 (2021)
6. Hajny, J., Malina, L., Martinasek, Z., Tethal, O.: Performance evaluation of primitives for privacy-enhancing cryptography on current smart-cards and smartphones. In: Garcia-Alfaro, J., Lioudakis, G., Cuppens-Boulahia, N., Foley, S., Fitzgerald, W.M. (eds.) DPM/SETOP -2013. LNCS, vol. 8247, pp. 17–33. Springer, Heidelberg (2014). https://doi.org/10.1007/978-3-642-54568-9_2
7. Janovsky, A., Jancar, J., Svenda, P., Chmielewski, Ł., Michalik, J., Matyas, V.: sec-certs: examining the security certification practice for better vulnerability mitigation. arXiv:2311.17603 (2023)
8. Konstantinos Manikas and Klaus Marius Hansen: Software ecosystems-a systematic literature review. J. Syst. Softw. **86**(5), 1294–1306 (2013)
9. Mavroudis, V., Svenda, P.: JCMathLib: wrapper cryptographic library for transparent and certifiable javacard applets. In: 2020 IEEE European Symposium on Security and Privacy Workshops, pp. 89–96, IEEE. Genoa, Italy (2020)
10. Oracle. Java card^{TM} platform, application programming interface, classic edition version 3.2 (2023). https://docs.oracle.com/en/java/javacard/3.2/jcapi/api_classic/index.html. Accessed 29 Sept 2023
11. Martin Paljak. GlobalPlatformPro (2023). https://github.com/martinpaljak/GlobalPlatformPro. Accessed29 Sept 2023
12. Pawlak, R., Monperrus, M., Petitprez, N., Noguera, C., Seinturier, L.: Spoon: a library for implementing analyses and transformations of java source code. Softw. Pract. Experience **46**, 1155–1179 (2015)
13. Svenda, P., Kvasnovsky, R., Nagy, I., Dufka, A.: JCAlgTest: robust identification metadata for certified smartcards. In: 19th International Conference on Security and Cryptography, pp. 597–604. INSTICC. Lisabon (2022)

PQ.V.ALU.E: Post-quantum RISC-V Custom ALU Extensions on Dilithium and Kyber

Konstantina Miteloudi[1,2(✉)], Joppe W. Bos[3], Olivier Bronchain[3], Björn Fay[4], and Joost Renes[2]

[1] Radboud University, Nijmegen, The Netherlands
konstantina.miteloudi@ru.nl
[2] NXP Semiconductors, Eindhoven, The Netherlands
joost.renes@nxp.com
[3] NXP Semiconductors, Leuven, Belgium
{joppe.bos,olivier.bronchain}@nxp.com
[4] NXP Semiconductors, Hamburg, Germany
bjorn.fay@nxp.com

Abstract. This paper explores the challenges and potential solutions of implementing the recommended upcoming post-quantum cryptography standards (the CRYSTALS-Dilithium and CRYSTALS-Kyber algorithms) on resource constrained devices. The high computational cost of polynomial operations, fundamental to cryptography based on ideal lattices, presents significant challenges in an efficient implementation. This paper proposes a hardware/software co-design strategy using RISC-V extensions to optimize resource utilization and speed up the number-theoretic transformations (NTTs). The primary contributions include a lightweight custom arithmetic logic unit (ALU), integrated into a 4-stage pipeline 32-bit RISC-V processor. This ALU is tailored towards the NTT computations and supports modular arithmetic as well as NTT butterfly operations. Furthermore, an extension to the RISC-V instruction set is introduced, with ten new instructions accessing the custom ALU to perform the necessary operations. The new instructions reduce the cycle count of the Kyber and Dilithium NTTs by more than 80% compared to optimized assembly, while being more lightweight than other works that exist in the literature.

Keywords: CRYSTALS-Dilithium · CRYSTALS-Kyber · NTT · RISC-V · ISA extension

1 Introduction

Research in the field of quantum computing has advanced tremendously in recent years, which has brought significant changes to the field of cryptography. The security of traditional public-key cryptography (PKC) is based on hard mathematical problems such as the factorization of large integers and the calculation

The original version of the chapter has been revised. the second and third authors' affiliation has been corrected. A correction to this chapter can be found at https://doi.org/10.1007/978-3-031-54409-5_14

S. Bhasin and T. Roche (Eds.): CARDIS 2023, LNCS 14530, pp. 190–209, 2024.
https://doi.org/10.1007/978-3-031-54409-5_10

of discrete logarithms, which are difficult to solve on a classic computer (i.e., in (sub-)exponential time). However, these problems would be solved in polynomial time on a large-scale quantum computer, using Shor's algorithm [22]. This has led to a growing interest in post-quantum cryptography (PQC). Post-quantum cryptography refers to algorithms that run on the same classical hardware as traditional PKC, but are deemed secure against quantum adversaries.

Since 2016, PQC algorithms are in the process of being standardized by the National Institute of Standards and Technology (NIST) [18]. In July 2022, NIST recommended two primary algorithms to be implemented for most use cases: CRYSTALS-Kyber [21] and CRYSTALS-Dilithium [16]. Dilithium is a digital signatures algorithm, for example used to verify the authenticity and integrity of a message or document and to secure digital transactions. Kyber is a Key-Encapsulation Mechanism (KEM) and it is used to securely establish keys for a variety of applications, such as secure web browsing and email encryption. Both Dilithium and Kyber are based on hard problems coming from the theory of *ideal* lattices.

Lattice-based cryptographic algorithms have gained significant attention due to their security, efficiency and versatility. Although their efficiency is comparable to their classical counterparts on many platforms, their implementation can present challenges on resource-constrained systems. Looking at the performance numbers on the pqm4 benchmarking framework [13] that compares software implementations results on Arm Cortex-M4 cores, it can, for example, be seen that more than 80% of the runtime of Dilithium signature verification is spent in Keccak (SHAKE-256). For Kyber decapsulation more than 75% of the time is due to Keccak. Although this does not include side-channel protections beside executing in a constant running time, and while these numbers may differ for RISC-V, it shows that Keccak is a dominant factor in both algorithms. Because (hardened) Keccak accelerators are already well-researched (e.g., [6,23]), we do not investigate them in this work. Instead we focus on the next bottleneck that arises, namely polynomial operations. The Number-Theoretic Transform (NTT), a specialized form of the Discrete Fourier Transform (DFT), is used by many lattice-based algorithms to reduce the time complexity of polynomial multiplication. Despite this optimization, the polynomial arithmetic remains time-consuming, making efficient hardware implementations crucial, especially in constrained settings.

Resource-constrained devices, characterized by their limited computational capabilities, energy resources, and memory, pose unique challenges for the implementation of cryptographic algorithms. Typical examples include Internet of Things (IoT) devices, such as sensors, healthcare devices, automotive processors, and other embedded systems in vehicles and industrial control systems. The importance of implementing PQC algorithms with low area in mind lies in the need to accommodate the constraints of these devices. However, these implementations are expensive due to the complexity of the operations involved. This has led to exploration of hardware/software (HW/SW) co-design strategies, which aim to combine the high-speed performance of hardware implementations with the flexibility of software designs.

In the context of HW/SW codesign, RISC-V presents a promising approach. RISC-V is an open-source instruction set architecture (ISA) that supports custom extensions, providing a flexible platform for the implementation of various algorithms, including lattice-based algorithms. The support for custom extensions allows developers to add specialized instructions tailored to specific applications. These instructions can operate on the typical processor datapath or on a co-processor connected to the main processor as a peripheral, or they can operate on a modified processor datapath that integrates custom accelerators.

Contributions. The objective of this work is to design and implement a custom hardware accelerator for the NTT and other polynomial operations of Dilithium and Kyber. The primary aim is to optimize resource utilization of this accelerator and to minimize the number of changes to the RISC-V core, while increasing the speed of NTT computations for both Dilithium and Kyber. In short, in this work we introduce PQVALUE and summarize our contributions as follows:

- A lightweight custom ALU that is tailored to the requirements of the NTT computations for both schemes. The ALU is seamlessly integrated into a 4-stage pipeline 32-bit RISC-V processor and it supports modular operations (addition, subtraction and multiplication) as well as the NTT and inverse NTT butterfly operations (Cooley-Tukey and Gentleman-Sande).
- An extension to the RISC-V ISA, by introducing ten new instructions, five for each scheme. These instructions access the custom ALU and perform the operations mentioned above. Each instruction takes one clock cycle to execute.
- We show that our ALU is the smallest to appear in the literature and only increases the resource utilization of the RI5CY core by 13–17% (depending on the operating frequency). We measured our design for specific clock frequencies (100–400 MHz), similar to those used in real-world microprocessors, and confirmed that our custom RISC-V core with the new tailored ALU operates at these frequencies without any degradation compare to the original RISC-V core.
- We implement and benchmark the various operations and confirm that PQVALUE decreases the cycle count of the Kyber and Dilithium NTTs by more than 80% compared to optimized assembly, and significantly improves the runtime of the full algorithms.

2 Preliminaries

2.1 Number-Theoretics Transformations (NTTs)

The introduction of Ring- and Module-LWE has moved the arithmetic operations from general lattices to polynomial rings of the form $R = \mathbb{Z}_q[X]/(X^n + 1)$ for a prime q and integer n. The critical operations are polynomial addition and subtraction of linear complexity of n, and polynomial multiplication. The latter is a more involved operation that naïvely has quadratic complexity using schoolbook multiplication, but can be sped up using a variety of techniques such as Karatsuba, Toom-Cook, and *Number-Theoretic Transformations* (NTTs). The latter is not possible in generic polynomial rings, which is why the lattice-based constructions choose R to be of this specific form. The prime q is selected to be small, e.g., 12-bit for Kyber and 23-bit for Dilithium, while n is selected to be a power of 2 (typically 256). Moreover, q is selected such that it has a root of unity ζ of sufficiently large order. That is, of order n or $2n$.

An NTT can come in various forms or shapes, depending on the choice of root of unity and the implementation. For example, the Kyber ring contains only n-th roots of unity while the Dilithium ring contains $2n$-th roots of unity. This leads to slightly different strategies for implementing the NTT. For Kyber, given an input polynomial f, the NTT computes (up to re-ordering)

$$\begin{aligned} R &\rightarrow \prod_{i=0}^{127} \mathbb{Z}_q[X]/(X^2 - \zeta^{2i+1}) \\ f &\mapsto (f \bmod (X^2 - \zeta), \ldots, f \bmod (X^2 - \zeta^{255})) \ , \end{aligned}$$

which is an isormophism via the Chinese Remainder Theorem (CRT). The inverse NTT computes the inverse isomorphism. For Dilithium, the NTT instead computes (up to re-ordering)

$$\begin{aligned} R &\rightarrow \prod_{i=0}^{255} \mathbb{Z}_q[X]/(X - \zeta^{2i+1}) \\ f &\mapsto (f \bmod (X - \zeta), \ldots, f \bmod (X - \zeta^{511})) \ , \end{aligned}$$

and the inverse computes the inverse isomorphism according to the CRT.

The actual implementation of the operation closely mimics the strategies of Discrete Fourier Transforms (DFTs). The NTT can be constructed as a sequence of $\log_2(n)$ layers containing exactly $n/2$ *butterflies*, where each butterfly consists of a base field multiplication, addition and subtraction. These are known as Cooley-Tukey (CT) butterflies for the forward NTT, named after the authors that first introduced the butterfly structure. It is graphically described in Fig. 1a. The total complexity of the algorithm is therefore $n \log(n)$, which is significantly faster than other existing methods. The sequence of butterflies in the forward NTT introduces a re-ordering of the coefficients that deviates from the CRT isomorphism. This is not a problem as long as this is included in the inverse NTT operation (typically without performance loss). For example, this can be done by changing the butterfly structure from Cooley-Tukey to Gentleman-Sande (GS) for the inverse NTT. The GS butterfly is described in Fig. 1b. Eventually, we note that Kyber NTT only uses $\log_2(n) - 1 = 7$ layers while Dilithium uses $\log_2(n) = 8$ layers because of the different isomorphism described above.

2.2 Barrett Reduction

Barrett reduction [4] is an efficient method to compute modular reductions using precomputed data which only depends on the modulus used. Given a positive odd modulus q such that $2^{n-1} < q < 2^n$ and some input value $0 \leq c < q^2$ then one can compute $c' = c \bmod q$ as

$$c' = c - m \cdot q, \text{ where } m = \left\lfloor \frac{c}{q} \right\rfloor .$$

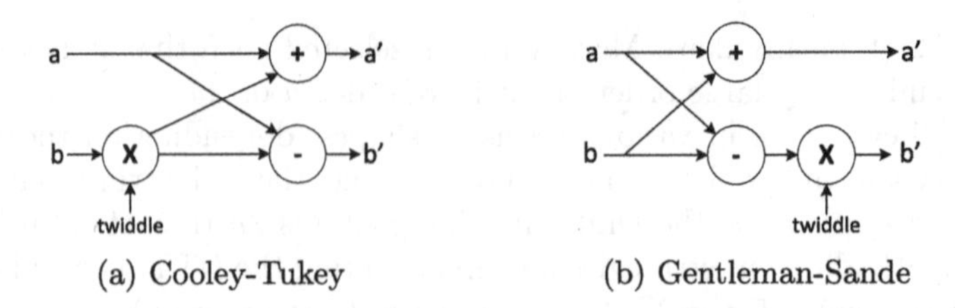

(a) Cooley-Tukey (b) Gentleman-Sande

Fig. 1. Butterflies blockdiagrams

Algorithm 1. Barrett Reduction in Dilithium

Input: $0 \leq x < 8\,380\,417^2$,
Output: $z = x \mod 8\,380\,417$
1: $t \leftarrow (x \ll 23) + (x \ll 13) + (x \ll 3) - x$
2: $t \leftarrow t \gg 46$
3: $t \leftarrow (t \ll 23) - (t \ll 13) + t$
4: $z \leftarrow x - t$
5: **if** $z \geq 8\,380\,417$ **then**
6: $\quad z \leftarrow z - 8\,380\,417$
7: **return** z

Algorithm 2. Barrett Reduction in Kyber

Input: $0 \leq x < 3329^2$,
Output: $z = x \mod 3329$
1: $t \leftarrow 5039 \cdot x$
2: $t \leftarrow t \gg 24$
3: $t \leftarrow (t \ll 11) + (t \ll 10) + (t \ll 8) + t$
4: $z \leftarrow x - t$
5: **if** $z \geq 3329$ **then**
6: $\quad z \leftarrow z - 3329$
7: **return** z

The idea behind Barrett reduction is inspired by a technique of emulating floating point data types with fixed precision integers: namely, *approximate* $m = \lfloor c/q \rfloor$ by

$$m_1 = \left\lfloor \frac{c}{2^{2n}} \cdot \left\lfloor \frac{2^{2n}}{q} \right\rfloor \right\rfloor = \left\lfloor \frac{c \cdot \mu}{2^{2n}} \right\rfloor,$$

where $\mu = \lfloor 2^{2n}/q \rfloor$ is a pre-computed constant. Since $m - 1 \leq m_1 \leq m$ this approximation is almost correct while only efficient integer divisions by powers of two (i.e., right shifts) are required.

Barrett modular reduction for the prime moduli used in Dilithium and Kyber is shown in Algorithm 2 and Algorithm 1, respectively. Using the special form of the modulus was also used in [3]. The Barrett reduction for Kyber is shown in Algorithm 2. One has $q = 3329$, $2^{11} < q < 2^{12}$ and $\mu = \lfloor 2^{24}/q \rfloor = 5039$. The computation of $\lfloor c \cdot \mu/2^{24} \rfloor$ is performed in Line 1 and 2. Line 3 and 4 compute $c' = c - m_1 \cdot q$ where the multiplication by $q = 3329 = 2^{11} + 2^{10} + 2^8 + 2^0$ is performed as a sequence of shifts and additions.

The Barrett reduction for Dilithium is shown in Algorithm 1. One has $q = 8\,380\,417$, $2^{22} < q < 2^{23}$ and $\mu = \lfloor 2^{46}/q \rfloor = 8\,396\,807 = 2^{23} + 2^{13} + 2^3 - 1$. The computation of $\lfloor c \cdot \mu/2^{46} \rfloor$ is performed in Line 1 and 2 (where the multiplication by μ makes use of the special form). Line 3 and 4 compute $c' = c - m_1 \cdot q$ where the multiplication by $q = 8\,380\,417 = 2^{23} - 2^{13} + 1$ is done as a series of shifts and additions.

31 25	24 20	19 15	14 12	11 7	6 0
funct7	rs2	rs1	funct3	rd	opcode

Fig. 2. RISC-V R-type instruction format

2.3 RISC-V

RISC-V defines a flexible and extensible Instruction Set Architecture (ISA). The ISA defines small instruction sets that must be implemented by all the RISC-V cores along with additional extensions that can optionally be implemented depending on the target application. This makes this ISA portable to a large variety of applications ranging from embedded system to high-performance computers. The ISA does not define how instructions must be implemented, or their latency. Compared to Arm architectures, cryptographic implementations on RISC-V generaly take advantage of its large register file, but suffer from the limited instruction sets when the appropriate extensions are not implemented [23].

For the purpose of this work, we focus on the 32-bit RISC-V architecture where both the datapath and the instructions operate on 32-bit words. These architectures have 32 registers from which 27 can be freely used. We have a closer look at the R-type instructions that are highlighted in Fig. 2. The instruction is uniquely defined by the three fields `opcode`, `funct3` and `funct7`, that combine to a total of 17 control bits. Each instruction has a specific combination of control bits and the CPU will use them to execute the corresponding operation. The R-type instructions have 2 source and 1 destination registers that are the registers that the instruction must operate on. These are encoded as `rs1`, `rs2` and `rd` respectively in Fig. 2. Each of these registers are encoded on 5 bits as 32 different registers could be accessed.

2.4 Related Work

Research into PQC hardware accelerators has increased in recent years, with a growing number of published works addressing its various aspects. There have been hardware implementations of the full scheme of Dilithium [5,12,15,20, 25,26] and Kyber [7,19,24]. They use various optimizations in order to have small and fast designs while also exploring how these designs can evolve into co-processors capable of running the full schemes. Some studies have also focused on designing co-processors that can accelerate more than one scheme. Aikata, Mert, Imran, Pagiliarini, and Roy [1] create a unified architecture for both Dilithium and Kyber by designing common building blocks for polynomial multiplication and SHAKE. Moreover, Banerjee, Ukyab and Chandrakasan [3] create custom RISC-V instructions for polynomial arithmetic and sampling to support a co-processor for Frodo, NewHope, qTESLA, Kyber and Dilithium. All publications mentioned above connect their hardware accelerators as peripherals to the main processor. However, these peripherals, while beneficial, can be slowed down by the bus of the processor and the transfer of data between the different components. An alternative approach is to integrate the hardware accelerator directly into the datapath of the processor. For example Karl, Schupp, Fritzmann and Sigl [14] accelerate the NTT transformation and pointwise multiplications as well as the functions SHAKE128 and SHAKE256 for Dilithium and Falcon. The NTT accelerator is connected as peripheral and it needs its own memory in order to reduce the transaction cost, but this increases also the resources, while the Keccak round function for SHAKE is incorporated into RISC-V datapath.

There are two publications where a custom ALU with ISA extensions is implemented to accelerate Dilithium and Kyber that can be directly compared with our own. Fritzmann, Sigl and Sepúlveda [8] present an enhanced RISC-V architecture that integrates tightly coupled accelerators directly into the processor's pipeline datapath to speed-up NewHope, Kyber, and Saber while extending the RISC-V ISA by twenty-nine new instructions. Nannipieri, Di Matteo, Zulberti, Albicocchi, Saponara and Fanucci [17] introduce an extension to the RISC-V ISA to facilitate the NTT operations of the Dilithium and Kyber cryptographic schemes.

3 Extensions for Polynomial Arithmetic

In this section, we describe our dedicated ALU for polynomial arithmetic including adders, subtractors, multipliers and butterflies. The main challenge to support polynomial arithmetic for both Dilithium and Kyber is the difference in prime size. For Dilithium we have the 23-bit prime $q = 8\,380\,417$ while for Kyber, we have the 12-bit prime $q = 3\,329$. As our goal is to have a small hardware design, we re-use the same base adders and multipliers for both schemes.

Representation: When implementing arithmetic modulo q, the question that arises is how to represent the values. Optimized (embedded) software typically use the signed interval $[-(q-1)/2, (q-1)/2\,]$, as it may have advantages for the number of reductions that take place in the butterflies [10,11,13]. Instead, we choose the canonical interval $[0, q-1]$ as it leads to slightly simpler hardware. Of course, the choice does not have a large impact as transforming between representations is trivial and cheap.

3.1 Single-Cycle Modular Arithmetic

First, we describe the modular addition and the modular subtraction units. Both have a similar structure as detailed in Fig. 3. We start by describing the addition for Dilithium, then its subtraction and finally we describe how this can be re-used for Kyber.

The modular adder is based on an addition and a subsequent subtraction approach as shown in Fig. 3a. Concretely, both inputs a and b are first added together with a 24-bit unsigned integer adder to obtain $\texttt{c'} = \texttt{a} + \texttt{b}$. Next, we perform a 24-bit unsigned subtraction such that $\texttt{c''} = \texttt{c'} - q$. The final result c is then selected conditionally: $\texttt{c} = \texttt{c'}$ if $\texttt{c'} < q$ (i.e., a borrow occurred in

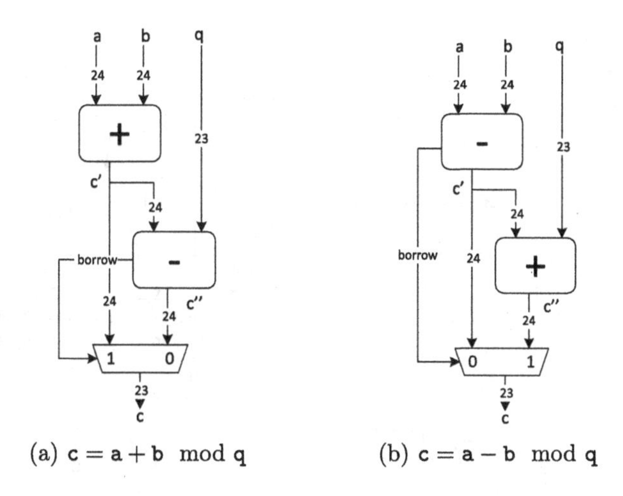

(a) $\mathtt{c} = \mathtt{a} + \mathtt{b} \mod \mathtt{q}$ (b) $\mathtt{c} = \mathtt{a} - \mathtt{b} \mod \mathtt{q}$

Fig. 3. Modular addition and subtraction circuits.

the subtraction), and c = c'' otherwise. After the selection, only the 23 least significant bits are needed, as it guarantees that $\mathtt{c} \in [0, q-1]$.

The modular subtraction is described in Fig. 3b and works in a similar fashion. Namely, first compute c' = a − b using a 24-bit unsigned subtractor. The output signal c is then set to c = c' + q if c' < q (i.e., a borrow occurred in the subtraction) and c = c' otherwise. The selection between both cases is performed based on the borrow bit and a multiplexer. Again, only the 23 least significant bits are needed afterwards.

Exactly the same circuit can be used for all $q < 2^{23}$, and hence in particular for both Dilithium and Kyber. The only difference is in the position of the borrow bit. Interestingly, as additions and subtractions are based on similar designs, the same circuit could be shared also between modular addition and subtraction. However, as detailed in Sect. 3.2, one addition and one subtraction needs to be performed in parallel to have butterflies in a single cycle. Hence, we use independent circuits for modular addition and subtraction.

As detailed in Sect. 2.2, modular multiplication $\mathtt{c} = \mathtt{a} \cdot \mathtt{b} \mod q$ is performed in two steps. The first step is an integer multiplication where both inputs a and b are multiplied. In order to reduce the cost of the integer multiplier, we share it between the modular multiplications used in both Dilithium and Kyber. This means adding a 23 × 23-bit multiplier with 46-bit output. As Kyber requires only a 12 × 12-bit multiplication with 24-bit output, this is far from an optimal multiplier size. We discuss possible alternatives at the end of this section.

The second step for modular multiplication is to perform the Barrett reduction. The Barrett reduction circuits tailored for both Dilithium and Kyber are shown in Fig. 4. These circuits are the direct implementation of Algorithm 1 and Algorithm 2. While several reduction methods exist, we chose Barrett reduction. It avoids expensive division operations and the special form of the modulus in Dilithium and Kyber allows the multiplication to be performed as a sequence

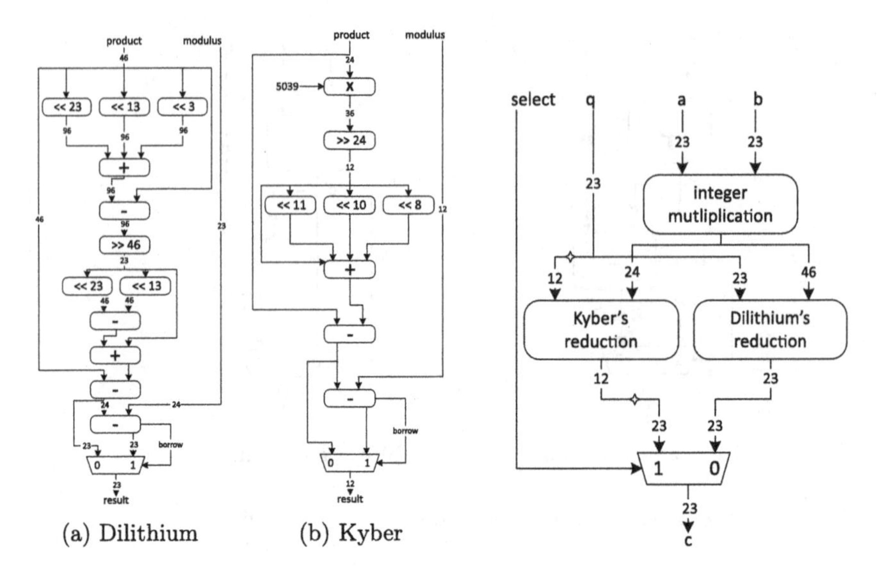

Fig. 4. Dedicated Barrett reductions hardware from [3].

Fig. 5. Modular multiplication unit. Computes $\texttt{c} = \texttt{a} \cdot \texttt{b} \bmod q$ for Dilithium and Kyber primes q.

of shifts and additions, reducing the cost in time and area. For Dilithium, the Barrett reduction has a logical depth of 7 additions/subtractions. For Kyber, a total for 13 additions/subtractions is needed, of which 8 are for the constant multiplication by 5039. Indeed, it can be decomposed into a sum of powers of two as follows: $5039 = 2^{12} + 2^9 + 2^8 + 2^7 + 2^5 + 2^3 + 2^2 + 2^1 + 2^0$. We rely on the synthesizer's optimization strategies to efficiently implement this, allowing it to choose between structures like adder trees or utilizing carry chains. In case of an adder tree is used, the total logical depth of 8 adders is needed for the reduction (with 4 for the constant multiplication). We note that other options to perform the modular reduction are also possible (see [15]), which target FPGA implementations.

Finally, as the 23×23 base multiplier is shared for both Dilithium and Kyber, both Barrett reductions are connected to it. The result of the modular multiplication is then chosen via a select signal and a multiplexer as illustrated in Fig. 5. As mentioned above, this base multiplier is not optimal for Kyber, leading to the underutilization of one DSP block when Kyber is executed. While a potential solution could be a vector instruction that executes two multiplications for the Kyber prime in parallel, thereby doubling the throughput and fully utilizing the DSP block, we chose the simpler shared multiplier design without the additional logic for parallel multiplications. This decision prioritizes area savings and design simplicity, especially given the differing performance requirements between Kyber and Dilithium. Another interesting approach would be to re-use the 32×32-bit multiplier from the original ALU if available, leading

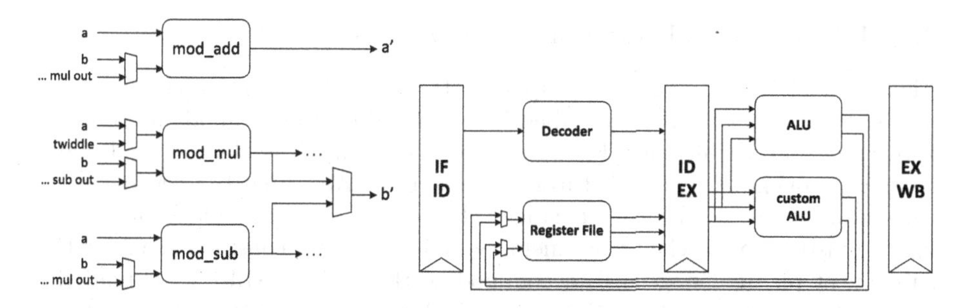

Fig. 6. Custom ALU unit.

Fig. 7. Modified datapath.

to an even smaller design. As the performance would depend strongly on the availability and latency of the instructions in the RISC-V core, it is not clear what the benefit would be. Therefore, we take the general approach and include our own multiplier.

3.2 Single-Cycle Butterflies

From the hardware for modular arithmetic, one can efficiently implement both Cooley-Tukey and Gentleman-Sande butterflies by connecting them together as illustrated in Fig. 1. The simplest option is to use a sequence of assembly instructions for the addition, subtraction, and multiplication (see Sect. 3.4 and Fig. 8c) with a latency of 3 cycles. In this section, we also describe the option to directly connect these operations within the custom ALU in order to perform a butterfly in a single cycle.

From Fig. 1, we observe that the Cooley-Tukey butterfly first performs the multiplication followed by two independent addition and subtraction operations. Conversely, the Gentleman-Sande butterfly performs the addition and subtraction before the multiplication with the twiddle factor. As a result, in order to use a single unit for all three operations, multiplexers are needed. The resulting connections between the different operations are detailed in Fig. 6. For the Cooley-Tukey butterfly, the multiplier is connected to `b` and `twiddle`, and both the subtraction and addition are connected to `a` and the output of the multiplication. For Gentleman-Sande, the output of the subtraction is directly connected to the multiplier that takes the twiddle factor as its second input.

Finally, we note that to perform a single-cycle butterfly, three read and two write ports on the register files are required. A typical register file has two read ports and one write port, resulting in two clock cycles for reading the inputs and two clock cycles for writing the outputs of the butterfly operations. However, PULPino's register file has three read and two write ports, which allows us to read and write in a single clock cycle.

3.3 Integration and Additional Instructions

Figure 7 depicts the modified datapath with our ALU at the execution stage. There, the *custom ALU* is described in Fig. 6 and contains a modular adder, a subtractor and a multiplier. For simplicity, the datapath is abstract and many of its components are excluded from the diagram. We edited the decoder unit, the decode stage and the execute stage. The decoder is responsible for interpreting the instruction fetched from memory and generating control signals for the subsequent pipeline stages. To accommodate the custom ALU, we extend the decoder to recognize additional custom opcodes representing the new operations supported by the custom ALU. This includes adding logic to the decoder to generate new control signals that will be used to steer data to and from the custom ALU. In the decode stage, the instruction is further analyzed, and the operands are fetched from the register file. We modified this stage to handle the additional control signals generated by the decoder for the custom ALU. This ensures that, when an instruction targeting the custom ALU is encountered, the appropriate operands are fetched and routed to the custom ALU. We integrated the custom ALU into the execution stage. The custom ALU is connected in parallel with the existing ALU, and both share the same input interfaces. A multiplexer is used to select between the output of the custom ALU and the regular ALU based on the control signals generated by the decoder.

Table 1 lists the new instructions which are of R-type format and they have the same opcode. Between the two schemes, the instructions are distinguished by the funct7 and vary by a single bit. The Least Significant Bit (LSB) of funct7 is utilized to control the modulus to be used and the effective width of the signals. The operations themselves are distinguished by the funct3 field. Modular addition, subtraction, and multiplication operations are analogous to their integer counterparts performed by the classic ALU. The operands for these operations are read from two source registers, and the result is written back to a single destination register. For the butterfly instructions, three read and two write ports are needed, which are already available in the PULPino core we consider. There, the coefficients `a` and `b` are read from the first and second source registers, while the `twiddle` is read from the destination register. The new coefficients `a'` and `b'` are then computed and written back to the first and second source registers. Furthermore, we managed to maintain a low number of sources by creating only the bare minimum control signals for our multiplexers. In addition, by ensuring that the same operands that go to the regular ALU also go to our custom ALU, we were able to avoid increasing the pipeline registers. This approach allowed us to maintain the efficiency of our design while expanding its capabilities.

3.4 PQVALUE: NTTs

We have described hardware blocks to support various modular arithmetic operations with specific moduli, as well as their integration into a RISC-V microcontroller. In this section, we elaborate on how software can use the added instructions to improve the performance of polynomial arithmetic. The instructions

Table 1. Custom arithmetic instructions for RISC-V ISA

opcode	funct3	funct7	operation name
1110111	001	0000000	pq.mod_add_dil
1110111	010	0000000	pq.mod_sub_dil
1110111	011	0000000	pq.mod_mul_dil
1110111	100	0000000	pq.ct_btrfly_dil
1110111	101	0000000	pq.gs_btrfly_dil
1110111	001	0000001	pq.mod_add_kyb
1110111	010	0000001	pq.mod_sub_kyb
1110111	011	0000001	pq.mod_mul_kyb
1110111	100	0000001	pq.ct_btrfly_kyb
1110111	101	0000001	pq.gs_btrfly_kyb

detailed in Sect. 3.3 can be used to build the butterflies as detailed in Fig. 8c and Fig. 8e. We consider the case where no single-cycle butterfly is available (2 read and 1 write ports) and where it is available (3 read and 2 write ports), respectively. We refer to the first design as PQVALUE^1, and the latter as PQVALUE^2. For reference, we also include optimized RISC-V software butterflies in Fig. 8a for comparison.

As can be seen from Sect. 2.1, NTTs and inverse NTTs can be implemented in place with 8 separate layers, each containing 128 independent butterflies. A straightforward implementation could process the layers of butterflies one by one, in which case for each butterfly the input coefficients are read, the butterfly is computed, and the results are stored back in memory. Hence, the polynomial is read and stored at every layer: this leads to signification number of memory access that can become a bottleneck. Hence, a common technique in software implementations is to merge multiple layers in order to reduce these overheads [10,11]. More specifically, multiple coefficients are loaded into the register file such that multiple layers can be (partially) computed without need for loading/storing for every butterfly. To merge ℓ layers one would load and store 2^ℓ (in-place) inputs/outputs and apply all $\ell \cdot 2^{\ell-1}$ butterflies before loading the next batch of coefficients. As this is repeated $256/2^\ell = 2^{8-\ell}$ times, the total cost of the ℓ merged layers is $2^{8-\ell} \cdot (2 \cdot 2^\ell\, \texttt{IO} + \ell \cdot 2^{\ell-1}\, \texttt{BFLY})$, where `IO` is the cost of a read/write to memory and `BFLY` is the cost of a single butterfly. For example, with single-cycle read/writes and using a single-cycle butterfly the cost of 4 merged NTT layers is (at least) $2^4 \cdot (32 + 4 \cdot 8) = 1024$ cycles. Since our hardware designs affect `BFLY` but not `IO`, it is worthwhile merging layers even with our extensions as it reduces the overheads of memory accesses. Finally, the twiddle factors need to be loaded once at a cost of 256 `IO`. Note that an actual implementation will have some additional overhead for stack and control flow operations: we give more precise results in Sect. 4.

As the register file on RISC-V has 32 general purpose registers, we can merge at most 4 layers. Although we have 16 inputs/outputs, some registers are reserved for other values such as twiddle factors. The Dilithium forward NTT can be constructed by merging 4 layers twice and loading the twiddle factors once throughout the layers. The Kyber forward NTT has an early-abort strategy that achieves a cheaper NTT at the cost of a more expensive base multiplication. It can be constructed by merging 4 layers, then merging 3 layers, loading the twiddle factors once throughout the layers. The inverse NTT functions in the same way using Gentleman-Sande butterflies, and with the addition of a final `poly_mul` to remove the constant factor n. Note that this can also help to remove any factors of the Montgomery domain that might remain. For example, the values in the matrix $\mathbf{A}$ are sampled into the Montgomery domain for Kyber, which remains as we only use Barrett reduction in our implementation. The additional inversion/multiplication makes the inverse NTT more expensive than the forward NTT, but as it is used less frequently than the forward NTT, the overall impact on Kyber or Dilithium is minimal. In that case, the final division by 256 can include the Montgomery constant as well. Finally, the base multiplication is a direct application of `mod_mul` on each of the coefficients.

```
.macro montgomery al, ah, qi, q
    mul \al, \a, \qi
    mulh \al, \al, \q
    sub \al, \ah, \al
.endm

.macro ct_butterfly a, b, qi, q, zeta,
    tmp
    mul \tmp, \zeta, \b
    mulh \b, \zeta, \b
    montgomery \tmp, \b, \qi, \q
    sub \b, \a, \tmp
    add \a, \a, \tmp
.endm
```

(a) Cooley-Tukey, RV32

```
.macro montgomery al, ah, qi, q
    mul \al, \a, \qi
    mulh \al, \al, \q
    sub \al, \ah, \al
.endm

.macro gs_butterfly a, b, qi, q, zeta,
    tmp
    sub \tmp, \a, \b
    add \a, \a, \right
    mul \b, \zeta, \tmp
    mulh \tmp, \zeta, \tmp
    montgomery \b, \tmp, \qi, \q
.endm
```

(b) Gentleman-Sande, RV32

```
.macro ct_butterfly a, b, z, tmp
    pq.mod_mul \tmp, \z, \b
    pq.mod_sub \b, \a, \tmp
    pq.mod_add \a, \a, \tmp
.endm
```

(c) Cooley-Tukey, PQVALUE[1]

```
.macro gs_butterfly a, b, zeta, tmp
    pq.mod_sub \tmp, \a, \b
    pq.mod_add \a, \a, \b
    pq.mod_mul \b, \zeta, \tmp
.endm
```

(d) Gentleman-Sande, PQVALUE[1]

```
.macro ct_butterfly a, b, zeta
    pq.ct_btrfly \a, \b, \zeta
.endm
```

(e) Cooley-Tukey, PQVALUE[2]

```
.macro gs_butterfly a, b, zeta, tmp
    pq.gs_btrfly \a, \b, \zeta
.endm
```

(f) Gentleman-Sande, PQVALUE[2]

Fig. 8. Butterfly with custom assembly and two read, one write ports.

4 Results

In this section we present area and performance results of our hardware design. We conduct synthesis for PQVALUE, and compare the unmodified RI5CY core with the modified core that integrates our ALU. We target two different platforms: Application-Specific Integrated Circuits (ASICs) and Field-Programmable Gate Arrays (FPGAs) in order to assess their performance and resource utilization. We also measure the cycle counts of our HW/SW co-design implementations and compare to existing literature.

4.1 Setup

For integration and benchmarking, we use the open-source PULPino platform with its RI5CY core that has a 4-stage pipeline implementing the RISC-V ISA [9]. It has support for the integer (I), compressed (C) and multiplication (M) instruction set extensions. Notably, two 32-bit words are multiplied in 6 cycles where 1 cycle is spent in the `mul` instruction, and 5 cycles are spent in `mulh` (to compute the top half). The memory accesses are all single-cycle. PULPino provides a comprehensive framework and a set of scripts that streamline the process of running automated tests and simulations. This infrastructure simplifies the task of compiling and running simulations, collecting and analyzing results, and comparing these results against expected outputs. To verify the correct functionality of our modified RI5CY core and measure the cycle counts, we utilize the PULPino framework in conjunction with ModelSim HDL Simulator.

For the synthesis on ASICs we use Cadence Genus Synthesis Solution and the TSMC 28 nm standard cell library. We set timing constraints at frequencies of 100 MHz, 200 MHz, 300 MHz, and 400 MHz. We choose these values to mimic closely the typical operating ranges of modern microprocessors. We have confirmed that up to 600 MHz, our ALU is not in the critical path of the processor. This approach facilitates potential integration of our design in existing computational systems, while also accounting for the characteristics and limitations of real-world microprocessor architectures. For the synthesis on FPGAs we used Vivado 2019.2 and we target the ZedBoard Zynq-7000 development board. The timing constraint for the FPGAs is already set by PULPino team at 20 Mhz. From the whole PULPino system, we excluded the communication peripherals and the memories and we measured the overhead in the RI5CY core only.

4.2 Hardware Cost of PQVALUE

We report the utilization on ASICs for both of our designs, comparing three parameters: cells count, total cell area, and gate equivalents. The first two metrics are reported by the tool. The gate equivalents are calculated by dividing the total cell area with the size of the two-input NAND gate of the specific library. For the TSMC 28 nm library, the size of a two-input NAND gate is $0.378\,\mu m^2$.

We provide a comparison between the two designs: the unmodified RI5CY core and the RI5CY core modified with a custom ALU (referred to as RI5CY +

Table 2. Resource utilization of PQVALUE in ASICs. The Total Cell Area is measured in μm^2. All numbers are rounded to the nearest integer.

Freq.	Metric	RI5CY + PQVALUE	RI5CY	Difference	
				Abs	Rel
100 MHz	Total Cell Area	15 848	18 315	2 467	16%
	Cell Count	19 044	22 204	3 160	17%
	Gate Equivalent	41 926	48 452	6 526	16%
200 MHz	Total Cell Area	15 844	18 256	2 412	15%
	Cell Count	19 026	21 951	2 925	15%
	Gate Equivalent	41 915	48 297	6 382	15%
300 MHz	Total Cell Area	16 087	18 496	2 409	15%
	Cell Count	19 756	22 723	2 967	15%
	Gate Equivalent	42 559	48 932	6 373	15%
400 MHz	Total Cell Area	16 532	18 834	2 302	14%
	Cell Count	20 912	23 670	2 758	13%
	Gate Equivalent	43 737	49 825	6 088	14%

PQVALUE). The Total Cell Area represents the physical area occupied by the design on the silicon. The Cell Count represents the number of logic cells used in the design. Table 2 shows that as the operating frequency increases, the Total Cell Area, Cells Count, and Gate Equivalents for both designs increase. This is expected as meeting higher frequency requirements often necessitates the use of more resources. However, we notice that at 200 MHz, all the metrics have values slightly lower than the ones in 100 MHz. This is probably due to the variability involved in the synthesis process. At 200 MHz, it's possible that the synthesis tool found a combination of logic cells that resulted in lower area utilization and cell count. At 300 Mhz and 400 Mhz, the resource utilization significantly increases. Notably, the percentage difference between the two designs decreases as the timing constraint increases. This suggests that the unmodified RI5CY core scales its resources more aggressively to meet higher frequency requirements compared to the modified RI5CY + PQVALUE design. The relative increase of our design compared to the unmodified RI5CY core ranges from 13% to 17% for all metrics across the different frequencies that we measured with.

Table 3 provides a comparison between the two designs, when targeting the same FGPA with the same operating frequency. Slice Look-Up Tables (LUTs) are a fundamental resource in FPGAs used to implement combinational logic functions. The modified RI5CY + PQVALUE shows a percentage increase of 5.29% in LUTs, a percentage increase of 0.72% in Registers, and a percentage increase of 33.3% in Digital Signal Processors (DSPs). This increase was expected as our ALU is pure combinational logic and we did the minimum modification in the registers of RI5CY core. The 2 extra DSPs are from our base 23×23 integer multiplier.

Table 3. Resource utilization of PQVALUE in the Zynq-7000 FPGA.

Resources	RI5CY	RI5CY + PQVALUE	% Difference
Slice LUTs	6 797	7 157	5.3%
Slice Registers	2 212	2 228	0.7%
DSPs	6	8	33.3%

4.3 PQVALUE in Pulpino: NTTs

In this section, we discuss the concrete improvements that PQVALUE has on the NTTs on Pulpino. We focus on Dilithium for simplicity, but very similar numbers can be achieved for Kyber. Recall that Pulpino requires 1 cycle for memory accesses, 1 cycle for `mul`, `add`, and `sub`, and 5 cycles to execute `mulh`. Therefore the full Cooley-Tukey and Gentleman-Sande butterflies as shown in Fig. 8a can be performed in 15 clock cycles. Looking at the estimates from Sect. 3.4 for 4 merged layers with $\texttt{IO} = 1$ and $\texttt{BFLY} = 15$, we expect the forward NTT to require about 16 640 cycles. In this case 15 360 cycles are spent on butterflies, and 1 280 on reads/writes. We confirm this by implementing and measuring the cycle count on the Pulpino platform, resulting in 17 041 cycles in total. The additional 401 cycles are expected as they come from stack and control flow operations.

By reducing the butterfly operations to 3 cycles in the case of PQVALUE[1], we reduce the of the butterfly operations from 15 360 to 3 072 cycles. As the cost of reads/writes and stack/control flow operations remains at 1 280 and 401 respectively, the expected cost is 4 753 cycles. The total impact of butterflies is reduced from 90% for a full software implementation, to 65% in PQVALUE[1]. We confirm this precise cycle count by implementing the code from Fig. 8c and benchmarking. For the single-cycle butterfly, the butterflies require 1 024 cycles and the forward NTT in total 2 705 cycles. The butterflies only contribute to 38% of the runtime, and are now dominated by the overhead for reads and writes. Again, we confirm by implementing the code from Fig. 8e and measuring the number of cycles.

Similarly, we implement the operations for the inverse NTT, polynomial multiplication and addition. We provide an overview of the result in Table 4. Overall, we see that PQVALUE[1] leads to a performance improvement of 72% (70%) for the (inverse) NTT, while PQVALUE[2] leads to a reduction by 84% (80%) for the (inverse) NTT. Both improve the polynomial multiplication by 71%, and have no impact on polynomial addition or subtraction.

4.4 PQVALUE in Pulpino: Dilithium

In order to evaluate the impact of the polynomial arithmetic on Dilithium, we integrate them into the reference implementation and compare its run time for various configurations in Table 5. The reported number are the minimal run-time for each of the algorithms. This choice has an influence for the Sign results,

Table 4. Cycle counts of polynomial operations in Dilithium for various instructions. PQVALUE^2 is with single-cycle butterfly, PQVALUE^1 is without single-cycle butterfly.

	RV32	PQVALUE^1	PQVALUE^2
`poly_NTT`	17 041	4 753 *(−72%)*	2 705 *(−84%)*
`poly_iNTT`	20 372	6 027 *(−70%)*	3 979 *(−80%)*
`poly_mul`	4 346	1 274 *(−71%)*	1 274 *(−71%)*
`poly_add`	1 274	1 274 *(−00%)*	1 274

because of its non-deterministic execution time. The reported numbers include a single execution of the rejection loop. Average run-time can be obtained by multiplying by the average number of aborts reported in [2]. For all the algorithms the public matrix is computed on-the-fly in order to save stack.

The two left-most columns compare Dilithium-3 with and without PQVALUE^2 entirely in software. Concretely, we consider the RISC-V optimized Keccak-1600 software implementation proposed in [23] and assembly optimized polynomial addition, subtraction, Montgomery and base multiplication. In this context, we observe that PQVALUE improves the performances by a small factor. Indeed, KeyGen is sped up by a factor 1.09, Sign by 1.25 and Verify by a factor 1.13. This is because Keccak is the bottleneck as usual in pure software implementations [13]. In the two right-most columns of Table 5, we consider a configuration where a Keccak co-processor with 24 cycles per permutation is available. In that context, the bottleneck operations are polynomial operations and therefore the impact of PQVALUE is larger. Concretely, Keygen is sped up by a factor 1.72, Sign by 2.32 and Verify 1.91. We note that in this case, the rest of the operations such as packing, rejection sampling, norm checks, and hint manipulations take a large portion of the overall execution time as they are not assembly optimized. The impact of NTT extensions will be even larger on a fully optimized implementation.

4.5 Comparison

As mentioned in Sect. 2.4, we compare our custom ALU to two other works that target RISC-V extensions [8,17]. Fritzmann, Sigl and Sepúlveda [8] include an arithmetic unit PQR-ALU for vectorized modular arithmetic and NTT operations, a vectorized modular multiply accumulate unit, a Keccak accelerator for pseudo-random bit generation, and a binomial sampling unit for the generation of distributed samples. The arithmetic units for modular arithmetic and NTT operations are added to the decode stage of RISC-V leading to a *decrease* in the overall processor clock frequency (see [8, Section 5.3]). This is a high price to pay, as all functionality (possibly more critical than PQC) that runs on the RISC-V core is impacted. In comparison, we have added our ALU to the execution stage, while we managed to maintain the original clock frequency ensuring optimal performance of our design.

Table 5. Minimum cycle count for Dilithum-3 for various available instructions. Available Keccak co-processor is marked with ^{k}. PQVALUE^2 has a single cycle butterfly.

	RV32		PQVALUE^2		RV32^k		PQVALUE^{2,k}	
	kCycles	%	kCycles	%	kCycles	%	kCycles	%
KeyGen	4 316	100.0	3 970	100.0	825	100.0	479	100.0
Poly	518	12.0	172	4.3	518	62.8	172	36.0
Keccak	3 514	81.4	3 514	88.5	23	2.8	23	4.9
Others	283	6.6	283	7.1	283	34.3	283	59.1
Sign	5 253	100.0	4 218	100.0	1 817	100.0	782	100.0
Poly	1 469	28.0	434	10.3	1 469	80.9	434	55.6
Keccak	3 459	65.8	3 459	82.0	22	1.3	22	2.9
Others	324	6.2	324	7.7	324	17.9	324	41.5
Verify	4 178	100.0	3 712	100.0	976	100.0	511	100.0
Poly	697	16.7	232	6.3	697	71.4	232	45.5
Keccak	3 223	77.1	3 223	86.8	21	2.2	21	4.2
Others	257	6.2	257	6.9	257	26.4	257	50.3

Table 6. Size and efficiency comparison of post-quantum ALUs.

	Resources				Kyber perf.		
	LUT	Reg.	DSP	BRAM	Core	NTT	NTT^{-1}
PQR-ALU [8]	2 908	170	9	0	RI5CY	1 935	1 930
PQ ALU [17]	555	0	15	1	CVA6	18 448	18 448
PQVALUE^2	459	0	2	0	RI5CY	2 577	3 851

Nannipieri, Di Matteo, Zulberti, Albicocchi, Saponara and Fanucci [17] introduce an extension to the RISC-V ISA to facilitate the NTT operations of the Dilithium and Kyber cryptographic schemes. They implement two distinct custom ALUs, each tailored to a specific scheme. This approach, however, leads to a significant increase in resource overhead compared to our design, which employs a unified ALU for both schemes. The resource utilization and performance of PQR-ALU and PQ ALU are detailed in Table 6.

As shown in Table 6, our proposed PQVALUE^2 demonstrates a significant advantage in terms of resource utilization compared to the other two designs. The PQVALUE^2 uses fewer LUTs, DSPs, and Block RAMs (BRAMs) than the other two designs. In terms of performance for Kyber, PQVALUE^2 performs the NTT operation in 2 577 cycles and the inverse NTT operation in 3 851 cycles. This is slower than the PQR-ALU [8] which performs both operations in under 2 000 cycles, meaning that we present a different area/performance trade-off that puts the focus on small designs. We believe this is worthwhile as the efficiency is sufficiently high to move the bottlenecks elsewhere (see Table 5 for Dilithium), and we have the additional benefit that we are able to maintain operating frequency of the RI5CY core. We require significantly fewer cycles than the PQ ALU [17], which requires 18 448 cycles for both operations. However, direct comparison is difficult as they benchmark on the CVA6 core with external DDR4 memory. For [17] a similar conclusion can be made for Dilithium, while the efficiency of the NTT is not reported for Dilithium by Fritzmann et al. [8].

References

1. Aikata, A., Mert, A.C., Imran, M., Pagliarini, S., Roy, S.S.: KaLi: a crystal for post-quantum security using Kyber and Dilithium. IEEE Trans. Circ. Syst. I: Regular Pap. 1–12 (2022)
2. Bai, S., et al.: CRYSTALS-Dilithium algorithm specifications and supporting documentation (Version 3.1) (2021). https://pq-crystals.org/dilithium/
3. Banerjee, U., Ukyab, T.S., Chandrakasan, A.P.: Sapphire: a configurable crypto-processor for post-quantum lattice-based protocols. IACR Trans. Cryptogr. Hardw. Embed. Syst. **4**, 17–61 (2019)
4. Barrett, P.: Implementing the Rivest Shamir and Adleman public key encryption algorithm on a standard digital signal processor. In: Odlyzko, A.M. (ed.) CRYPTO 1986. LNCS, vol. 263, pp. 311–323. Springer, Heidelberg (1987). https://doi.org/10.1007/3-540-47721-7_24
5. Beckwith, L., Nguyen, D.T., Gaj, K.: High-performance hardware implementation of crystals-Dilithium. In: 2021 International Conference on Field-Programmable Technology (ICFPT), pp. 1–10 (2021)
6. Bertoni, G., Daemen, J., Peeters, M., Van Assche, G.: Building power analysis resistant implementations of Keccak. In: Second SHA-3 Candidate Conference, vol. 142 (2010)
7. Bisheh-Niasar, M., Azarderakhsh, R., Mozaffari-Kermani, M.: A monolithic hardware implementation of Kyber: comparing apples to apples in PQC candidates. In: Longa, P., Ràfols, C. (eds.) LATINCRYPT 2021. LNCS, vol. 12912, pp. 108–126. Springer, Cham (2021). https://doi.org/10.1007/978-3-030-88238-9_6
8. Fritzmann, T., Sigl, G., Sepúlveda, J.: RISQ-V: tightly coupled RISC-V accelerators for post-quantum cryptography. IACR Trans. Cryptogr. Hardw. Embed. Syst. **2020**(4), 239–280 (2020)
9. Gautschi, M., et al.: Near-threshold RISC-V core with DSP extensions for scalable IoT endpoint devices. IEEE Trans. Very Large Scale Integr. (VLSI) Syst. **25**(10), 2700–2713 (2017)
10. Greconici, D.O.C., Kannwischer, M.J., Sprenkels, A.: Compact Dilithium implementations on cortex-M3 and cortex-M4. IACR TCHES **2021**(1), 1–24 (2021). https://doi.org/10.46586/tches.v2021.i1.1-24, https://tches.iacr.org/index.php/TCHES/article/view/8725
11. Güneysu, T., Oder, T., Pöppelmann, T., Schwabe, P.: Software speed records for lattice-based signatures. In: Gaborit, P. (ed.) PQCrypto 2013. LNCS, vol. 7932, pp. 67–82. Springer, Heidelberg (2013). https://doi.org/10.1007/978-3-642-38616-9_5
12. Gupta, N., Jati, A., Chattopadhyay, A., Jha, G.: Lightweight hardware accelerator for post-quantum digital signature CRYSTALS-Dilithium. IEEE Trans. Circ. Syst. I: Regular Pap. 1–10 (2023)

13. Kannwischer, M.J., Petri, R., Rijneveld, J., Schwabe, P., Stoffelen, K.: PQM4: post-quantum crypto library for the ARM Cortex-M4. https://github.com/mupq/pqm4
14. Karl, P., Schupp, J., Fritzmann, T., Sigl, G.: Post-quantum signatures on RISC-V with hardware acceleration. ACM Trans. Embed. Comput. Syst. (2023)
15. Land, G., Sasdrich, P., Güneysu, T.: A hard crystal - implementing Dilithium on reconfigurable hardware. In: Grosso, V., Pöppelmann, T. (eds.) CARDIS 2021. LNCS, vol. 13173, pp. 210–230. Springer, Cham (2022). https://doi.org/10.1007/978-3-030-97348-3_12
16. Lyubashevsky, V., et al.: CRYSTALS-DILITHIUM. Technical report, National Institute of Standards and Technology (2022). https://csrc.nist.gov/Projects/post-quantum-cryptography/selected-algorithms-2022
17. Nannipieri, P., Di Matteo, S., Zulberti, L., Albicocchi, F., Saponara, S., Fanucci, L.: A RISC-V post quantum cryptography instruction set extension for number theoretic transform to speed-up CRYSTALS algorithms. IEEE Access **9**, 150798–150808 (2021)
18. National Institute of Standards and Technology: Post-Quantum Cryptography Standardization. https://csrc.nist.gov/Projects/Post-Quantum-Cryptography/Post-Quantum-Cryptography-Standardization
19. Ni, Z., Khalid, A., e Shahwar Kundi, D., O'Neill, M., Liu, W.: Efficient pipelining exploration for a high-performance CRYSTALS-Kyber accelerator. Cryptology ePrint Archive, Paper 2022/1093 (2022)
20. Ricci, S., et al.: Implementing CRYSTALS-Dilithium signature scheme on FPGAs. ARES 21, Association for Computing Machinery, New York (2021)
21. Schwabe, P., et al.: CRYSTALS-KYBER. Technical report, National Institute of Standards and Technology (2022). https://csrc.nist.gov/Projects/post-quantum-cryptography/selected-algorithms-2022
22. Shor, P.: Algorithms for quantum computation: discrete logarithms and factoring. In: Proceedings 35th Annual Symposium on Foundations of Computer Science, pp. 124–134. IEEE Computer Society Press (1994)
23. Stoffelen, K.: Efficient cryptography on the RISC-V architecture. In: Schwabe, P., Thériault, N. (eds.) LATINCRYPT 2019. LNCS, vol. 11774, pp. 323–340. Springer, Cham (2019). https://doi.org/10.1007/978-3-030-30530-7_16
24. Xing, Y., Li, S.: A compact hardware implementation of CCA-secure key exchange mechanism CRYSTALS-KYBER on FPGA. IACR Trans. Cryptogr. Hardw. Embed. Syst. **2021**(2), 328–356 (2021)
25. Zhao, C., et al.: A compact and high-performance hardware architecture for CRYSTALS-Dilithium. IACR Trans. Cryptogr. Hardw. Embed. Syst. **2022**(1), 270–295 (2021)
26. Zhou, Z., He, D., Liu, Z., Luo, M., Choo, K.K.R.: A software/hardware co-design of crystals-Dilithium signature scheme. ACM Trans. Reconfigurable Technol. Syst. **14**(2) (2021)

22. [illegible]
23. [illegible]

Side-Channel and Neural Networks

Keep It Unsupervised: Horizontal Attacks Meet Simple Classifiers

Sana Boussam[2,3](✉) and Ninon Calleja Albillos[1,4](✉)

[1] CEA, LETI, MINATEC Campus, 38054 Grenoble, France
ninon.callejaalbillos@cea.fr
[2] INRIA and LIX, Institut Polytechnique de Paris, Palaiseau, France
sana.boussam@inria.fr
[3] Thales ITSEF, Toulouse, France
[4] Univ. Grenoble Alpes, 38000 Grenoble, France

Abstract. In the last years, Deep Learning algorithms have been browsed and applied to Side-Channel Analysis in order to enhance attack's performances. In some cases, the proposals came without an in-depth analysis allowing to understand the tool, its applicability scenarios, its limitations and the advantages it brings with respect to classical statistical tools. As an example, a study presented at CHES 2021 [16] proposed a corrective iterative framework to perform an unsupervised attack which achieves a 100% key bits recovery. In this paper we analyze the iterative framework and the datasets it was applied onto. The analysis suggests a much easier and interpretable way to both implement such an iterative framework and perform the attack using more conventional solutions, without affecting the attack's performances.

1 Introduction

The Rise of DL-SCA. Side-Channel Attacks (SCA) are a type of physical attacks in which an attacker exploits implementation weaknesses of a cryptographic algorithm in embedded systems in order to extract secret information, by targeting all types of physical leaks (*e.g.* execution time, power consumption, electromagnetic radiation ...) generated during the execution of this algorithm. More generally, SCA fall into two categories: *profiled attacks* and *non-profiled attacks*. *Profiled attacks* are one of the most powerful SCA and require a device similar to the targeted one in order to build a learning database that models device behavior (*i.e.* leakage model) according to the values of sensitive variables that depend on the secret. One of the recent strategies, actively explored by the side-channel community since 2017, consists in using deep learning techniques to enhance the classical profiled attack scenario while mitigating the impact of some countermeasures (*e.g.* masking) [13]. On the other side, *non-profiled attacks* are less restrictive than *profiled attacks*. In this scenario, the attacker only captures physical traces and has possibly some details about the implementation of the targeted algorithm to carry out the attack. Therefore, one of the first difficulty he encounters consists in finding an adequate approach to extract sensitive information from the targeted embedded system. While this restrictive scenario is more

S. Bhasin and T. Roche (Eds.): CARDIS 2023, LNCS 14530, pp. 213–234, 2024.
https://doi.org/10.1007/978-3-031-54409-5_11

realistic than the profiled attacks from a practical point of view, the application of machine learning and deep learning techniques are less pronounced.

Motivations of Our Work. Recently, Perin *et al.* published in CHES 2021 [16] a *non-profiled* horizontal attack where they introduce a deep learning framework to correct errors after their clustering process. Correction of noisy labels is an open problem in machine learning. The methods used in the literature follow different approaches like using robust learning algorithms to mitigate the impact of noisy labels, noise identification and reduction techniques or even noise correcting methods [17]. Perin *et al.* are one of the first to present an attack followed by a deep learning correction framework, used as a black-box, which permits to retrieve 100% of correct bits for at least one key. While such contribution sounds promising for the community, the black-box property provided by the use of deep learning techniques can be problematic to assess the benefits and the limitations of such contribution. This raises the question of more appropriate and less costly solutions which can be beneficial from an interpretation point of view.

Contributions. This paper reduces this issue through the proposal of the following contributions:

- First, we identify that the datasets targeted by Perin *et al.* [16] are linearly separable. Then, we bring out the limitations of the corrective approach by showing that iterative framework's performance is strongly related to the neural network used and thus to the dataset considered. This illustrates the need of understanding how machine learning or deep learning techniques should be refined depending on use-case scenarios. Consequently, we adapt the iterative framework by justifying the design of a dedicated neural network structure. This results in an improvement in terms of network complexity, accuracy and computation time compared to the proposed attack. While the average accuracy is improved by up to 8%, the attack time is divided by 10. This contribution is reported in Sect. 4.
- To make our observation more consistent with a real use-case scenario, we assess the suitability of the corrective framework under a noisy environment and observe that such restriction highly impacts its performance. This contribution is reported in Sect. 5.
- As a consequence, we finally question the ability of a more conventional approach to conduct such attack scenario. In that purpose, we propose a *non-profiled* attack which is composed of two steps: a feature extraction based on principal components analysis and a k-means clustering performed onto the projected data. This attack is performed on the same datasets as those introduced in [16] and results are compared to those obtained by Perin *et al.*. Similarly, the average accuracy is improved by up to 8% with this deep learning free approach. This contribution is reported in Sect. 6.

Sections 2 and 3 provide the background and a succinct overview of the unsupervised attack proposed by Perin *et al.* [16].

2 Background

In this section we recall well-known terms and summarize well-known notions that we will use in the rest of the paper.

2.1 Supervised vs Unsupervised Learning

Machine learning is a vast domain. The problems it addresses may be categorized in many ways. Here, we are interested in distinguishing problems addressed by the so-called *supervised* and *unsupervised* learning algorithms. Supervised learning algorithms construct solutions to problems asking to provide predictions on the basis of labeled examples (*i.e. training dataset*), *e.g.* binary or multi-class classification or regression. On the contrary, unsupervised learning algorithms are not provided with labeled examples and are asked to construct a function that highlights certain data structures. Examples of problems addressed by unsupervised learning are density estimation, clustering and dimensionality reduction.

2.2 T-Test and Sum of Squared T-Statistics (SOST)

The Welch's t-test is an adaptation of the Student's t-test that aims at determining if two populations with unequal variances have equal means. In the context of side-channel analysis, this statistical tool is used to capture the data dependency with the physical signal (*e.g.* EM emanations). Indeed, a large t-value is interpreted as a difference in behavior in terms of side-channel signal which is induced by the value of the sensitive variable handled during the execution of a cryptographic algorithm. Samples with large t-value are thus points of interest (PoIs) (*i.e.* samples that potentially contain the most of relevant information related to the secret). In 2006, Gierlichs *et al.* [8] introduced a PoI selection method based on the so-called Sum of Squared T-statistics (SOST), which is a multi-class generalisation of the t-test. Such a PoI selection method is a supervised dimensionality reduction tool.

2.3 Principal Components Analysis (PCA)

In the context of dimensionality reduction, Principal Components Analysis (PCA) is a well-established unsupervised technique. Instead of performing a point selection, it performs a *feature extraction*, orthogonally projecting the original data onto a lower-dimensional space. PCA requires to determine the Principal Components (PCs), which are the projecting vectors that maximizes the variance of the projected data. Interpreting the side-channel traces as realizations of a d-dimensional zero-mean random vector X, it can be shown that the PCs are the eigenvectors $\alpha_1 ... \alpha_d$ of the covariance matrix of X. As X is an unknown distribution, the empirical covariance matrix estimated from acquisition is used. The estimated variances of each projected component $Y_i = \alpha_i X$ is its corresponding eigenvalue λ_i. Thus, the PCs are typically ordered in decreasing order with respect to their corresponding eigenvalue, such that a natural choice to maximize the variance of the projected data consists in projecting onto the first PCs.

In SCA context, it has been discussed that a major issue with PCA application consists in the choice of the PCs so that significant information is conserved. Various techniques and criteria have been proposed to solve this issue. In [4], the authors introduced a method using the Explained Global Variance (EGV).

This method recommends using natural principal component order to select components. In [3], an alternative selection method, is proposed: the Explained Local Variance (ELV) selection method consists in computations of local variance for each time sample j and component α_i. The sum of such variance, $ELV(\alpha_i, j)$, over all samples j gives $EGV(\alpha_i)$. Thus, this component selection method, based on the globally-accepted assumption that side-channel information is localised in few PoIs, strives for a balance between the EGV of a given component and the number of samples required to achieve consistent part of it.

Definition 1. *Given a principal component α_i and a sample j, the Explained Local Variance is defined by:*

$$ELV(\alpha_i, j) = \frac{\lambda_i \times \alpha_i[j]^2}{\sum_{k=1}^{r} \lambda_k} = EGV(\alpha_i)\alpha_i[j]^2$$

with $EGV(\alpha_i) = \frac{\lambda_i}{\sum_{k=1}^{r} \lambda_k}$, r the rank of the covariance matrix and λ_i the eigenvalue of the ith principal component.

The ELV can be sorted in descending order for each component and summed in a cumulative way. From this, we can observe trends followed by components to achieve their EGV and deduce a ranking of all PCs.

2.4 Clustering Algorithms, Voronoi Diagrams and *k*-Means

Clustering algorithms are used to find a structure in a unlabeled dataset. It is a partitioning process which gathers similar objects into groups according to defined metrics. There are different types of clustering algorithms. If all clusters are determined simultaneously, the algorithm is called partitional. Otherwise, if the clusters are established using previous clusters, the algorithm is hierarchical.

When the number of clusters is known, which is the case in this work, partitional algorithms are preferred. One of the most well-established partitional clustering algorithm is the k-means. The k-means clustering consists in determining a set of k points in $\mathbb{R}^d$ called centroids that minimize the distance between each data point and its nearest centroid. Finding the optimal solution of a k-means clustering problem is NP-hard even when $k = 2$ (see [1] proof). One of the best known heuristics for solving k-means clustering problem is Lloyd's algorithm [11] and its generalizations [10]. This algorithm gives an approximate solution to this problem by seeking a local minimum instead of the global one. To do so, k-means clustering problem is approximated by an iterative computation of

Voronoï diagrams (or *tessellations*) of k regions until a convergence criterion is satisfied. This results in a *centroidal Voronoï tessellation* of the input [6]. Let $\mathcal{E}$ be an Euclidean space of finite dimension and $P = \{\mathbf{p}_1, ..., \mathbf{p}_k\}$ be a finite set of k points called *generators* in $\mathcal{E}$.

Definition 2 (Voronoï diagrams). *A Voronoï diagram $\mathcal{V}_P$ is a partition of $\mathcal{E}$ into a set of k regions V_i such that $\bigcup_{i=1}^{k} V_i = \mathcal{E}$ and $\forall\,(i,j) \in \{1,...,k\}^2$, $i \neq j$, $V_i \cap V_j = \emptyset$. The Voronoï region V_i corresponding to the generator $\mathbf{p}_i$ is defined as:*

$$V_i = \{\mathbf{x} \in \mathcal{E} \quad | \quad |\mathbf{x} - \mathbf{p}_i| < |\mathbf{x} - \mathbf{p}_j| \quad \text{for}\, j = 1, ..., k,\, j \neq i\}$$

Definition 3 (Centroidal Voronoï tesselation). *A centroidal Voronoï tessellation (CVT) is a particular type of Voronoï diagram in which the generators $\mathbf{p}_i$ of Voronoï regions V_i are their mass centroids μ_i.*

Considering the fact that k-means algorithm generates a CVT, the decision boundaries between the k clusters found by a k-means are therefore linear since they correspond to the boundaries between the k Voronoï regions of the generated CVT. Indeed, given a Voronov region V_i, $i \in \{1, ..., k\}$, the decision boundary between V_i and $V_j \in \mathcal{A}(V_i) \subset \mathcal{V}_P$, with $\mathcal{A}(V_i)$ the set of the adjacent Voronoï regions of V_i, is an affine hyperplane defined by the set of points equidistant from μ_i and μ_j [7].

2.5 Perceptrons and Deep Neural Networks

A so-called *perceptron* is the simplest neural networks since it consists of an artificial neuron. This learning algorithm is a binary linear classifier. Let consider $\Theta = \{W, b\}$ as a set of parameters. Given an input x, a perceptron learns a parametric linear function $f_\Theta : \mathbb{R}^d \to \mathbb{R}$ of the form $f_\Theta(x) = W^T x + b$ where $W \in \mathbb{R}^d$ is a vector which represents the *weights* associated with x and $b \in \mathbb{R}$ is a *bias*. Its output is determined by the application of a non-linear function σ called *activation function.* Indeed, given a threshold $\tau \in \mathbb{R}$, $y = 1$ if $\sigma(f_\Theta(x)) > \tau$ and 0 otherwise. In binary classification, the *activation function* used is the *Sigmoid function* $\sigma : \mathbb{R} \to [0,1]$; also named *logistic function.* This function is defined as $\sigma(f_\Theta(x)) = \frac{1}{1+e^{-f_\Theta(x)}}$ and aims at converting, given the input x, $f_\Theta(x)$ into a probability for x to belong to class 1. Since the output of the *Sigmoid function* is a probability, the threshold τ is set to 0.5 (*i.e.* the case when x lies on the decision boundary). The decision boundary characterized by a perceptron is thus an affine hyperplane defined by $W^T x + b = 0$.

Deep neural networks (DNN) are neural networks with multiple layers of artificial neurons that can learn more complex parametric functions (*i.e.* non linear functions). Among the several typology of DNN existing in literature, Multi-Layer Perceptrons (MLP) and Convolutional Neural Networks (CNN) are the most widely used in SCA domain in order to perform classification task. The MLP consist in concatenations of layers (also called dense layers) comprising several artificial neurons and activation functions. The CNN are similar to MLP, but adds some special layers (convolutional and pooling layers) that allows them to extract geometrically local features and to be robust to geometrical deformations, such as misalignment.

3 Overview of [16]

In this section we report a succinct summary of the unsupervised attack proposed in 2021 by Perin *et al.*

Target. The attack's targets are the protected ECC software implementations on Curve25519 from μNaCl [5], a cryptographic library for ARM Cortex-M[1]. The library includes two regular and constant-time elliptic curve scalar multiplication (ECSM) implementations based on the Montgomery ladder. In [14], the authors strengthen side-channel resistance of these ECSM by adding coordinate re-randomization to both implementations. Although scalar randomization countermeasure is not implemented, Perin *et al.* collect side-channel traces using a random scalar for each ECSM execution, in such a way that the proposed attack could be considered effective also against implementations protected with scalar randomization.

Vulnerability. The vulnerability the attack exploits is the leakage coming from a conditional swap operation (*CSWAP*) [14] during the ECSM. Indeed, during the ith round of the main multiplication loop, the *CSWAP* condition depends directly on the ith bit of the secret scalar.[2] Hence, an attacker that can classify leakages from *CSWAP* operations with respect to their condition can recover the 255 secret bits of the key bit per bit.

Datasets and Scenario. Perin *et al.* conducted their attack on two datasets corresponding respectively to *CSWAP*'s implementation using pointer or arithmetic swapping: `cswap_pointer` and `cswap_arith`. Each dataset is composed of $N = 300$ complete traces. Each trace is split in 255 subtraces, corresponding to the execution of a single *CSWAP* operation. Thus the dataset contains $300 \times 255 = 76,500$ subtraces of $1,000$ time samples for `cswap_pointer` and $8,000$ time samples for `cswap_arith`.

The complete attack proposed in [16] comprises two phases, namely a prelabeling and a corrective phase. The **prelabeling phase** is decomposed into two steps. The first one uses an adapted unsupervised version of SOST [15] in order to select a set of POIs. Then, the k-means algorithm is applied on the selected POIs in order to group the traces corresponding to the same value of *CSWAP* condition. After this prelabeling phase, the resulting accuracies found in [16]

[1] https://munacl.cryptojedi.org/curve25519-cortexm0.shtml.

[2] More specifically, the swap condition value at the ith iteration is equal to the XOR between the ith bit and $(i-1)^{th}$ bit of the secret scalar.

are respectively of 52.44% and 52.24% on `cswap_arith` and `cswap_pointer` datasets. The **corrective phase** takes the prelabeling results as input and applies a corrective framework[3] in order to correct wrong labels. Based on `Co-teaching` strategy [9], it consists in iteratively correcting mislabeled traces by training twice[4] a neural network *ANN* (which is a CNN in [16]) to perform a classification task on two disjoints parts $\mathcal{D}_1$ and $\mathcal{D}_2$ of the complete dataset $\mathcal{D}$. After each training, the neural network is used to relabel the subdataset on which it did not train. The label correction provided by this framework relies on the ability of *ANN* to extract discriminative features despite the presence of noise in labels. The interested readers may refer to [16, section 4] for deeper details on iterative framework.

Performance Metric. For each dataset, training is done using 250 complete traces (*i.e.* $250 \times 255 = 63,750$ subtraces). Attack is performed using the 50 remaining traces (i.e. $50 \times 255 = 12,750$ subtraces). To estimate iterative framework's performance, the authors defined two metrics called *Average and Maximum single trace accuracy metric.* The maximum (resp. average) single trace accuracy is the maximum (resp. average) accuracy obtained using the iterative framework on these 50 remaining traces. An attack is considered to be successful if a maximum single trace accuracy of 100% is reached by the framework for at least one trace at any epoch.

Experimental Results Achieved in [16]. Perin *et al.* assess iterative framework's correction capability, using the previous defined metrics, following different scenarios including or not regularization[5]. In this paper, only the results combining several regularization techniques are reported as a particular focus is proposed on those scenarios (see Table 1 which summarizes results achieved for fixed CNN hyperparameters and 50 framework iterations[6]).

Table 1. Average and maximum single trace accuracies obtained in [16] for fixed CNN hyperparameters (50 framework iterations).

	`cswap_pointer`		`cswap_arith`	
	Average	Maximum	Average	Maximum
After prelabelling phase	52.24%	59.22%	52.44%	59.22%
Scenario without regularization	85%	97.64%	52%	76.07%
Scenario with regularization	91%	100%	83%	100%

[3] Source code: https://github.com/AISyLab/IterativeDLFramework.
[4] At each training phase, trainable paramaters are reset.
[5] Regularization consists in applying a technique to avoid overfitting.
[6] These results are those reported in [16] as we were not able to reproduce it using the source code.

The fact that Perin *et al.* achieve a successful attack only when applying regularization techniques suggests that their neural network overfits. The overfitting phenomenon can be due to several reasons [18] such as a lack of training data or a high model complexity (in terms of number of trainable parameters). For the latter reason, the function induced by the neural network, does not fit with the underlying classification problem. Here, in the case of [16], we assume that overfitting is induced by the model complexity, which was not considered in the original paper. In the following, we propose in Sect. 4.1 to simplify the approach introduced by Perin *et al.* through the analysis of the decision boundary's complexity. Based on the results found, we propose a new neural network that is more suitable for the classification problem inherent in both datasets.

4 Corrective Framework Fitting the Decision Boundaries

In this section, we aim at providing a better understanding of the attack scenario proposed in [16]. To do so, we conduct an in-depth analysis of this work and propose a new version of this attack with more suitable tools for `cswap_pointer` and `cswap_arith` datasets.

4.1 Simplification of the Neural Network

As mentioned above, we assume that overfitting is due to a high model complexity. To confirm our assumption, we first conduct additional investigations in order to characterize the decision boundaries of both datasets. Then, based on the results found, we simplify the new neural network such that we obtain a classifier more suitable for the classification problems.

Linear Decision Boundaries. The characterization of the decision boundaries is done in two steps. First, we conduct a (supervised) PoIs selection. Then, we determine the shape of the decision boundary by applying simple tools such as k-means and examining their performance (*i.e.* accuracy obtained). Figure 1 shows Welch's t-tests (recalled in Sect. 2.2) obtained on both datasets, using their true labels. As we can see, both datasets have significant side-channel leakages. Samples that maximize the t-value seem to be sampled around 600-650 for `cswap_pointer` and around $3,300$-$3,350$ for `cswap_arith`.

By applying the same methodology as Perin *et al.*, we first compute a k-means clustering on the PoIs of the two datasets (*i.e.* samples 600-650 for `cswap_pointer` and samples $3,300$-$3,350$ for `cswap_arith`) using 250 of the 300 complete traces (*i.e.* $63,750$ subtraces). Then we predict cluster index for the $50 \times 255 = 12,750$ remaining subtraces. We obtain an maximum and average single trace accuracy of 96.1% and 98.03% (resp. 80.26% and 87.05%) for `cswap_pointer` dataset (resp. `cswap_arith` dataset). These results highlight that a k-means is sufficient to discriminate subtraces processing bit 0 and bit 1. Therefore, we deduce that the classification problem is not so complex to solve as both classes can be separable by a k-means decision boundary. According to

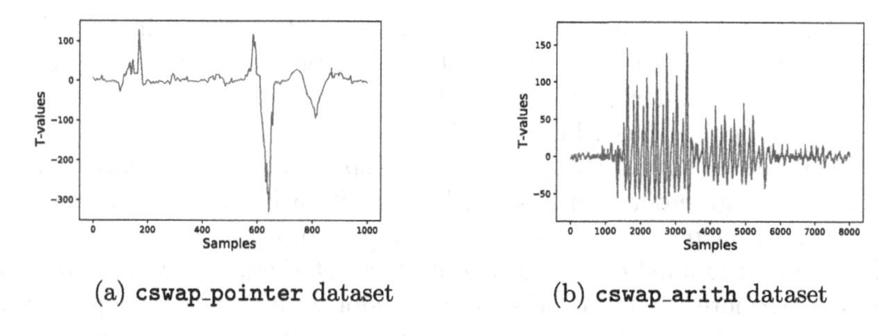

(a) `cswap_pointer` dataset (b) `cswap_arith` dataset

Fig. 1. Welch's t-tests obtained on the two datasets using true labels.

Sect. 2.4, a hyperplane seems to be sufficient to discriminate the two clusters (*i.e.* subtraces processing bit 0 and bit 1). This means that a linear classifier would be more suitable for the underlying classification problem of both datasets.

Linear Classifier. Starting from the observations above, we simplify the neural network proposed in [16] in order to adapt it to a linear classification problem. In fact, a linear classifier that models a linear function, such as a perceptron (see Sect. 2.5), should be sufficient to approximate the true unknown decision boundary. To verify this assumption, we reduce the complexity of the CNN architecture used in [16] until we obtain a neural network composed of a single layer of 1 neuron.

4.2 Experimental Results Obtained on Attacked Datasets

This section presents the results achieved on both datasets with this new neural network according to the setting (supervised/unsupervised context) considered.

Supervised Learning. In order to confirm our hypothesis that the data are linearly separable and to assess the suitability and effectiveness of our proposed architecture, we first try this neural network in a supervised context (*i.e.* training is done using the true labels). To do so, for both datasets, we define a single layer of one neuron and use a *sigmoid* activation function. We use *RMSprop* optimizer and set the learning rate to 10^{-4}. Training is done for 20 epochs with a batch size of 100. No regularization technique is applied. We use the same methodology as Perin *et al.*, namely we first train our neural network with 250 complete traces = 63,750 subtraces and then we attack the remaining 50 traces = 12,750 subtraces.

After training our neural newtwork, we achieve for both dataset a maximum single trace accuracy of 100%. The average single trace accuracies are respectively 99.77% and 98.33% for `cswap_pointer` and `cswap_arith` datasets. These results appear to clearly confirm our previous hypothesis: both datasets are linearly separable.

Unsupervised Learning. Once we validate our hypothesis in a supervised learning context, we assess the ability of our new neural network to correct the mislabeled data by following the methodology provided in [16]. As a remainder, after the prelabeling phase, Perin *et al.* retrieves 52.24% (resp. 52.44%) of the private key bits when the `cswap_pointer` (resp. `cswap_arith`) dataset is considered. To configure our neural network in the iterative framework, we use the same hyperparameters as those configured in the supervised context. Only the batch size and the number of epochs differ, we set them to 250 and 10 respectively. Furthermore, we keep the same methodology as the one adopted in [16] (*i.e.* training on 250 complete traces/attack on the 50 remaining traces).

Figure 2 shows the results we obtain using the same inputs (*i.e.* subtraces and labels) as Perin *et al.* and without regularization. As we can see, after 50 iterations, the framework achieved a maximum (resp. average) single trace accuracy of 100% (resp. 98.9%) for `cswap_pointer` dataset (see Fig. 2a). Hence, for this dataset, we conduct a successful attack using a perceptron. However, it seems to be more complicated for `cswap_arith` dataset. Indeed, for this dataset, the framework reached a maximum (resp. average) single trace accuracy of 79.22% (resp. 70.93%) (see Fig. 2b). These results are better than those obtained by Perin *et al.* (scenario without regularization) but still not sufficient to consider the attack as successful. As in [16], we decide to reduce the size of subtraces of `cswap_arith` dataset in order to see if when considering a smaller number of samples, the framework manages to correct mislabeled subtraces. Thus, we narrow down as in [16] the subtraces of `cswap_arith` from 8,000 to 2,000 samples by considering only the interval of samples 1,500-3,500. We call this dataset `cswap_arith_reduced`. Figure 2c shows the results achieved by the framework on this reduced dataset. By narrowing down the attacking sample interval, we now obtain a successful attack with a perceptron for `cswap_arith`. Indeed, when considering only the sample interval 1,500–3,500, the framework reached a maximum (resp. average) single trace accuracy of 100% (resp. 98%). Thus, these results, obtained without regularization and with a very simple neural network, highlight the fact that even in an unprofiled context, a linear classifier is sufficient to correct mislabeled subtraces.

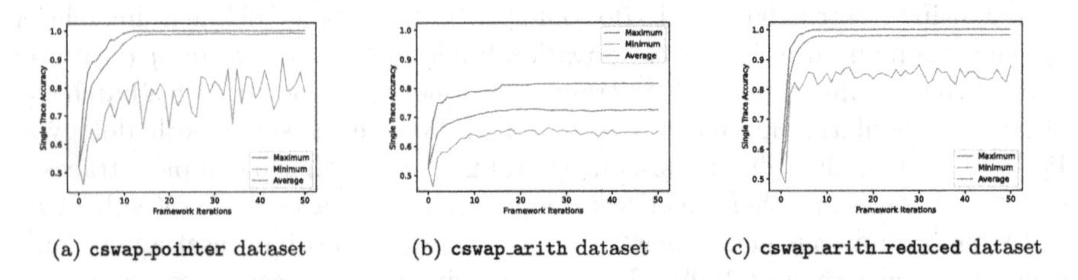

(a) `cswap_pointer` dataset (b) `cswap_arith` dataset (c) `cswap_arith_reduced` dataset

Fig. 2. Results obtained using iterative framework with our neural network after 50 iterations (no regularization).

Although we managed to carry out successful attacks for both datasets, we do not explain why, when considering the 8,000 samples (*i.e.* considering entire subtraces and not a restricted sample interval), we are unable to conduct a successful attack even though the neural network we consider is adapted to the classification problem inherent to `cswap_arith` dataset.

Since changing the hyperparameters (number of epochs and batch size) did not solve the problem, we decide to investigate the impact of increasing the initial accuracy on the framework to see if it is still possible to achieve a successful attack on `cswap_arith` dataset. To do so, we create new labels by applying a random uniform noise on the true labels. Figure 3 shows the results achieved by the iterative framework for `cswap_arith` dataset according to the accuracy related to the prelabeling phase. This accuracy will be called the "initial accuracy" in the rest of the paper. Indeed, when we increase the initial accuracy by a small amount *i.e.* from 52.44% (initial accuracy found by Perin *et al.*) to 53.66%, we still have the same behavior as in the previous experiments (see Fig. 3a). However, in Fig. 3b, we can notice that from an initial accuracy of about 54.57%, we can finally achieve a successful attack for `cswap_arith` dataset. With an even higher accuracy, the iterative framework seems to find more quickly all the bits of at least one trace, as we can see in Fig. 3c. The obtained results illustrate that the performance of the iterative framework highly depends on the quality of the prelabeling process. Even if the decision boundary to approximate is simple (*e.g.* linear decision boundary), the inappropriate use of the prelabeling phase can highly impact the ability of the iterative framework to correct the mislabeled subtraces.

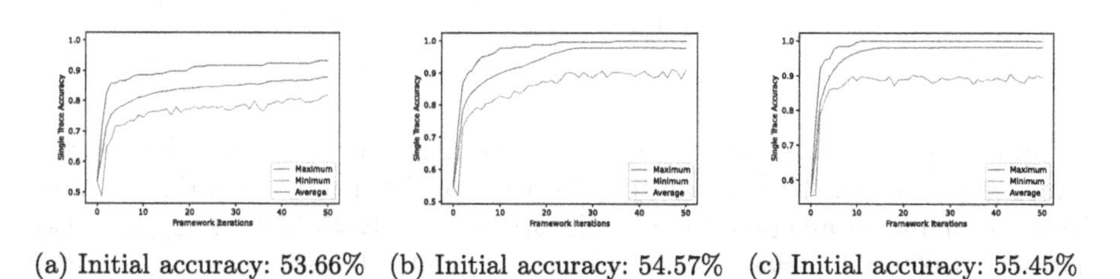

(a) Initial accuracy: 53.66% (b) Initial accuracy: 54.57% (c) Initial accuracy: 55.45%

Fig. 3. Results achieved for `cswap_arith` dataset using iterative framework with our neural network after 50 iterations (no regularization).

4.3 Comparison of the Two Approaches and Discussion

Table 2 compares the results achieved for the same inputs and methodology according to the neural network considered in this work and in [16]. We observe that our proposition achieves better results than the CNN on both datasets.

For the `cswap_pointer` dataset, our perceptron without regularization retrieves on average 98.9% of the bits while the CNN used in [16] reaches an average accuracy of 91% when regularization techniques are applied. This enhancement allows us, on average, to complete the attack using brute force regardless of the trace considered. Indeed, an average single trace accuracy of 98.9% means that there are on average 3 remaining bits to correct in order to recover the secret scalar. This correction can be done by performing $n = \sum_{i=0}^{3} \binom{255}{i} \sim 2^{22}$ tests. In comparison with the results reported in [16], an average single trace accuracy of 91% corresponds to a 23-bit correction *i.e.* $\sim 2^{109}$ tests to recover the secret key. In this case, the use of brute force to complete the attack is not feasible. For `cswap_arith_reduced` dataset, our neural network without regularization obtains similar results as [16] when they use regularization techniques. Here again, on average, the secret scalar can be retrieved using brute force (correction of 6 bits $\sim 2^{39}$ tests). Although our proposition does not obtain better results than the scenario with regularization for `cswap_arith` dataset, we still achieve better results than the CNN used without regularization. More generally, in a scenario without regularization, an improvement of average accuracies of 13.9% (`cswap_pointer` dataset), 18.93% (`cswap_arith` dataset) and 40% (`cswap_arith_reduced` dataset) can be observed by finding a neural network that suitably approximates the linear decision boundary.

Table 2. Comparison of average and maximum single trace accuracy obtained after 50 framework iterations for both datasets.

	`cswap_pointer`		`cswap_arith`		`cswap_arith_reduced`	
	Average	Maximum	Average	Maximum	Average	Maximum
1 neuron, no regularisation	98.9%	100%	70.93%	79.22%	98%	100%
CNN [16], no regularisation	85%	97.64%	52%	76.07%	58%	86.27%
CNN [16], dropout + data augmentation	91%	100%	83%	100%	98%	100%

Moreover, using a suitable neural network allows us to reduce the network complexity. Indeed, through the simplification of the neural network, the number of trainable parameters decreased from respectively 45, 978 (`cswap_pointer` dataset), 3, 069, 820 (`cswap_arith` dataset) and 669, 820 (`cswap_arith_reduced` dataset) to 1, 001, 8, 001 and 2, 001. This enhancement highlights the high complexity of the neural network used by Perin *et al.* and thus confirms the assumption we made about the overfitting that occurs in [16]. Finally, this enhancement allows us also to significantly reduce the computational time. Using a 1.8 GHz Intel Core i7-10510U processor, the attack (*i.e.* 50 framework iterations) duration drops from respectively 1 h and 15 min (`cswap_pointer` dataset), 18 h (`cswap_arith` dataset) and 5 h (`cswap_arith_reduced` dataset) down to respectively 8 min, 35 min and 10 min.

Finally, through this section, we saw that the performance of the iterative framework is strongly related to the neural network used, which is itself very dependent on the dataset considered. As a general fact, deep-learning techniques provide advantages with respect to conventional techniques when one has to tackle with noisy data or complex problems, *e.g.* non-linear decision boundaries. Datasets used in [16] do not have such characteristic, thus do not allow to assess the suitability of the iterative corrective framework in interesting cases. To get a full overview of the corrective framework capability, next section identifies its benefits and its limitations under a noisy environment.

5 Correction Capability of the Iterative Framework

In this section, we question the ability of the corrective approach to perform correctly when the clusters are entangled. To do so, we detail our methodology in Sect. 5.1 and introduce the results we obtained in Sect. 5.2.

5.1 Experiments

In order to assess the correction capability of the iterative framework on more noisy datasets we consider `cswap_pointer` and `cswap_arith_reduced` datasets, with an additional artificial Gaussian noise following $\mathcal{N}(0, \sigma^2)$, with $\sigma \in \{15, 25, 30, 50, 75, 90\}$. This is beneficial to generate different clusters entanglement levels. Figure 4 shows examples of clusters entanglement on `cswap_pointer` traces for different σ values. As expected, the higher the noise level, the more entangled the clusters become. Therefore, they can no longer be separated by a linear decision boundary. For each noise configuration, we consider two kind of neural networks architectures, namely perceptron and CNNs. Those two structures will produce different hypothesis space $\mathcal{F}$. We perform a supervised training in order to first ensure the presence of leakages when we entangle the clusters, and then, to obtain a neural network $f_{\text{sup}} \in \mathcal{F}$ that we considered as a correction upper bound. Then we apply the unsupervised iterative corrective strategy (using the same prelabels provided in [16]) and compare the obtained neural network f to f_{sup}. If f performs similarly to f_{sup}, we deduce that the unsupervised

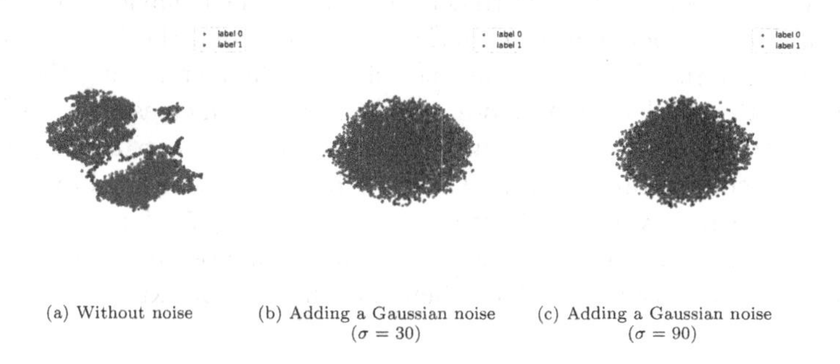

(a) Without noise (b) Adding a Gaussian noise ($\sigma = 30$) (c) Adding a Gaussian noise ($\sigma = 90$)

Fig. 4. Visualization of `cswap_pointer` dataset's clusters using t-SNE [12] property of the iterative framework does not affect the attack performance, and thus that the iterative approach is suitable to deal with entangled clusters. If f has poorer performance than f_{sup}, we deduce that the iterative framework is not able to extract an efficient predictive model under the assumption of entangled clusters.

5.2 Results

Supervised Learning. For both perceptron and CNN, training is done for 20 epochs and a batch size of 100. As in [16], we use *RMSprop* as optimizer with a learning rate set to 10^{-4}. As we can see in Table 3, adding Gaussian noise with standard deviation up to $\sigma = 50$ on traces results in non-linear separable clusters and leads the performance of the perceptron to be slightly lower than the CNN. Beyond $\sigma = 50$, performances of both neural networks are deteriorated by noise for both datasets.

Table 3. Average and maximum single trace accuracy achieved for `cswap_pointer` (1) and `cswap_arith_reduced` (2) datasets in a supervised setting.

		$\sigma = 15$		$\sigma = 25$		$\sigma = 30$		$\sigma = 50$		$\sigma = 75$		$\sigma = 90$	
		Average	Maximum	Average	Maximum	Average	Maximum	Average	Maximum	Average	Maximum	Average	Maximum
(1)	CNN [16]	99.87%	100%	99.32%	100%	98.83%	100%	93.11%	96.47%	84.66%	89.41%	79.43%	87.84%
	Perceptron	98.87%	100%	97.8%	99.6%	96.66%	98.82%	90.19%	95.29%	79.62%	84.7%	77.1%	83.13%
(2)	CNN [16]	97.46%	99.21%	92.13%	96.07%	89.84%	94.12%	73.52%	81.96%	67.78%	75.29%	64.56%	72.94%
	Perceptron	93.29%	96.86%	87.45%	91.76%	84.83%	89.01%	72.76%	84.31%	67.63%	74.11%	65.3%	74.11%

Unsupervised Learning. For the unsupervised setting, we adopt for CNN the configuration in which regularization techniques are employed (see Table 1). Regarding hyperparameters for training, we pick the same ones used in supervised context. The only difference lies in choosing the batch size for the perceptron and the number of epochs for both neural networks. We set them respectively to 250 and 10. For all σ values, we summarize in Table 4 accuracies obtained after 50 framework iterations for `cswap_pointer` and `cswap_arith_reduced` datasets.

As we can see in Table 4, for both neural networks the behavior of the iterative framework differs depending on noise level. As expected, the higher the noise, the more the decision boundaries are non-linear. Therefore, using the iterative framework with a CNN seems more suitable when σ increases. Indeed, when σ equals up to 30 (resp. 50) for `cswap_pointer` (resp. `cswap_arith_reduced`) dataset, the results obtained with a CNN are better than those obtained with a perceptron. However, when considering $\sigma \geq 50$ (resp. $\sigma \geq 75$) for `cswap_pointer` (resp. `cswap_arith_reduced`) dataset, CNN performances are on average lower than perceptron performances. This phenomenon could be explained by the fact

Table 4. Average and maximum single trace accuracy achieved for `cswap_pointer` (1) and `cswap_arith_reduced` (2) datasets after 50 framework iterations.

		$\sigma = 15$		$\sigma = 25$		$\sigma = 30$		$\sigma = 50$		$\sigma = 75$		$\sigma = 90$	
		Average	Maximum	Average	Maximum	Average	Maximum	Average	Maximum	Average	Maximum	Average	Maximum
(1)	CNN [16]	97.11%	99.22%	98.62%	100%	71.25%	81.96%	50.73%	65.49%	50.68%	63.92%	50.78%	60.78%
	Perceptron	98.21%	100%	64.03%	78.43%	67.15%	76.08%	62.73%	72.55%	57.99%	71.76%	56.43%	66.27%
(2)	CNN [16]	90.26%	97.65%	80.6%	89.8%	66.28%	80.39%	71.87%	82.75%	50.26%	61.96%	50.13%	63.92%
	Perceptron	88.07%	95.67%	68.81%	81.96%	63.03%	75.29%	58.84%	70.2%	54.74%	67.06%	54.22%	65.88%

that deep neural networks have a high capacity to memorize noisy labels (*i.e.* they can easily fit random labels) [2,19]. Thus it seems to appear that, in an unsupervised setting, the more the clusters are entangled, the less correction is provided by the framework.

Indeed, for `cswap_pointer` dataset, average single trace accuracy decreases from 97.11% (resp. 98.21%) when $\sigma = 15$ for CNN (resp. perceptron) to 50.78% (resp. 56.43%) when $\sigma = 90$. Based on the results obtained from Table 3, note that a huge performance gap is observed between the supervised and the unsupervised settings. While the predictive model f_{sup} can retrieve at least 90% of the private key bits when the standard deviation of the applied Gaussian noise is up to $\sigma = 50$, the unsupervised approach affects the selection of the predictive model $f \in \mathcal{F}$ by reducing for CNN (resp. perceptron) the correction process by 43% (resp. almost 30%) when $\sigma = 50$. Therefore, the noisy configuration blurred the iterative framework to correct the wrong bit predictions even if the prelabeling phase retrieves 52.44% of the private key bits (same prelabeling configuration as in [16]). This observation suggests that some improvement can be provided to the iterative framework in order to reduce this performance gap while preserving its unsupervised property. In addition, depending on the noise level, if the prelabeling phase retrieves 52.44% of the private key bits, the use of the iterative framework does not improve the bit correction such that, in the worst use-case scenario (*i.e.* $\sigma = 90$), the iterative framework degrades the attack performance. This observation illustrates the limitation of this machine learning tools. In particular, it cannot extract a (non-)linear separation between the entangled clusters if it does not exist in the first place.

The same observations hold true for `cswap_arith_reduced` dataset: the higher the noise level, the harder it becomes for the framework to correct wrong bits regardless of the neural network used. As mentioned in Sect. 5.1, this suggests that the unsupervised property of the iterative framework affects the selection of the predictive model $f \in \mathcal{F}$. Therefore, even if the prelabeling phase retrieves 52% of the private key bits for both datasets, the suitability of the tool proposed in [16] should be shaded when it deals with noisy traces which is a classical setting in side-channel context.

Through this section, we identified that the efficiency of the proposed framework can be altered under a noisy environment. Indeed, this experiment illustrates the limitation of the iterative approach to correctly deal with entangled clusters in unsupervised context as the performance gap between profiled and unprofiled setting increases with the amount of noise. Therefore, we question the need of using advanced statistical tools against the one classically introduced in the side-channel community (*e.g.* dimensionality reduction techniques, clustering algorithms...). In that purpose, the following section proposes an easier alternative and introduces a new strategy based on the combination of PCA and k-means.

6 Deep Learning Free Approach

Now that we have highlighted the limitations of the methodology proposed in [16], we want to show that nor correction techniques nor deep learning are required for this attack to be effective.

6.1 Alternative Attack: Overview

The alternative DL-free strategy that we propose consists in two steps:

- Extracting a single linear combination of time samples via PCA (see Sect. 2.3).
- Clustering with k-means algorithm to directly conclude the attack.

In other words, we propose to settle for the prelabeling phase of the attack proposed in [16], simply performing it in a more opportune way.

PCA is performed on all the 76,500 subtraces of each dataset with 20 components. Then, we select, by visual selection or by Explained Local Variance (ELV, recalled in Sect. 2.3), a single component to project data onto.

6.2 Alternative Attack: Experimental Results

Visual Component Selection Methodology

By visually observing the PCs, we made the following observation on both datasets: at least one PC is very similar to the t-test traces depicted in Fig. 1. For the `cswap_pointer` dataset, exactly one component, the 8th, depicted in Fig. 5a, presents similarities with such t-test trace depicted in Fig. 1a. We observe, on both traces, peaks at the same samples. Based on this observation, we would have chosen this component for our attack.

Similarly, in `cswap_arith`, several components are similar to the t-test analysis but one in particular, the 6th, depicted in Fig. 5b, seems to be less impacted by noise. It presents the main of the two groups of peaks that can be observed in the t-test outcome depicted in Fig. 1b. Thus, this component would visually be the best candidate for our attack.

When the two datasets are projected onto the chosen PCs, each subtrace only consists in one sample. These datasets are then used as inputs of the k-means algorithm. We obtain a satisfying result on `cswap_pointer` for k-means

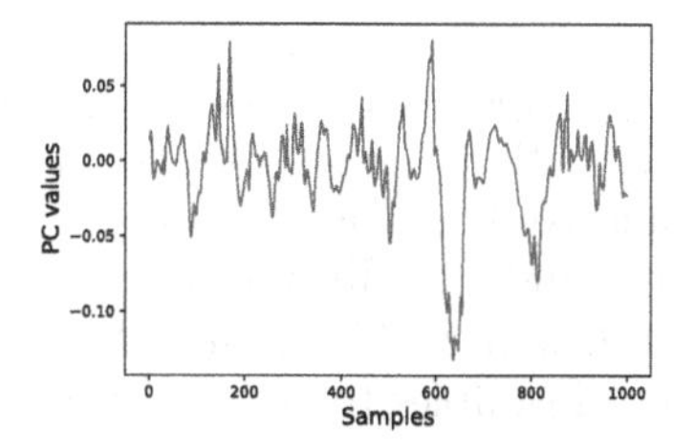

(a) 8th PC of the Principal Component Analysis for the `cswap_pointer`

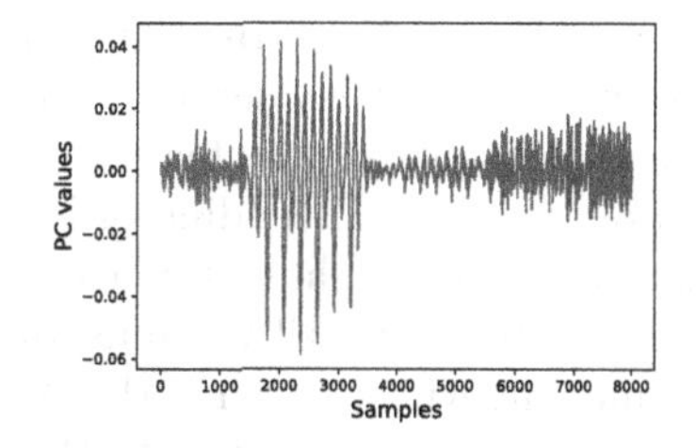

(b) 6th PC of the Principal Component Analysis for `cswap_arith`

Fig. 5. PCs visually identified as similar to t-test outcomes (see Fig. 1).

on the 8th component. Indeed, the algorithm results in a 94.1% (resp. 97.7%) average (resp. maximal) single trace accuracy. However, for `cswap_arith`, the accuracy is lower than expected. k-means results in a 60.8% (resp. 72.2%) average (resp. maximal) single trace accuracy for 6th component. We suppose the k-means algorithm did not perform well onto `cswap_arith` because significant information is spread within multiple components and the chosen one has a higher noise ratio.

These results are nonetheless more satisfying than those presented in [16] and reported in the first row of Table 1. Nevertheless, this methodology requires an access to the t-test results which is in contradiction with the unsupervised context. One possibility is to relax this hypothesis such that the attacker can conduct leakage assessment analysis. In this case, the t-test results can also be used to delete non significant intervals which might improve PCA results. For example, for `cswap_arith`, we could conduct the same attack on samples 1,000 to 6,000 of traces and obtain 80% as the average single trace accuracy on the 6th PC.

If we do not want to alter unsupervised context, an alternative would be to test the 20 PCs independently with k-means. For `cswap_pointer`, only the 8th PC gives an accuracy far from random clustering. In this case, visual selection was more efficient but the success rate is equivalent. However, in `cswap_arith`, several components give an average single trace accuracy over 60% with k-means. As presented above, the 6th component results in a 60.8% average single trace accuracy and the 12th and 18th PC respectively result in a 61.4% and 76.1% average single trace accuracy. In this case, brute force is slower than visual selection but more effective because the final success rate is higher. This increase in time complexity might be mitigated by the use of a component selection method. In the following section, we use the Explained Local Variance in order to limit the brute force complexity of the principal component's choice.

Explained Local Variance Technique
After computing ELV values for each sample and component, we obtain 20 traces showing significant samples for each component. For the analysis, samples of ELV traces are sorted in descending order and cumulative traces are computed.

Figure 6 presents the cumulative ELV trends individually normalized such that their cumulative ELV values are between 0 and 1. The goal is to select the component which requires the least number of samples to achieve its global variance. We can clearly identify that the 8th principal component requires the least number of samples to achieve its global variance. For this dataset, both the visual observations and the ELV criteria resulted into the same conclusion.

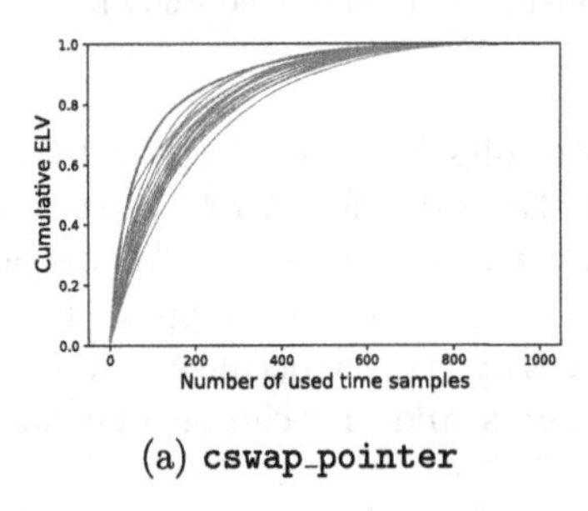

(a) `cswap_pointer`

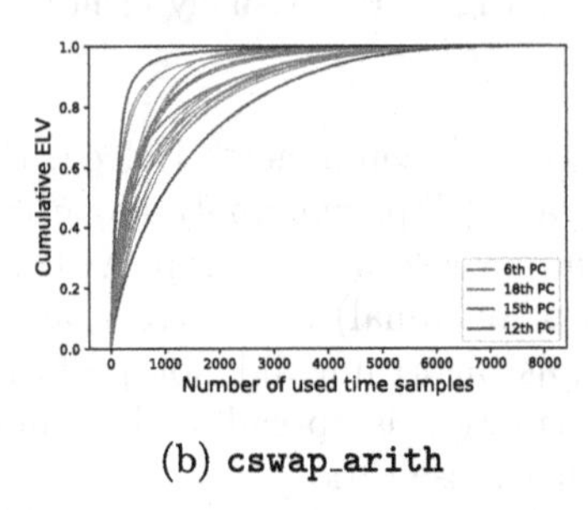

(b) `cswap_arith`

Fig. 6. Normalized cumulative ELV trend of principal components. In red, the components that we would have visually selected. In blue and magenta, for the `cswap_arith` dataset, the 2 PCs which gave k-means accuracies higher than 60%, and in green the PC that would be chosen by the ELV criterion. (Color figure online)

Figure 6b presents the cumulative ELV trends for `cswap_arith`. Here, the component suggested by the ELV criterion is not the same as in the visual selection method.

In this case scenario, the k-means results for both datasets are the following:

- For `cswap_pointer`, the selected PC is the same as the visually selected. Thus, 97.7% of maximal single trace accuracy are also achieved. Here, the ELV method permits to avoid the brute force onto the component selection.
- For `cswap_arith`, the clustering algorithm on the 15th PC, the one selected by ELV criterion, results in a 50.2% average single trace accuracy which is really close to the accuracy achieved by a random clustering. Observing Fig. 6b we realize that the most performing PCs would be select almost as fast by the ELV criterion than by their natural order. In fact, ELV would select the 12th PC as the last one. In that particular case the natural order is a better choice. Thus, brute force is still necessary and testing components in the natural order would require similar effort as using ELV criteria.

In [3] the ELV concept is also derived in an attempt to denoise components. In the case where ELV selection method is used, we also apply the described denoising method on the selected component.

For `cswap_pointer`, if we only keep the 80 samples where the 8th PC presents the highest ELV, we fix the others to zero and we use the k-means algorithm after projection, we obtain an average (resp. maximal) single trace accuracy of 98.0% (resp. 100%). Figure 7a presents the evolution of the k-means accuracy for the 8th component depending on the estimated number of POIs. It shows that there is a wide range of number of points that would bring an advantage to be fixed to zero.

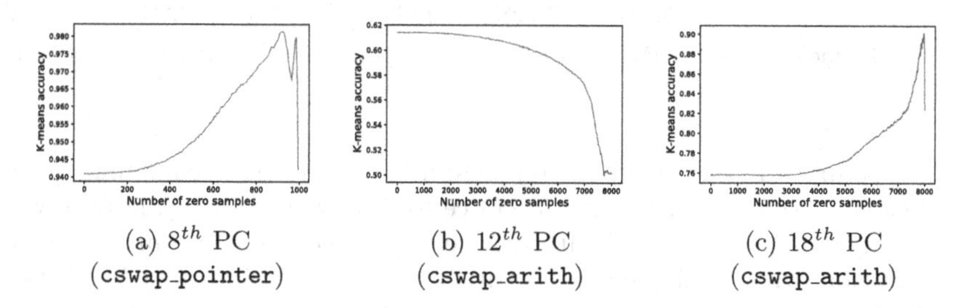

(a) 8th PC (`cswap_pointer`)

(b) 12th PC (`cswap_arith`)

(c) 18th PC (`cswap_arith`)

Fig. 7. k-means accuracy depending on the number of samples fixed to zero in the selected component.

For `cswap_arith`, if we only keep the 55 samples where the 18th PC presents the highest ELV, we fix the others to zero and we use the k-means algorithm after projection, we obtain an average (resp. maximal) single trace accuracy of 90.6% (resp. 96.1%). Figure 7c presents the evolution of the k-means accuracy for the 18th component depending on the estimated number of POIs. Here again, an attacker does not necessarily have to guess that 55 samples is the optimum. Indeed the trend of the accuracy raises continuously to such a maximum, while fixing to zero samples presenting increasing values of ELV. Interestingly, the behaviour of the 12th component is different. Indeed, the more we "denoise" by fixing to zero samples which does not contribute a lot to the global variance, the more the accuracy decreases. We can suppose that its global variance is mostly due to noisy signals and that, all in all, this PC is not that interesting.

Results Summary and Conclusion

In this section, we presented an unsupervised attack scenario which makes use of both PCA and k-means algorithms. Several solutions have been proposed to select a satisfying component depending on whether the primary objective is to keep it unsupervised or simple. In addition to that, a denoising method has been detailed, that allowed to significantly raise the attack performances.

The best results we could obtain for both datasets are summarized in Table 5, and compared with results obtained in [16].

In conclusion, we proposed an alternative unsupervised approach (namely PCA+ELV+Denoising) based on classical statistical tools for dimensionality reduction and clustering that still outperforms on average the solution proposed in [16] in all tested cases when comparing average single trace accuracy.

Table 5. Comparison of the average and maximal single trace accuracy between the different methods used to extract significant information.

	cswap_pointer			cswap_arith		
	Average	Maximum	PC/PoIs	Average	Maximum	PC/PoIs
PCA + Visual	94.1%	97.7%	8th	60.8%	72.2%	6th
PCA + Natural order	94.1%	97.7%	8th	76.1%	82.4%	18th
PCA + ELV + Denoising	98%	100%	8th	90.6%	96.1%	18th
SOST [16]	52.24%	55.22%	20 PoIs	52.44%	55.22%	20 PoIs
SOST + Framework [16]	91%	100%	20 PoIs	83%	100%	20 PoIs

This simple strategy provides complete attacks since the remaining entropy is sufficiently low to allow using brute force to retrieve the all key. Indeed, when considering the PCA equipped with ELV selection and denoising, in average a 5-bit correction (average single trace accuracy of 98%), which is done in 2^{33} tests, is sufficient to recover the entire secret scalar for the `cswap_pointer` dataset. Concerning the `cswap_arith` dataset, the average residual complexity is too high to be fulfilled (*i.e.* 2^{111} tests), but we remark that, even though we do not reach as in [16] a 100% maximal single trace accuracy, we can still conduct a successful attack since we obtain a minimal residual complexity of 2^{58} (corresponding to a maximal single trace accuracy of 96.1%), that allows a complete attack as well.

Moreover, the obtained results are in some cases close to the ones claimed in [16] after the application of the recursive neural-network-based correcting framework. In this section, we provide simple state-of-the-art solutions to defeat the implementations targeted in [16]. Thus, through those propositions, we question the real benefit of using deep-learning based solutions in the side-channel attacks context when datasets that are linearly separable are targeted.

7 Conclusion

Dealing with noisy labels is a generic machine learning problem which has a direct application in the side-channel domain. Many techniques have been proposed in the machine learning community to attempt to address this problem. In particular, one of the approaches studied is based on a co-teaching strategy: iterative correction of noisy labels [17]. Perin *et al.* applied such an approach in SCA domain [16].

In this paper, we provide an in-depth analysis of their iterative framework. We observed that the study was particularly biased by the simplicity of the used datasets (*i.e.* simple linear separability). Anyway, this particularity of the datasets was hidden by the huge complexity of the proposed deep-learning-based solution. This fact should raise a warning about the pertinence of using very complex machine-learning solution for general problems. Indeed, in the considered case, we show that a deep-learning-free solution, *i.e.* the use of a simple perceptron or the combination of PCA and k-means, results in a faster, more efficient and more interpretable attack than the CNN proposed in [16]. Through this contribution, we believe that focusing on the simplest possible attacks also provides a better insight on the actual threats and attack scenarios that one should take into account when implementing or designing cryptosystems. In particular, we consider that it is important to warn that the implementation studied is not only vulnerable to a somewhat involved deep learning framework but also to simpler and faster approaches. We acknowledge that machine-learning solution are more suitable in more complex context *e.g.* noisy or non-linearly separable data or datasets with desynchronization effects (for instance, a future work could investigate the property of convolutional layer to mitigate the desynchronization effect) but they should be sized carefully. Finally, we also question in this paper the relevance of this corrective approach when considering noisy traces. From these experiments, we conclude that this approach does not solve the problem of handling noisy labels, which implies that this problem is still an open problem in both side-channel and machine learning communities.

Acknowledgments. The authors would like to thank Guilherme Perin and the anonymous reviewers for their valuable comments which helped to improve this work. We would also like to thank Simon Abelard, Eleonora Cagli, Julien Eynard, Antoine Loiseau, Ange Martinelli, Guénaël Renault and Gabriel Zaid for fruitful discussions about this work. This work was financially supported by the defense innovation agency (AID) from the French ministry of armed forces.

References

1. Aloise, D., Deshpande, A., Hansen, P., Popat, P.: Np-hardness of Euclidean sum-of-squares clustering. Mach. Learn. **75**, 245–248 (2009). https://doi.org/10.1007/s10994-009-5103-0
2. Arpit, D., et al.: A closer look at memorization in deep networks (2017)
3. Cagli, E., Dumas, C., Prouff, E.: Enhancing dimensionality reduction methods for side-channel attacks. In: Homma, N., Medwed, M. (eds.) CARDIS 2015. LNCS, vol. 9514, pp. 15–33. Springer, Cham (2016). https://doi.org/10.1007/978-3-319-31271-2_2
4. Choudary, O., Kuhn, M.G.: Efficient template attacks. In: Francillon, A., Rohatgi, P. (eds.) CARDIS 2013. LNCS, vol. 8419, pp. 253–270. Springer, Cham (2014). https://doi.org/10.1007/978-3-319-08302-5_17
5. Düll, M., et al.: High-speed curve25519 on 8-bit, 16-bit, and 32-bit microcontrollers. Des. Codes Cryptography **77** (2015). https://doi.org/10.1007/s10623-015-0087-1
6. Du, Q., Faber, V., Gunzburger, M.: Centroidal Voronoi tessellations: applications and algorithms. SIAM Rev. **41**(4), 637–676 (1999)
7. Gallier, J.: Dirichlet–Voronoi diagrams and Delaunay triangulations. In: Gallier, J. (ed.) Geometric Methods and Applications. Texts in Applied Mathematics, vol. 38, pp. 301–319. Springer, New York (2011). https://doi.org/10.1007/978-1-4419-9961-0_10
8. Gierlichs, B., Lemke-Rust, K., Paar, C.: Templates vs. stochastic methods. In: Goubin, L., Matsui, M. (eds.) CHES 2006. LNCS, vol. 4249, pp. 15–29. Springer, Cham (2006). https://doi.org/10.1007/11894063_2

9. Han, B., et al.: Co-teaching: robust training of deep neural networks with extremely noisy labels (2018)
10. Linde, Y., Buzo, A., Gray, R.: An algorithm for vector quantizer design. IEEE Trans. Commun. **28**(1), 84–95 (1980). https://doi.org/10.1109/TCOM.1980.1094577
11. Lloyd, S.: Least squares quantization in PCM. IEEE Trans. Inf. Theory **28**(2), 129–137 (1982). https://doi.org/10.1109/TIT.1982.1056489
12. van der Maaten, L., Hinton, G.: Viualizing data using t-SNE. J. Mach. Learn. Res. **9**, 2579–2605 (2008)
13. Maghrebi, H., Portigliatti, T., Prouff, E.: Breaking cryptographic implementations using deep learning techniques. In: Carlet, C., Hasan, M.A., Saraswat, V. (eds.) SPACE 2016. LNCS, vol. 10076, pp. 3–26. Springer, Cham (2016). https://doi.org/10.1007/978-3-319-49445-6_1
14. Nascimento, E., Chmielewski, L.: Applying horizontal clustering side-channel attacks on embedded ECC implementations. In: Eisenbarth, T., Teglia, Y. (eds.) CARDIS 2017. LNCS, vol. 10728, pp. 213–231. Springer, Cham (2017). https://doi.org/10.1007/978-3-319-75208-2_13
15. Perin, G., Chmielewski, Ł: A semi-parametric approach for side-channel attacks on protected RSA implementations. In: Homma, N., Medwed, M. (eds.) CARDIS 2015. LNCS, vol. 9514, pp. 34–53. Springer, Cham (2016). https://doi.org/10.1007/978-3-319-31271-2_3
16. Perin, G., Chmielewski, U., Batina, L., Picek, S.: Keep it unsupervised: horizontal attacks meet deep learning. IACR Trans. Cryptographic Hardware Embed. Syst. **2021**(1), 343–372 (2020). https://doi.org/10.46586/tches.v2021.i1.343-372. https://tches.iacr.org/index.php/TCHES/article/view/8737
17. Song, H., Kim, M., Park, D., Shin, Y., Lee, J.G.: Learning from noisy labels with deep neural networks: a survey (2022)
18. Ying, X.: An overview of overfitting and its solutions. In: Journal of Physics: Conference Series, vol. 1168, no. 2, p. 022022 (2019). https://doi.org/10.1088/1742-6596/1168/2/022022
19. Zhang, C., Bengio, S., Hardt, M., Recht, B., Vinyals, O.: Understanding deep learning requires rethinking generalization (2017)

Deep Stacking Ensemble Learning Applied to Profiling Side-Channel Attacks

Dorian Llavata[1,2,3](✉), Eleonora Cagli[1,2], Rémi Eyraud[3], Vincent Grosso[3], and Lilian Bossuet[3]

[1] Univ. Grenoble Alpes, 38000 Grenoble, France
[2] CEA, LETI, MINATEC Campus, 38054 Grenoble, France
{dorian.llavata,eleonora.cagli}@cea.fr
[3] Univ. Jean Monnet Saint-Etienne, CNRS, Institut d Optique Graduate School, Lab. Hubert Curien UMR 5516, 42023 Saint-Etienne, France
{remi.eyraud,vincent.grosso,lilian.bossuet}@univ-st-etienne.fr

Abstract. Deep Learning is nowadays widely used by security evaluators to conduct side-channel attacks, especially in profiling attacks that allow a supervised learning phase. However, designing an efficient neural network model in a side-channel attack context can be a difficult task that may require a laborious hyperparameterization process. Hyperparameter selection is known to be a challenging problem in Deep Learning, while being a crucial factor for neural networks performances. Recent works investigate the so-called Deep Ensemble Learning in the side-channel context. It consists in using multiple neural networks in a single predictive task and aggregating the several predictions in an opportune way. The intuition behind is to use the power of numbers to improve the attack performance. In this work, we propose to use Stacking as an aggregation method, in which a meta-model is trained to learn the best way to combine the output class probabilities of the ensemble networks. Our proposal is supported by several experimental results, that allow to conclude that the use of Stacking can relieve the security evaluator from performing a fine hyperparameterization.

Keywords: Side-Channel Attacks · AES · Neural Networks · Ensemble Learning · Stacking

1 Introduction

Embedded cryptography on constrained electronic devices like smart cards can be vulnerable to Side-Channel Attacks (SCA). These attacks exploit physical leaks collected on a device during the execution of cryptographic operations, such as energy consumption [16] or electromagnetic emission [10]. The analysis of these physical leaks can allow an attacker or an evaluator to retrieve sensitive data and compromise the devices security. Depending on the level of access and control of the target device, SCA can be categorized as profiling

S. Bhasin and T. Roche (Eds.): CARDIS 2023, LNCS 14530, pp. 235–255, 2024.
https://doi.org/10.1007/978-3-031-54409-5_12

(*e.g.* template attacks [7]), or non-profiling attacks (*e.g.* DPA [16]/CPA [5]). The profiling scenario works on the principle that the evaluator has full control over a clone device identical to the target device. In this configuration, the evaluator splits the process into two phases. First, a profiling or characterization phase in which he uses the clone device to determine when the sensitive variable is leaking and to design an accurate model of the physical leakage of the clone device. Second, an attack phase in which the characterized leakage is used to attack the real target device. Since a profiling SCA may be viewed as a classical supervised learning problem, various machine learning methods have been investigated, such as Support Vector Machines [14] and Random Forests [17]. In recent years, profiling SCA based on deep learning have proved to be very efficient [3,6,18]. However, deep learning algorithms have much more tunable hyperparameters than other techniques. Their correct configuration is essential to obtain a good attack performance and it is very difficult to precisly know which hyperparameters influence the attack performance. In particular, too complex neural network architectures may be prone to the overfitting phenomenon, that is, when a model learns the training data by heart and is no more able to generalize on unseen data.

1.1 Related Works

A few papers discuss about methodologies to build neural networks for SCA. Zaid *et al.* [27] proposed a methodology to generate robust convolutional neural network architectures. The authors used visualization tools to try to understand the impact of each convolutional hyperparameter. Robissout *et al.* [23] explore several regularization techniques (Batch Normalization, Weight decay and Dropout) to improve the attack performance of a neural network. Some work has investigated methods to automate the search of hyperparameters, Wu *et al.* [26] proposed to use Bayesian optimization to find optimal hyperparameters for neural network architectures. In the same way, Rijsdijk *et al.* [22] propose the use of Reinforcement Learning techniques. More recently, the SCA community has begun to experiment with neuroevolution and genetic algorithms for neural network design [1]. On the other hand, to improve the generalization and to limit the efforts of hyperparameterization, several works propose to use Ensemble Learning. Destouet *et al.* [8] use an ensemble of models for approximate a leakage model by targeting different sensitive values. Gao *et al.* [11] explore different ensemble methods based on decision tree to improve the attack success, including RusBoost, Bagging, and Adaboost methods. Recent work has begun to investigate the use of ensemble learning with deep neural network as weak model in SCA context. Perin *et al.* [20] provided experimental results on implementations of symmetric algorithms which show that combining predictions from multiple neural networks of different architectures with a bagging method allows to gain in attack performance. An extension of this work has been proposed by Zaid *et al.* [28], on asymmetric algorithms, with the proposal of a new loss function named *Ensembling loss* which aims to maximizing diversity between weak models during the training process.

1.2 Contributions and Paper Organization

Contributions. This paper extends the preliminary results of deep ensemble learning applied in profiling SCA context by proposing to use stacking as an aggregation method [25]. We report an experimental exploration of the stacking method, whose goals are to assess the soundness of such a technique in profiling SCA context to highlight its adventages and inconvenients and to compare it with the bagging. The performance optimization is out of the scope of the experimental campaign.

Paper Organization. The paper is organized as follows. Section 2 provides background about profiling SCA and ensemble learning. Section 3 presents our experimental results, together with a discussion about how stacking can significantly improve generalization and attack performance by comparing it with bagging. In Sect. 4 we discuss about results, highlight the cost and the effectiveness of such an approach to reduce the hyperparameterization effort and try to deduce a way to construct suitable architectures for a meta-model. Finally, in Sect. 5 we conclude and discuss possible future research directions.

2 Introduction to Profiling Side-Channel Attacks and Ensemble Learning

2.1 Notations

Let capital letters X denote random variables (random vectors if in bold $\boldsymbol{X}$), and the corresponding lowercase letters x denote their realizations. (resp. $\boldsymbol{x}$ for vectors). During their acquisition, each trace is associated with a target sensitive variable $Z = f(K, P)$, where P denotes some public variable, e.g. a plaintext, and K the part of secret key the evaluator aims to retrieve.

2.2 Profiling Side-Channel Attacks and Evaluation Metrics

Profiling SCA. Side-channel attacks are typically performed using a divide-and-conquer strategy to independently attack chunk of the secret key called *subkeys.* For example, in the case of AES-128, instead of directly attacking the entire 16-byte key (which is computationally infeasible), the attack is divided into 16 parts and attempts to recover each 1-byte subkey separately. In the profiling phase, the evaluator aims to characterize the leakage from the clone device. The issue of such a characterization may be, for example in the case of classification-task-inspired deep learning attacks, a model $F(z, t)$ that provides an estimate of the posterior probability $Pr[Z = z|\boldsymbol{T} = \boldsymbol{t}]$, being Z the target sensitive variable and $\boldsymbol{T}$ the random vector representing side-channel traces. In this work, the model F is assumed to be either a Multilayer Perceptron (MLP) or a Convolutional Neural Network (CNN). Once the profiling phase is done, and the model is able to establish the relationship between the leakage and the corresponding value of the sensitive variable (which is linked to the target subkey), the evaluator applies the model on additional traces from the real target device. Finally to retrieve the subkey value, the evaluator will need to match the predicted sensitive values to an estimation of the subkey value. To do this, all possible values of a subkey are enumerated, and for each of them, all resulting sensitive variables are computed. Then, for all the subkey candidates, the evaluator exploits the set of traces by summing the logarithms of the output probabilities of their respective labels. This gives the following logarithmic probability vector $\boldsymbol{g} = (g_1, \ldots, g_C)$ used to determine the likelihood that each of the C candidates is the correct subkey:

$$g_k^Q = \sum_{i=1}^{Q} log(F[f(k, p_i), t_i]), \tag{1}$$

where Q is the number of attack traces and $f(\cdot)$ is the sensitive operation, The value $F[f(k, p_i), t_i]$ denotes the $f(k, p_i)$-th compotent of output of the neural network model, given the trace t_i as an input. It is interpreted as the probability assigned by the profiling model of obtaining the sensitive variable $f(k, p_i)$ corresponding to leakage trace t_i with a subkey hypothesis k and a plaintext p_i.

Evaluation Metrics. Accuracy (defined as the successful classification rate) is the most common metric to evaluate a deep learning model. Nevertheless, while this metric is perfectly suitable for a general classification problem, it may not be suitable for SCA, as discussed by Cagli *et al.* [6]. Indeed, in SCA context it only corresponds to the success rate of a *simple attack*, *i.e.* a single-trace attack. When the evaluator can exploit several traces for varying plaintexts, the accuracy metric is not sufficient to evaluate the attack performance. In this case, it is worth considering SCA specific metric like the Empirical Guessing Entropy (GE) [24]. After the attack, the evaluator exploits the sorted logarithmic vector of candidate subkeys called the rank vector $\boldsymbol{r} = (r_1, r_2, \ldots, r_C) = Sort(\boldsymbol{g})$. In the vector, r_1 is considered the most likely subkey and r_C as the least likely. The position of the good subkey k^* in the vector is called the rank of the subkey.

$$Rank_Q(k^*) = \text{i such that } r_i^Q = k^* \tag{2}$$

Guessing entropy is defined as the expected rank of the correct subkey:

$$GE_Q = \mathbb{E}[Rank_Q(k^*)] \tag{3}$$

It may be estimated by the empirical average rank of the subkey k^*, among all the subkey hypothesis. This metric is estimated empirically, by performing the attack several times from different subsets of traces. In this work, we randomly select subsets from all available attack traces and set the number of attacks to 100. To compare the attack performance of our models, we will also look at another metric derived from the GE computation: N_a will denote the number of traces required for a successful attack. In other words, an attack is considered successful using N_a traces if the Guessing entropy is stably equal to 1.

$$N_a^* = min(Q|GE_Q(k^*) = 1) \tag{4}$$

2.3 Ensemble Learning

In ensemble learning, the models used within the ensemble usually perform poorly individually (slightly better than random guessing) and are called *weak models*. Ensemble methods combine the individual predictions of weak models via an aggregation method. The principle is to synergistically use an ensemble of several different weak models and to correctly combine their predictions in order to reduce their variance and their biases to obtain a more accurate and robust model called the *ensemble model*. As explained in [2], if the weak models errors are uncorrelated, the ensemble can be more efficient than any individual weak model. Combining the predictions of complementary weak models can thus improve the generalization, *i.e.* the ability of a trained model to perform on an unseen test set as well as on the training set. Concerning how to combine the weak models and build the ensemble model, in this work we use two types of meta-algorithms[1] called bagging [4] and stacking [25].

Bagging. The principle of bagging (depicted in Fig. 1a) is to build several weak models (usually models of homogeneous types) independently, then aggregate them by an averaging or voting process to obtain the final predictions. The training set is usually subsampled to train the weak models on different subsamples of the training data. However in the SCA context the profiling phase requires a high number of traces. Thus, as the previous work of Perin *et al.* [20], we considered in this work the same training set for all weak models. The aggegation method that we used is the method proposed by Perin *et al.* [20] which consists of summing the log-probabilities during the attack phase. The new sum of the log-probabilities e based on the Eq. 1 is calculated for each key byte hypothesis k:

$$e_k = \sum_{m=1}^{W} \sum_{i=1}^{Q} log(F[f(k, p_i), t_i]_m), \tag{5}$$

where W is the number of weak models.

Stacking. The principle of stacking (depicted in Fig. 1b) differs mainly from bagging in three ways. First, in stacking, it is common to consider weak models of heterogeneous types. Second, weak models are usually trained on the same training set. Finally, to aggregate the weak models predictions, stacking uses a higher-level model, called *meta-model*, which is trained to produce new predictions.

2.4 Datasets

For all datasets, the experiments are implemented in Python 3.9 using the Keras 2.8 library and are run on a workstation equipped with 32 GB RAM and a NVIDIA Quadro P4000 with 8 GB memory.

[1] There are other ensemble methods, in particular the Boosting [9], which we have experimented without obtaining good enough performance for the considered datasets.

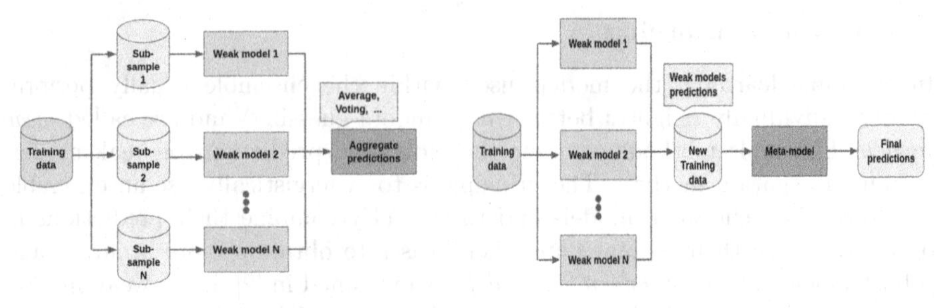

(a) Bagging ensemble learning. (b) Stacking ensemble learning.

Fig. 1. Different ensemble learning methodologies.

ASCADv1. This dataset contains traces of an 8-bit AVR microcontroller running a masked AES-128 implementation. There are actually two versions of the dataset that we will name ASCADF and ASCADV. ASCADF has a fixed key for training traces and consists of 50,000 traces for profiling and 10,000 for attack. Traces contain 700 time samples, a.k.a. features. ASCADV has variable keys for training traces and consists of 200,000 traces for profiling and 100,000 traces for attack. Traces contain 1400 features. Each of these versions also includes 3 variants to add a desynchronization type countermeasure of 0, 50 and 100 desynchronization samples respectively. The dataset was introduced in [3] and is publicly available at https://github.com/ANSSI-FR/ASCAD.

AES HD. This dataset was introduced in [21]. It contains traces from an unprotected AES-128 hardware implementation. The AES HD dataset does not include any countermeasure but has the particularity to be very noisy. 50,000 traces are used for profiling and 25,000 are used for attack. Traces contain 1250 features. The dataset is publicly available at https://github.com/AESHD/AES_HD_Dataset.

3 Experiments

3.1 Weak Models

Number of Weak Models. Some works has investigated the question of how many weak models should be used in an ensemble [12,13,19]. To summarize, it appears that the optimal ensemble size depends on the problem and the performance of the weak models, so generally the ensemble size can be considered as another hyperparameter that can be searched by experimental analysis. Notably, Hansen and Salamon [12] suggested that ensemble with as few as 10 weak models were in general adequate to sufficiently reduce test error and improve generalization. Therefore in this work we have chosen to limit the scope of analysis to ensemble of up to 10 weak models.

Weak Models Training and Results. To evaluate bagging and stacking ensemble learning methods, we trained for each dataset 10 neural networks,

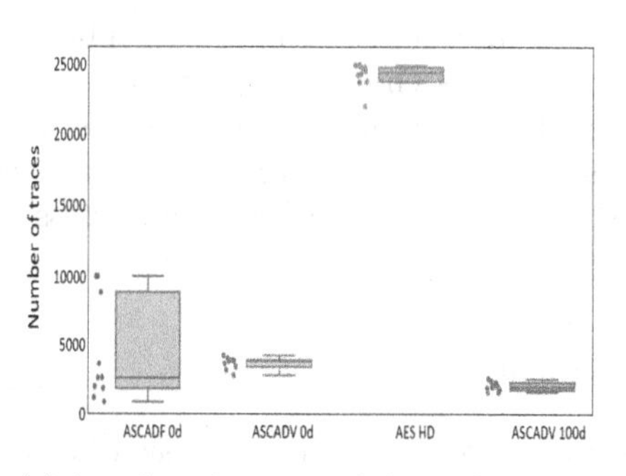

(a) Attack performance of the weak models on each datasets.

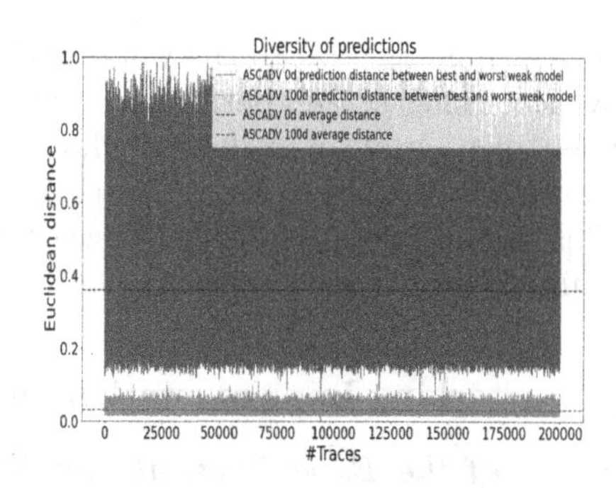

(b) ASCADV 0d and ASCADV 100d diversity of predictions.

Fig. 2. Weak models informations.

including MLPs and CNNs with some hyperparameters randomly selected from defined ranges. Varying models architectures will allow to learn different features from the same training set. Appendix A provides the search spaces of the weak models. An exception to such a random selection of hyperparameters is done in the case of ASCADV 100d dataset. Here, the convolutional part of the CNNs has been fixed once for all in order to deal with the high desynchronization and obtain weak models able to perform successfull (even if poorly performning) attacks independently. However, in Sect. 3.2.3 we explore some experimental results on ASCADV 50d where the convolutional part has not been fixed. Each network was trained with a stop criterion by monitoring the validation loss. The training and validation sets are obtained from the labeled data with a split ratio of 80%/20%. We focused onto a single byte of the AES secret key, and chose to target directly the corresponding first round Sbox output value as sensitive target variable. It may assume 256 possible values. The attack performance of the weak models are depicted in Fig. 2a. It may be remarked that on ASCADF 0d we have a significant performance gap between the weak models. On the other hand on AES HD our weak models are all very poorly performing due to the very high noise level of the dataset. Figure 2b) depicts the Euclidean distance between the predictions of the best and worst weak models for the ASCADV 0d and ASCADV 100d datasets. A striking difference in diversity can be observed through the fact that fixing the convolutional part on ASCADV 100d resulted in weak models with very close predictions.

Choice of Weak Models. Once obtained 10 weak models, and in order to use a posteriori ensemble learning, we have to choose the models to use in the ensemble. Intuitively, a good ensemble is one where the individual weak models are both accurate and make their errors on different parts of the input space. Nevertheless as with the ensemble size, the choice of weak models to combine must often be found experimentally. Due to the performance gap and the lack of diversity on some datasets, we ranked our 10 weak models from the best performing in the attack (minimal Na) to the worst performing in order to increase ensemble size by decreasing attack performance.[2] We made this choice in order to take the point of view of an evaluator who knows the attack performance of his weak models and wishes to use a posteriori ensemble learning to improve his attack. This also allowed us to consider the ideal conditions for bagging and assess its limitations in this context.

3.2 Stacking Implementation and Results

Scope of the Experimentations. Neural network stacking approach in SCA context has been mentioned by Perin *et al.* [20], but to the best of our knowledge it has never been explored in more detail in the literature. Our intuition is that a meta-model should perform better than an average process resulting from bagging. However, the complexity is increased by the addition of meta-model training, and we believe that this approach may be sensitive to the meta-model configuration. Therefore, rather than using a single hyperparameterized meta-model, our experiment consists in training 30 meta-models with random hyperparameter configurations. By analyzing the variability of the results for the 30 meta-models, we are able to verify if the stacking approach is robust or if it requires a careful meta-model hyperparameterization. We chose to use MLP as meta-models. Our 30 MLPs from the search space shown in Table 1 are the same for all ensemble sizes, in order to check the meta-models behavior when the ensemble size increases. In order to study the impact of stacking on our weak models, we compared the performance for different ensembles size with the performance of bagging and the best trained weak model. Our performance criterion is the convergence of Guessing Entropy. We trained the meta-models with the same training data of the weak models.[3]

Concatenation Method. For the sake of completeness, we provide a description of the way we stacked the weak models predictions in our stacking experiments. We stack the weak models predictions in depth-wise sequence: if we have N weak models and each of them produces 256 value per prediction, we finally get a stacked prediction c of shape $256 * N$ to train the meta-models:

$$c = [P_0^0, P_0^1, ..., P_0^N, ..., P_{256}^0, P_{256}^1, ..., P_{256}^N], \tag{6}$$

where P_i^j denotes the i-th compotent of output of the weak model j. Anyway, we remark that this choice should have no impact a priori on the learning process, since we use MLP as meta-models and the input layer of the MLP is fully connected (the order of the input thus not affect the possible functions it could model). Interestingly, we believe that this choice may have an important impact

[2] Other criteria has been tested during the experimental campaign, but the obtained results were less performant and uninteresting in our opinion. Thus, they have been omitted.

[3] We also tried to train on the validation dataset, but the results were generally worse due to the lack of data. Results have thus been omitted.

in cases where the meta-model is a different kind of model, for example a CNN. Indeed, CNN extracts information locally from the input, thus the proximity of the probability predictions for a same class could be a benefit for this kind of meta-model. An analysis of these impacts is left for future works.

Table 1. Hyperparameter search space for meta-models.

Hyperparameter	min	max	step
Number of layers	2	8	1
Number of neurons	100	1000	100
Activation	Relu, Elu, Selu, Gelu, Tanh		
Epoch	Early stopping: Val loss Patience 20		
Learning Rate	0.0001		
Mini Batch	100		
Optimizer	RMSprop		
Loss	Categorical Crossentropy: metric accuracy		

3.2.1 Results on ASCADF 0d and ASCADV 0d

Stacking Results. The results of our experiments are summarized in Table 2. On ASCADF 0d, the best meta-model was trained on the predictions of the 4 best weak models. This meta-model successfully performed the attack in 203 traces, reducing the number of traces required by 81.69% (compared to 1109 traces for the best weak model). A similar performance improvement was obtained on ASCADV 0d with a meta-model trained on the predictions of the 5 best weak models that successfully performed the attack in 582 traces, reducing the number of traces needed to succeed in the attack by 80.42% (compared to 2973 traces for the best weak model). If we look at the performance of the best meta-models obtained on each ensemble size, we see that stacking improved the overall attacks performance on both datasets by 60% and 70% respectively. We also notice on both datasets that by increasing the ensemble size, the number of meta-models that improved attack performance tends to decrease. This is probably due to the fact that the addition of weak models and their knowledge make the meta-model learning task easier and on these two datasets the meta-models appeared to be too complex for the learning task. Thus the meta-models overfit immediately without learning relevant information. This may be confirmed by observing the behaviour of the early-stopping mechanism: on ASCADF 0d up to an ensemble size of 5 weak models, the meta-models began to overfit in an average of 25 epochs, but from an ensemble size of 6, the average learning epoch before overfitting is only 2 epochs. The phenomenon is even more impressive on ASCADV 0d, where the meta-models began to overfit in an average of 3 epochs for the ensemble size of 2 weak models and 2 epochs for the other ensemble sizes.

Table 2. Stacking on ASCADF 0d and ASCADV 0d datasets. *Nb success* column refers to the number of meta-models that improved attack performance compared to the best weak model. Min, Max and Mean Na values are estimated considering only such *Nb success* meta-models. The best result is highlighted by a green cell.

Size of Ensemble	Nb success (Na <1109)	Na			Improvement in number of traces
		Min	Max	Mean	
2	30/30	371	853	576	66.54%
3	23/30	368	1098	696	66.81%
4	24/30	203	1064	680	81.69%
5	23/30	342	1062	674	69.16%
6	14/30	452	1043	588	59.24%
7	13/30	450	1070	604	59.42%
8	18/30	357	1086	666	67.80%
9	17/30	377	814	589	66.00%
10	15/30	427	989	631	61.49%

(a) Stacking on ASCADF 0d.

Size of Ensemble	Nb success (Na <2973)	Na			Improvement in number of traces
		Min	Max	Mean	
2	21/30	673	2448	1306	77.36%
3	11/30	635	2306	1533	78.64%
4	11/30	626	2879	1509	78.94%
5	9/30	582	2601	1341	80.42%
6	8/30	789	2693	1427	73.46%
7	7/30	604	2143	1327	79.68%
8	6/30	678	2475	1412	77.19%
9	5/30	607	2909	1606	79.58%
10	3/30	745	2502	1385	74.94%

(b) Stacking on ASCADV 0d.

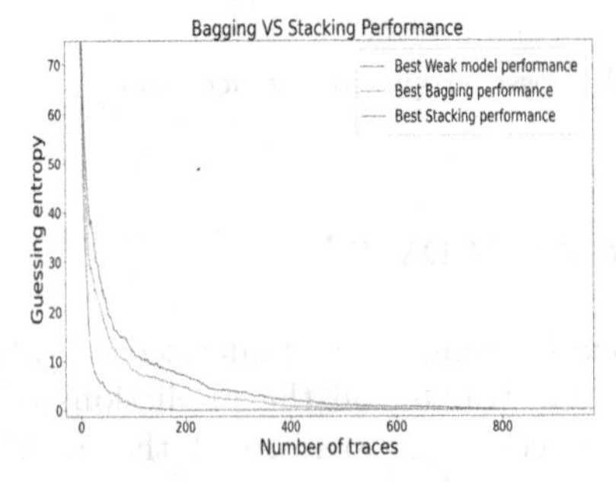

(a) ASCADF 0d GE of the best stacking and bagging ensemble.

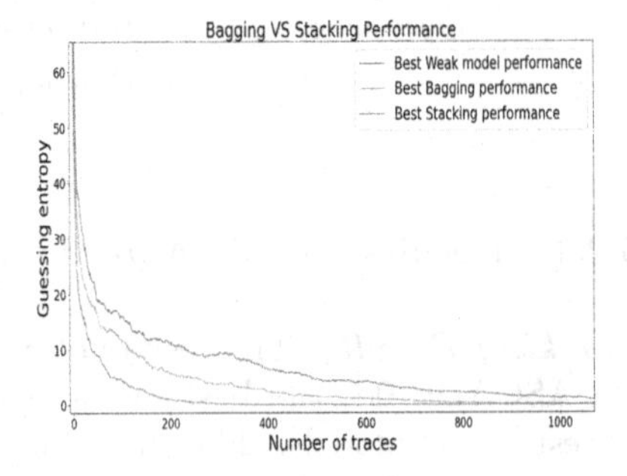

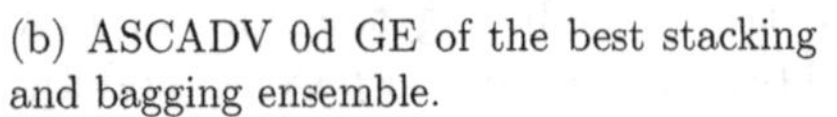

(b) ASCADV 0d GE of the best stacking and bagging ensemble.

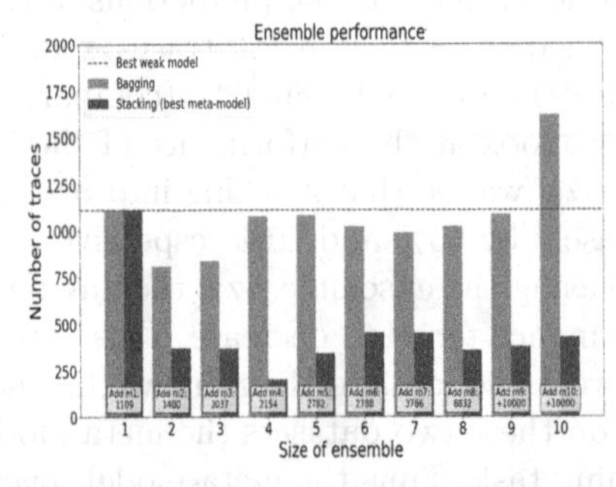

(c) ASCADF 0d Comparison of stacking and bagging across all ensemble sizes.

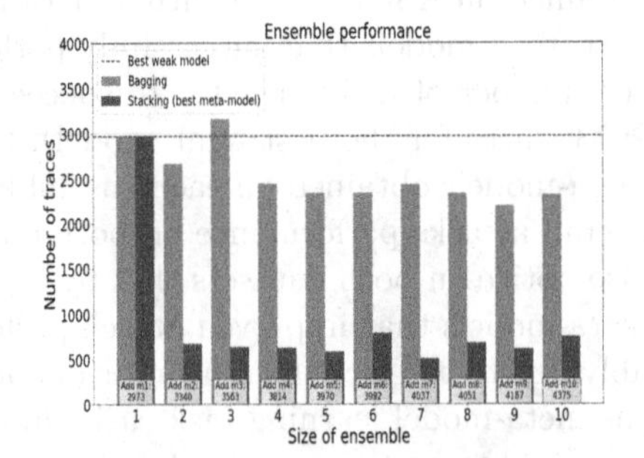

(d) ASCADV 0d Comparison of stacking and bagging across all ensemble sizes.

Fig. 3. Comparison of stacking and bagging ensemble on ASCADF 0d and ASCADV 0d. For Figures (c) and (d), each box lying at the bottom of the i-th column, informs about the performance (in terms of Na) of the i-th weak model.

Comparison with Bagging. The best stacking and bagging attack performance on both datasets are depicted in Fig. 3a and Fig. 3b. The behavior of the attack performance across all ensemble sizes are depicted in Fig. 3c and Fig. 3d. We can see that stacking converges faster and allows us to obtain higher attack performance than bagging. In particular, we can observe from the results of ASCADF 0d that the bagging process is strongly impacted when our ensemble contains weak models with a significant performance gap. We observe that by adding less and less performing weak models in the ensemble, bagging becomes less and less suitable, until it loses its interest by obtaining lower attack performance than the best individual weak model. Interestingly, stacking aggregation is less impacted by the high variability in weak model performance since the meta-model learns the relevance of each weak model. Our intuition is confirmed by results on ASCADV 0d, where, in absence of a performance gap between weak models, the bagging aggregation works properly. However, we find that stacking aggregation significantly improves generalization and attack performance regardless of ensemble size.

3.2.2 Results on AES HD and ASCADV 100d

Stacking Results. The results of our experiments are summarized in Table 3. On ASCADV 100d the best meta-model was trained on the predictions of the 7 best weak models that successfully performed the attack in 351 traces, reducing the number of traces needed to succeed in the attack by 80.41% (compared to 1792 traces for the best weak model). If we look at the performance of the best meta-models obtained on each ensemble size, we see that stacking improved the overall attacks performance by 80%. An even greater performance improvement was obtained on AES HD with a meta-model trained on the predictions of the 9 best weak models. This meta-model successfully performed the attack in 1220 traces, reducing the number of traces required by 94.46% (compared to 22034 traces for the best weak model). If we look at the performance of the best meta-models obtained on each ensemble size, we see that stacking improved the overall attacks performance by more than 90%. The use of stacking has shown significant interest for this dataset, allowing to obtain for all ensemble sizes performances below 2000 traces with ensemble of weak models that have independently attack performance above 20,000 traces. Unlike previous experiments, on these two datasets, stacking proved to be more robust, adding weak models did not decrease the number of meta-models that improved attack performance. This may be explained by the fact that the prediction tasks are more complex on these two datasets, due to the presence of desynchronization for ASCADV 100d and a very high level of noise for AES HD. Therefore the meta-models did not immediately overfit. On ASCADV 100d, we observe that all meta-models improved attack performance for all ensemble sizes.

Table 3. Stacking on ASCADV 100d and AES HD datasets. *Nb success* column refers to the number of meta-models that improved attack performance compared to the best weak model. Min, Max and Mean Na values are estimated considering only such *Nb success* meta-models. The best result is highlighted by a green cell.

Size of Ensemble	Nb success (Na <1792)	Na			Improvement in number of traces
		Min	Max	Mean	
2	30/30	429	1172	808	76.06%
3	30/30	423	1256	735	76.39%
4	30/30	362	1160	763	79.79%
5	30/30	369	1141	711	79.40%
6	30/30	352	1070	700	80.35%
7	30/30	351	1130	742	80.41%
8	30/30	351	1333	741	80.41%
9	30/30	369	1097	737	79.40%
10	30/30	369	1137	717	79.40%

(a) Stacking on ASCADV 100d.

Size of Ensemble	Nb success (Na <22034)	Na			Improvement in number of traces
		Min	Max	Mean	
2	25/30	1365	4179	2212	93.80%
3	27/30	1507	20542	2704	93.16%
4	28/30	1324	11394	2286	93.99%
5	28/30	1251	8014	2038	94.32%
6	27/30	1253	9641	1988	94.31%
7	29/30	1324	12604	2377	93.99%
8	26/30	1315	8947	1962	94.03%
9	27/30	1220	4556	1865	94.46%
10	27/30	1318	9092	2106	94.01%

(b) Stacking on AES HD.

Comparison with Bagging. The best stacking and bagging attack performance on both datasets are depicted in Fig. 4a and Fig. 4b. The behavior of the attack performance across all ensemble sizes are depicted in Fig. 4c and Fig. 4d. We can see that stacking converges faster and allows us to obtain higher attack performance than bagging. On ASCADV 100d, another limit of the bagging process can be visualized. As reported in Sect. 3.1, in this experiment we fixed the convolutional part of the CNNs in order to quickly obtain weak models able to perform attacks independently. This resulted in weak models with very close predictions. Since the bagging process draws its strength from the diversity of the weak models, the lack of diversity in this case leads to a poorly performing or even worthless ensemble attack. Interestingly, compared to bagging, we noticed that stacking was not affected by the lack of diversity between the weak models. On AES HD, the weak models present naturally a good diversity, thus bagging process is performant. However, as for ASCADV 0d, the performance of bagging is limited by the individual performance of the weak models. In comparison, meta-model training results in better association of weak model predictions and much greater improvement in attack performance.

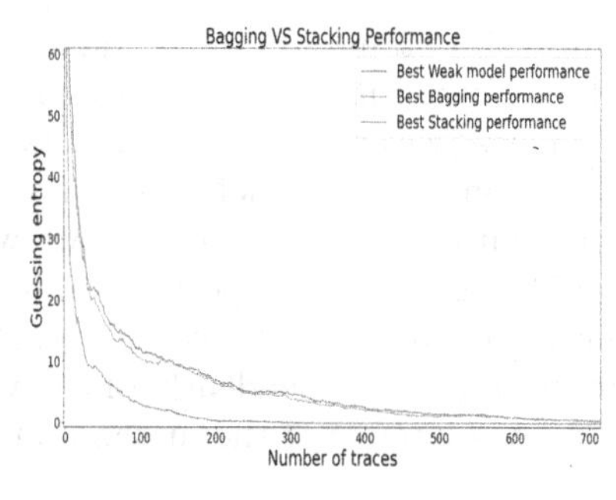

(a) ASCADV 100d GE of the best stacking and bagging ensemble.

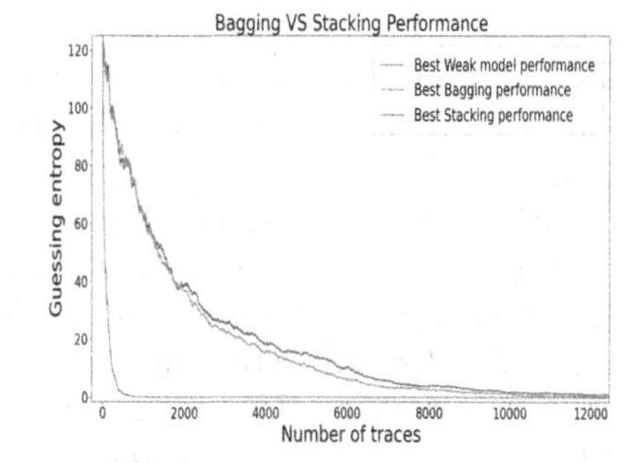

(b) AES HD GE of the best stacking and bagging ensemble.

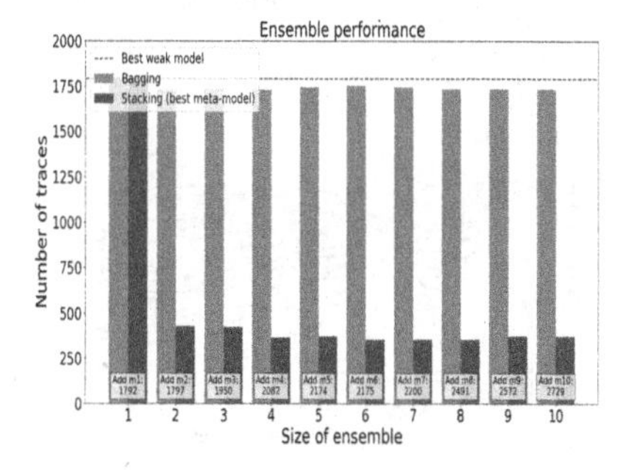

(c) ASCADV 100d Comparison of stacking and bagging across all ensemble sizes.

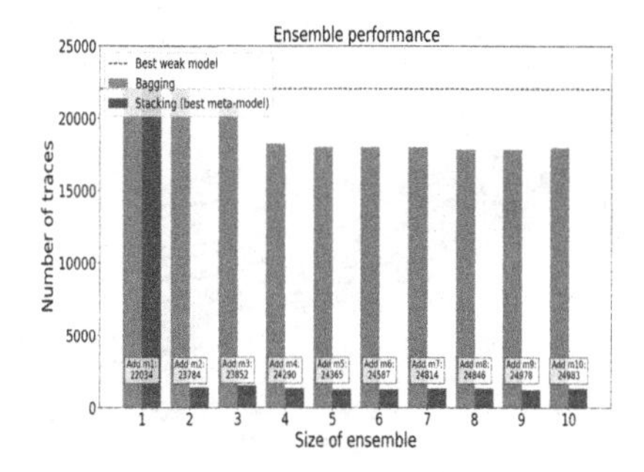

(d) AES HD Comparison of stacking and bagging across all ensemble sizes.

Fig. 4. Comparison of stacking and bagging ensemble on ASCADV 100d and AES HD. For Figures (c) and (d), each box lying at the bottom of the i-th column, informs about the performance (in terms of Na) of the i-th weak model.

3.2.3 Results on ASCADV 50d with Full Random CNNs

In previous experiments on the ASCADV dataset including desynchronization, in order to obtain individually performing weak models, we fixed the convolutional part of our weak models. In this experiment, in order to have a more objective analysis of the applicability of stacking in the presence of desynchronization, we trained new weak models in a completely random way by varying the convolutional part (*i.e.* including random hyperparameters selection for the convolutional part). As a result, we obtained very weak models that are not able to independently succeed in the attack with the 100,000 available attack traces. After applying stacking on different groups of two weak models, we found that in each case several meta-models were anyway able to improve GE convergence. The results of this experiment are shown in Fig. 5. In the first scenario, the two weak models show a converging trend, but had not enough traces to succeed in the attack. We observe that the best trained meta-model reaches the same performance as the best weak model in less than 50,000 traces (instead of 100,000 traces). Moreover, when we consider all traces, it obtains an average rank under 5 instead of an average rank of 25 for the best weak model. In the second scenario, stacking is particularly interesting: the two weak models are just starting to converge after 100.000 attack traces, while some meta-models reached a rank of 2 with the available traces. Finally, in the third scenario, one of the two weak models converged by placing the correct subkey guess in last position. Interestingly, stacking was able to correct this effect to provide a correct convergence of the GE. Therefore, stacking appears to be robust independently of the weak models performance. This is encouraging for the interest of the method and its application in a real attack cases.

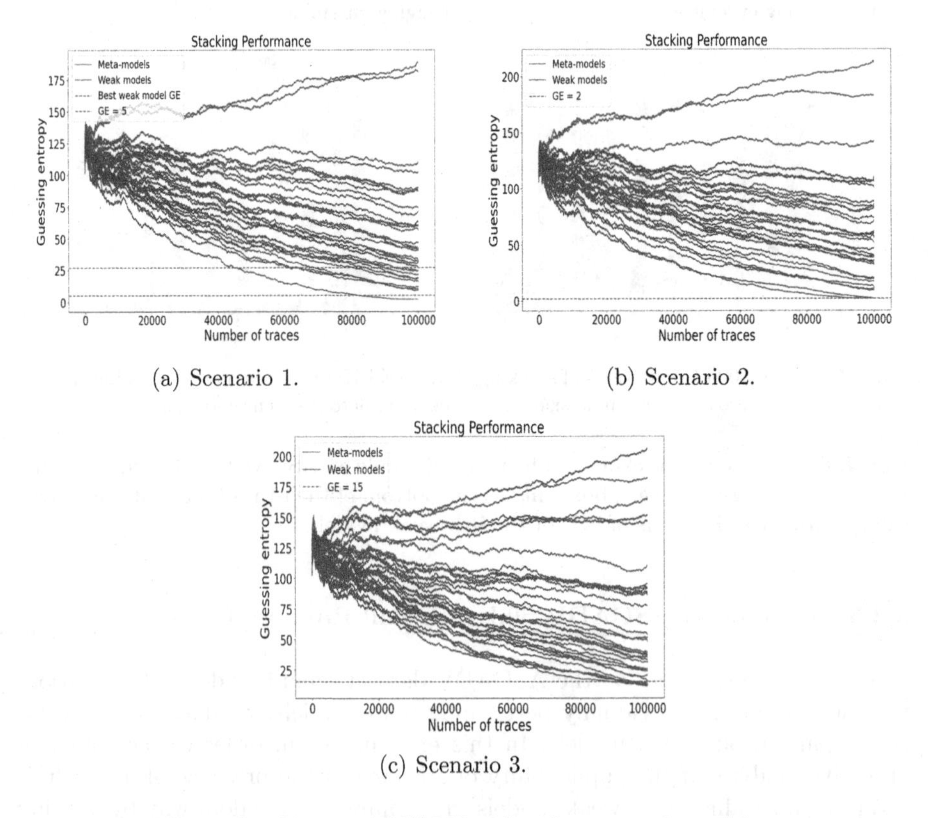

(a) Scenario 1.

(b) Scenario 2.

(c) Scenario 3.

Fig. 5. Stacking on ASCADV 50d with very weak models.

4 Discussion

4.1 Stacking Aggregation: Pros and Cons

The comparison of the attack performance improvements of bagging and stacking obtained on each datasets are summarized in Table 4. If we compare the best performance obtained with the two aggregation methods, we observe that for all datasets, stacking has always been a better choice, allowing to obtain significant attack performance improvements compared to bagging and non-ensemble models. This significant improvement is due to the fact that stacking allows more sophisticated combinations and transformations of the weak model predictions by the training of the meta-model. Indeed, the meta-model is able to capture non-linear relationships between weak model predictions, which allows for more granular and accurate ensemble predictions. When the bagging was constrained by the individual performance of the weak models or the lack of diversity within the ensemble, some meta-models proved to be robust and allowed to improve the attack performance. This can be explained by the fact that a properly trained meta-model is able to correctly learn the relevance of each weak model while being able to learn consistent information even over small variations in ensemble predictions. Thus, the choice as well as the number of weak models is less determining when considering the stacking ensemble, which makes it a more flexible method for evaluators interested in using a posteriori ensemble learning. Interestingly, our experiments revealed that even with very few weak models, significant performance gains can already be achieved. This suggests that stacking has the ability to extract relevant information from a small subset of diverse weak models and that there is no need to consider an overly complex ensemble model. On the other hand, stacking has some drawbacks. Since the meta-model training is determinant in the success of the ensemble, this adds a new constraint to the success of the attack. The ensemble model has a higher complexity than the weak ones due to the addition of the meta-model. Finally, we observed that the ensemble model often proved to be too complex for the problem. Therefore, the meta-models tends to overfit quickly. Furthermore, the meta-model needs a lot of data to generalize properly. For example, our omitted experiments in training the meta-model on the validation dataset did not work well due to the lack of data.

Table 4. Comparison of bagging and stacking results on all datasets.

Dataset	Best weak model	Bagging improvement in number of traces	Stacking (best meta-model) improvement in number of traces
AES HD	22034	17798 (20%)	1220 (94%)
ASCADF 0d	1109	709 (28%)	203 (81%)
ASCADV 0d	2973	2194 (26%)	582 (80%)
ASCADV 100d	1792	1730 (3%)	351 (80%)

Table 5. Comparison in terms of performance with state-of-the-art architectures.

Dataset	Reference	Hyperparameterization method	Na
ASCADF 0d	Arch. in [3]	–	1146
	Arch. in [22]	Reinforcement learning	202
	Our best Meta-model	4 random weak models with Na between [1109-2154]	203
ASCADV 0d	Arch. in [3]	–	1275
	Arch. in [22]	Reinforcement learning	490
	Our best Meta-model	5 random weak models with Na between [2973-3970]	582
ASCADV 100d	Arch. 1 in [23]	–	3333
	Arch. 2 in [23]	Regularization technique	347
	Our best Meta-model	7 random weak models with Na between [1792-2200]	351
AES HD	Arch. in [15]	–	25000
	Arch. in [27]	Visualization tools	1050
	Our best Meta-model	9 random weak models with Na between [22034-24983]	1220

4.2 Relieving Hyperparameterization Effort

In this section, we provide arguments to promote stacking as a technique to relieve the hyperparameterization effort for a security evaluator. To do so, we compare here for all datasets the performances of our best meta-model with different architectures, finely tuned, proposed in literature Table 5. Even if performance optimization was not at the core of the experiments, we observed that stacking ensemble can provide with less effort similar attack performance to rigorously hyperparameterized architectures. For ASCADF 0d and ASCADV 0d, we observe that with ensembles of weak models whose performance are similar (or worse) to those obtained with slightly hyperparameterized architectures [3], we obtain performance similar to high-performance architectures, which are hyperparameterized using reinforcement learning [22]. The interest of stacking is even more striking for AES HD, where with an ensemble of weak models (with individual performance higher than 20,000 traces), we obtain with less effort similar performance to high-performance architecture proposed by Zaid *et al.* [27] that are properly hyperparameterized. The main interest of stacking ensemble is to limit the hyperparameterization effort. The methodology proposed by Zaid *et al.* [27] make it possible to efficiently hyperparameterize its architecture and thus obtain a high-performance attack. However, they also assume much knowledge about datasets and an in-depth study of the impact of hyperparameters using data visualization tools (which can be time-consuming). Alternatively, the use of reinforcement learning proposed by Rijsdijk *et al.* [22] to automate hyperparameter search is effective, but the related process is extremely time-consuming.

Furthermore, the experiments described in Sect. 3.2.3 are very representative: stacking is able to improve the attack performance even with very weak models that had (almost) not started to converge. This indicates an interest in the method in a more realistic attack scenario. Therefore, stacking may be considered as a suitable approach to avoid the need for the security evaluator to perform a fine tuned hyperparameterization of the neural network architecture.

4.3 Generalizable Meta-model

In this section, we propose a quick analysis of the different meta-models we used in our experiment. The goal here was to deduce some general property to construct a good meta-model. If we take a look at a generalizable meta-model, we identified two meta-model architectures that proved to be suitable on all datasets. These two architectures, shown in Table 6, do not always achieve the best performance, but they always improved attack performance, on all datasets and all ensemble sizes. These appear to be the only two-layer architectures. We interpret this fact as a consenquence of the overfitting phenomenon: more complex meta-models often overfitted in our experiments. Especially on datasets that do not include desynchronization, where we can clearly see that increasing the number of layers in the meta-models degrades the attack performance (Fig. 6). Therefore, a small architecture with few layers is more appropriate for the meta-model.

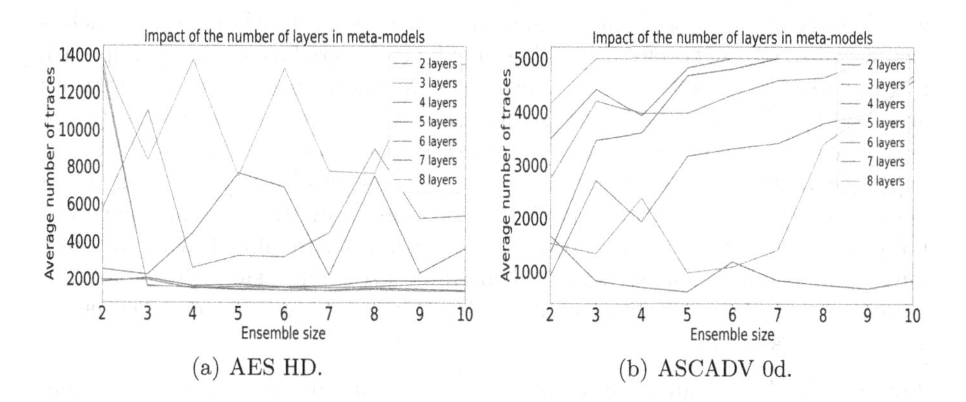

(a) AES HD. (b) ASCADV 0d.

Fig. 6. Impact of the number of layers on meta-model performance.

Table 6. Meta-models that always improve attack performance.

Hyperparameter	Architecture 1	Architecture 2
Number of layers	**2**	**2**
Number of neurons	600	300
Activation	elu	tanh
Epoch	Early stopping: Val loss Patience 20	
Learning Rate	0.0001	
Mini Batch	100	
Optimizer	RMSprop	
Loss	Categorical Crossentropy: metric accuracy	

5 Conclusion and Future Works

In this work, we propose a new study of Deep Ensemble Learning in the side-channel context. We extend the preliminary results of Perin *et al.* [20] who used bagging to aggregate the predictions of the weak models and we propose to use stacking as a more suitable choice in the aggregation method. Our experimental exploration on several publicly available datasets highlights some of the limitations of the bagging process and shows that stacking can significantly improve attack performance while providing a flexible solution to address these limitations. During our experiments, we observed that stacking ensemble can provide with less effort attack performance similar to those of rigorously hyperparameterized architectures. Therefore, stacking may be considered as a suitable approach to avoid the need for the security evaluator to finely tune the neural network architecture. However, stacking ensemble has proven to be extremely sensitive to overfitting, making it crucial to avoid using overly complex meta-models. In our experiments, two-layer meta-models have always succeeded in improving attack performance. We also noticed that the improvement in attack performance was not correlated with the number of weak models in the ensemble. Indeed, we often found similar improvements across all ensemble sizes. Thus, since the complexity increases with the addition of weak models, we recommend using an ensemble with few weak models.

Future Works. In our experimental campaign, our simplest meta-models consisted of two layers of 100 neurons. We think it would be interesting to further simplify the networks by experimenting with single-layer architectures with even fewer neurons. We also plan to extend this work by using a small CNN as a meta-model to take advantage of the concatenation in depth-wise sequence. Moreover, it would be interesting to study the applicability of the boosting ensemble methodology in SCA.

Acknowledgements. This work was financially supported by the Defense Innovation Agency (AID) from the french ministry of armed forces.

A Weak models

(See Fig. 7).

Hyperparameter	min	max	step
Learning Rate	0.0001	0.001	0.0001
Mini Batch	100	1000	100
Nb conv layers	2	8	1
Filters	8	32	4
Kernel size	10	20	2
Stride	5	10	5
Nb FC layers	2	3	1
Nb FC neurons	100	1000	100
Activations	Relu, Elu, Selu, Gelu, Tanh		

(a) Search space for ASCADF 0d / AES HD CNN hyperparameters.

Fixed Conv part			
Conv(32, 1) , AveragePooling(2, 2)			
Conv(64, 25), AveragePooling(25, 25)			
Conv(128, 3), AveragePooling(4, 4)			
Random Dense part			
Hyperparameter	min	max	step
Learning Rate	0.0001	0.001	0.0001
Mini Batch	100	1000	100
Nb FC layers	2	4	1
Nb FC neurons	500	4000	500
Activations	Relu, Elu, Selu, Gelu, Tanh		

(b) Search space for ASCADV 50d / ASCADV 100d CNN hyperparameters.

Hyperparameter	min	max	step
Learning Rate	0.0001	0.001	0.0001
Mini Batch	100	1000	100
Kernel size	16	128	16
Nb conv layers	1	4	1
Nb FC layers	1	3	1
Nb FC neurons	500	4000	500
Filters	1		
Strides	2		
Activations	Relu, Elu, Selu, Gelu, Tanh		

(c) Search space ASCADV 50d full random CNN.

Hyperparameter	min	max	step
Learning Rate	0.0001	0.001	0.0001
Mini Batch	50	1000	100
Nb FC layers	2	8	1
Nb FC neurons	100	1000	100
Dropout	0.0	0.4	0.1
Activations	Relu, Elu, Selu, Gelu, Tanh		

(d) Search space for ASCADF 0d/ AES HD/ ASCADV 0d MLP hyperparameters.

Fig. 7. Search space for weak models.

References

1. Acharya, R.Y., Ganji, F., Forte, D.: Information theory-based evolution of neural networks for side-channel analysis. IACR Trans. Cryptogr. Hardw. Embed. Syst. 401–437 (2023)
2. Ali, K.M., Pazzani, M.J.: On the link between error correlation and error reduction in decision tree ensembles (1995)
3. Benadjila, R., Prouff, E., Strullu, R., Cagli, E., Dumas, C.: Deep learning for side-channel analysis and introduction to ASCAD database. J. Cryptogr. Eng. **10**(2), 163–188 (2020)
4. Breiman, L.: Bagging predictors. Mach. Learn. **24**, 123–140 (1996)
5. Brier, E., Clavier, C., Olivier, F.: Correlation power analysis with a leakage model. In: Joye, M., Quisquater, J.J. (eds.) Cryptographic Hardware and Embedded Systems – CHES 2004. CHES 2004. LNCS, vol. 3156, pp. 16–29. Springer, Berlin, Heidelberg (2004). https://doi.org/10.1007/978-3-540-28632-5_2
6. Cagli, E., Dumas, C., Prouff, E.: Convolutional neural networks with data augmentation against jitter-based countermeasures: profiling attacks without pre-processing. In: Fischer, W., Homma, N. (eds.) Cryptographic Hardware and Embedded Systems – CHES 2017. CHES 2017. LNCS, vol. 10529, pp. 45–68. Springer, Cham (2017). https://doi.org/10.1007/978-3-319-66787-4_3

7. Chari, S., Rao, J.R., Rohatgi, P.: Template attacks. In: Kaliski, B.S., Koc, C.K., Paar, C. (eds.) Cryptographic Hardware and Embedded Systems – CHES 2002. CHES 2002. LNCS, vol. 2523, pp. 13–28. Springer, Berlin, Heidelberg (2003). https://doi.org/10.1007/3-540-36400-5_3
8. Destouet, G., Dumas, C., Frassati, A., Perrier, V.: Wavelet scattering transform and ensemble methods for side-channel analysis. In: Bertoni, G.M., Regazzoni, F. (eds.) Constructive Side-Channel Analysis and Secure Design. COSADE 2020. LNCS, vol. 12244, pp. 71–89. Springer, Cham (2021). https://doi.org/10.1007/978-3-030-68773-1_4
9. Freund, Y., Schapire, R.E.: A decision-theoretic generalization of on-line learning and an application to boosting. J. Comput. Syst. Sci. **55**(1), 119–139 (1997)
10. Gandolfi, K., Mourtel, C., Olivier, F.: Electromagnetic analysis: concrete results. In: Koc, C.K., Naccache, D., Paar, C. (eds.) Cryptographic Hardware and Embedded Systems – CHES 2001. CHES 2001. LNCS, vol. 2162, pp. 251–261. Springer, Berlin, Heidelberg (2001). https://doi.org/10.1007/3-540-44709-1_21
11. Gao, F., Mao, B., Wu, L., Wang, Z., Mu, D., Hu, W.: Leveraging ensemble learning for side channel analysis on masked AES. In: 2021 7th International Conference on Computer and Communications (ICCC), pp. 267–271. IEEE (2021)
12. Hansen, L.K., Salamon, P.: Neural network ensembles. IEEE Trans. Pattern Anal. Mach. Intell. **12**(10), 993–1001 (1990)
13. Hernández-Lobato, D., Martínez-Muñoz, G., Suárez, A.: How large should ensembles of classifiers be? Pattern Recogn. **46**(5), 1323–1336 (2013)
14. Heuser, A., Zohner, M.: Intelligent machine homicide: breaking cryptographic devices using support vector machines. In: Schindler, W., Huss, S.A. (eds.) Constructive Side-Channel Analysis and Secure Design. COSADE 2012. LNCS, vol. 7275, pp. 249–264. Springer, Berlin, Heidelberg (2012). https://doi.org/10.1007/978-3-642-29912-4_18
15. Kim, J., Picek, S., Heuser, A., Bhasin, S., Hanjalic, A.: Make some noise. unleashing the power of convolutional neural networks for profiled side-channel analysis. IACR Trans. Cryptogr. Hardw. Embed. Syst. 148–179 (2019)
16. Kocher, P., Jaffe, J., Jun, B.: Differential power analysis. In: Wiener, M. (eds.) Advances in Cryptology – CRYPTO' 99. CRYPTO 1999. LNCS, vol. 1666, pp. 388–397. Springer, Berlin, Heidelberg (1999). https://doi.org/10.1007/3-540-48405-1_25
17. Lerman, L., Bontempi, G., Markowitch, O.: A machine learning approach against a masked AES: reaching the limit of side-channel attacks with a learning model. J. Cryptogr. Eng. **5**, 123–139 (2015)
18. Maghrebi, H., Portigliatti, T., Prouff, E.: Breaking cryptographic implementations using deep learning techniques. In: Carlet, C., Hasan, M., Saraswat, V. (eds.) Security, Privacy, and Applied Cryptography Engineering. SPACE 2016. LNCS, vol. 10076, pp. 3–26. Springer, Cham (2016). https://doi.org/10.1007/978-3-319-49445-6_1
19. Opitz, D., Maclin, R.: Popular ensemble methods: an empirical study. J. Artif. Intell. Res. **11**, 169–198 (1999)
20. Perin, G., Chmielewski, L., Picek, S.: Strength in numbers: improving generalization with ensembles in machine learning-based profiled side-channel analysis. IACR Trans. Cryptogr. Hardw. Embed. Syst. 337–364 (2020)
21. Picek, S., Heuser, A., Jovic, A., Bhasin, S., Regazzoni, F.: The curse of class imbalance and conflicting metrics with machine learning for side-channel evaluations. IACR Trans. Cryptogr. Hardw. Embed. Syst. **2019**(1), 1–29 (2019)

22. Rijsdijk, J., Wu, L., Perin, G., Picek, S.: Reinforcement learning for hyperparameter tuning in deep learning-based side-channel analysis. IACR Trans. Cryptogr. Hardw. Embed. Syst. 677–707 (2021)
23. Robissout, D., Bossuet, L., Habrard, A., Grosso, V.: Improving deep learning networks for profiled side-channel analysis using performance improvement techniques. ACM J. Emerg. Technol. Comput. Syst. (JETC) **17**(3), 1–30 (2021)
24. Standaert, F.X., Malkin, T.G., Yung, M.: A unified framework for the analysis of side-channel key recovery attacks. In: Joux, A. (eds.) Advances in Cryptology – EUROCRYPT 2009. EUROCRYPT 2009. LNCS, vol. 5479, pp. 443–461. Springer, Berlin, Heidelberg (2009). https://doi.org/10.1007/978-3-642-01001-9_26
25. Wolpert, D.H.: Stacked generalization. Neural Netw. **5**(2), 241–259 (1992)
26. Wu, L., Perin, G., Picek, S.: I choose you: automated hyperparameter tuning for deep learning-based side-channel analysis. IEEE Trans. Emerg. Top. Comput. (2022)
27. Zaid, G., Bossuet, L., Habrard, A., Venelli, A.: Methodology for efficient CNN architectures in profiling attacks. IACR Trans. Cryptogr. Hardw. Embed. Syst. 1–36 (2020)
28. Zaid, G., Bossuet, L., Habrard, A., Venelli, A.: Efficiency through diversity in ensemble models applied to side-channel attacks:-a case study on public-key algorithms-. IACR Trans. Cryptogr. Hardw. Embed. Syst. 60–96 (2021)

Like an Open Book? Read Neural Network Architecture with Simple Power Analysis on 32-Bit Microcontrollers

Raphaël Joud[1,2], Pierre-Alain Moëllic[1,2(✉)], Simon Pontié[1,2], and Jean-Baptiste Rigaud[3]

[1] CEA Tech, Centre CMP, Equipe Commune CEA Tech - Mines Saint-Etienne, 13541 Gardanne, France
{raphael.joud,pierre-alain.moellic,simon.pontie}@cea.fr
[2] Univ. Grenoble Alpes, CEA, Leti, 38000 Grenoble, France
[3] Mines Saint-Etienne, CEA, Leti, Centre CMP, 13541 Gardanne, France
rigaud@emse.fr

Abstract. Model extraction is a growing concern for the security of AI systems. For deep neural network models, the architecture is the most important information an adversary aims to recover. Being a sequence of repeated computation blocks, neural network models deployed on edge-devices will generate distinctive side-channel leakages. The latter can be exploited to extract critical information when targeted platforms are physically accessible. By combining theoretical knowledge about deep learning practices and analysis of a widespread implementation library (ARM CMSIS-NN), our purpose is to answer this critical question: *how far can we extract architecture information by simply examining an EM side-channel trace?* For the first time, we propose an extraction methodology for traditional MLP and CNN models running on a high-end 32-bit microcontroller (Cortex-M7) that relies only on simple pattern recognition analysis. Despite few challenging cases, we claim that, contrary to parameters extraction, the complexity of the attack is relatively low and we highlight the urgent need for practicable protections that could fit the strong memory and latency requirements of such platforms.

Keywords: Side-Channel Analysis · Confidentiality · Machine Learning · Neural Network · Model Architecture

1 Introduction

Deployment of Deep Neural Network (DNN) models continues to gain momentum, typically with Internet of Things (IoT) applications with microcontroller-based platforms. However, their security is regularly challenged with works focused on availability, integrity and confidentiality threats [17]. Latter topic keeps gathering growing attention from the adversarial Machine Learning (ML) community with *Model Extraction* [2,7,8] becoming a major concern.

An attacker performs a model extraction attack either to steal a well-trained model performance or to precisely recover its characteristics to obtain a *clone*

S. Bhasin and T. Roche (Eds.): CARDIS 2023, LNCS 14530, pp. 256–276, 2024.
https://doi.org/10.1007/978-3-031-54409-5_13

model. Additionally, it would allow the adversary to enhance his level of control over the victim system to design more adapted and powerful attacks. Therefore, model extraction attacks have been the subject of a growing number of works in recent years with different adversarial scenarios regarding the level of knowledge related to the model and the training data distribution. In the literature, model architecture and parameters are usually studied separately. Many efforts have been brought to parameters recovery with milestones works relying on the assumption that model architecture is known by the attacker [2,5,7,13,18]. However, such an assumption is quite strong and can be questioned as architecture details are generally not disclosed. However, whatever the attacker's objectives, knowledge of victim model architecture significantly increases adversarial ability.

This work is focused on architecture extraction of models embedded on 32-bit microcontrollers thanks to ARM CMSIS-NN library. Our purpose is to know how far an adversary can extract information from a victim model architecture by jointly exploiting the knowledge of the deployment library and very limited side-channel (EM) traces (averaged trace from a single input). Surprisingly and worryingly, we show that (with methods and classical ML expertise) almost all the most important information and hyper-parameters are reachable because of the high *repetitiveness* of the underlying computations. More particularly, for convolutional neural networks, we demonstrate a *Russian doll* effect: one regular EM pattern is related to one hyper-parameter and zooming in presents other patterns related to others hyper-parameters, and so on.

Illustrations, Code and Data Availability. This work relies on the visual analysis of side-channel traces with many (colored) illustrations to explain our approach. For conciseness purpose, we select the most representative ones. We propose additional contents in a public repository[1] with codes and data (traces).

2 Background

2.1 Neural Network Models

We consider a supervised deep neural network M_W, with W its internal parameters. When fed with an input $x \in \mathbb{R}^d$, the model outputs a set of *prediction scores* $M_W(x) \in \mathbb{R}^K$ with K, the number of labels. Then, the output predicted label is $\hat{y} = \arg\max(M_W(x))$.

We note $\mathcal{A}_M$ the architecture of the model M that corresponds to the organisation of its *layers*. $\mathcal{A}_M$ is defined by the nature of each layer, their connections, size and hyper-parameters (non-trainable ones, e.g. the number of convolutional kernels). We note L the number of layers. In this work, we only consider feed-forward models with layers stacked horizontally.

MultiLayer Perceptron (MLP) are composed of several vertically stacked *neurons* (or *perceptron*) called *dense layers* (also named *fully-connected* or even *linear*). Each perceptron first processes a weighted sum of its trainable parameters w and b (called *bias*) with the input $x = (x_0, ..., x_{n-1}) \in \mathbb{R}^n$.

[1] https://gitlab.emse.fr/securityml/model-architecture-extraction.

Then, it non-linearly maps the output thanks to an *activation function* σ: $a(x) = \sigma(w_0x_0 + ... + w_{n-1}x_{n-1} + b)$, where a is the perceptron output. For MLP, a neuron from layer l gets inputs from all neurons belonging to previous layer $l-1$.

Convolutional Neural Network (CNN) process input data with a set of convolutional *kernels* (also called *filters*). The kernels are usually low-dimensional squared matrices (e.g. 3×3 for image classification). In addition, kernels third dimension matches the number of channels of the input tensor C_{in} (e.g. for RGB images, $C_{in} = 3$). As such, for a *convolutional layer* (hereafter shorted to conv. layer) composed of K kernels of size $Z \times Z \times C_{in}$ applied to an input tensor of size $H_{in} \times H_{in} \times C_{in}$ (we use square inputs for simplicity), the weight tensor W will have the shape $[K, Z, Z, C_{in}]$ (i.e. $K \cdot C_{in} \cdot Z^2$ parameters without bias, $(K+1) \cdot C_{in} \cdot Z^2$ otherwise). Additionally, The number of output tensor channels is $C_{out} = K$. Equation 1 is a convolution (without bias) expressed as dot-products between kernel and local regions of the input tensor, classically processed by a sliding window. With Y the output tensor, we have $\forall k, l, n \in [\![0, H_{in}[\![^2 \times [\![0, C_{out}[\![$:

$$Y_{k,l,n} = \sigma\Big(\sum_{m=0}^{C_{in}-1}\sum_{i=0}^{Z-1}\sum_{j=0}^{Z-1} W_{i,j,m,n} \cdot X_{k+i,l+j,m}\Big) \tag{1}$$

Other hyper-parameters are *Padding* P and *Stride* S that, respectively, adds extra dimensions (P) to handle the borders of the input tensor[2] and enables to downsample (by S) the sliding[3]. The output tensor shape is defined as in Eq. 2.

$$H_{out} = \frac{H_{in} - Z + 2 \cdot P}{S} + 1 \tag{2}$$

Usually, a conv. layer is followed by a **pooling layer** that aims at reducing the dimensions of the output tensor by locally reducing it with some statistics. A classical approach is to apply a *Max pooling* or an *Average pooling* with a 2×2 kernel (we note $Z_{pool} = 2$) over the output tensor Y of size $H_{out} \times H_{out} \times C_{out}$ so that the resulting tensor is half the size $(H_{out}/2) \times (H_{out}/2) \times C_{out}$.

Activation functions (σ) inject non-linearity through layers. Typical functions map the output of a dense or conv. layer into specific space like $[0, +\infty[$ (ReLU), $[-1, +1]$ (*Tanh*) or $[0, 1]$ (*Sigmoid*, *Softmax*). A widely used function is the Rectified Linear Unit defined as $ReLU(x) = max(0, x)$. The same activation function is applied to all units of a layer and thus can be functionally considered as an independent layer, as in this work.

2.2 Model Deployment on Cortex-M Platforms

Several tools are available to deploy DNN on Cortex-M platforms such as TF-LM[4], Cube.MX.AI[5] (STMicroelectronics), NNoM [12] or MCUNet [10]. Most of

[2] With $P = 0$, borders are not considered and the output tensor is shorter.
[3] $S = 1$ is the standard default sliding, one element at a time.
[4] https://www.tensorflow.org/lite/microcontrollers.
[5] https://www.st.com/en/embedded-software/x-cube-ai.html.

them are based on the CMSIS-NN library from ARM [9]. In this work, we study implementations based on NNoM with CMSIS-NN as back-end.

NNoM [12] (standing for *Neural Network on Microcontroller*) is an open-source, high-level neural network inference library dedicated to microcontrollers. It allows to easily implement DNN models previously trained on Keras-Tensorflow, while supporting complex structures (e.g. Inception or ResNet). Converted NNoM models are layer-wise quantized to reduce memory footprint. Scaling factor of the quantization scheme is restricted to powers of two, allowing to perform efficient shifting operations rather than divisions. Back-end operations can be performed by NNoM's eigenfunctions or by efficient CMSIS-NN ones when compatible.

CMSIS-NN [9] is a collection of optimized neural network basic operations, developed for Cortex-M processor cores. They enhance performance and reduce memory footprint with different optimisation techniques that depend on the target platform. They handle quantized variables on 8 or 16 bits. Implementations considered hereafter manipulate 8-bit variables. Using such variables enables to leverage on Single Instruction Multiple Data (SIMD).

2.3 Model Extraction

The goal of *model extraction* attacks is to steal a *victim model* M_W with different possible adversarial goals [7]. A *task-performance* objective [15,19] is to steal the performance of M_W to reach equal or better one at lower cost (e.g., save prohibitive training time). In that case, knowledge of (exact) victim model architecture or parameter values is not necessary. In a *fidelity* scenario, attacker wants to craft a *substitute model* M'_Θ that mimics M_W as close as possible. M' should provide the same predictions as M (correct and incorrect ones). This *similarity* between M and M' is typically defined by measuring the agreement at the label-level [7], i.e. $\arg\max(M'_\Theta(x)) = \arg\max(M_W(x))$ for every x sampled from a target distribution $\mathcal{X}$. A more complex and optimal objective (*Functionally Equivalent Extraction*) targets equal predictions ($M'(x) = M(x), \forall x \in \mathcal{X}$). Importantly, the strongest *Exact Extraction* attack ($\mathcal{A}_M = \mathcal{A}_{M'}$, $W = \Theta$) is impracticable by simply exploiting input/output pairs from victim model [7]. *Fidelity*-oriented scenarios receive a growing interest because of challenging extraction processes, more essentially for the parameters. When dealing with parameter extraction, a usual assumption is that the adversary knows $\mathcal{A}_M$. This is the case for cryptanalysis-like approaches [2,7], active learning techniques [16,19] and recent efforts relying on physical attacks such as side-channel (SCA) [1,8] or fault injection (FIA) [6,18] analysis. Interestingly, whatever the adversarial goal, victim model architecture is a crucial information: it is compulsory for fidelity scenarios, and its knowledge significantly strengthen attacker's abilities to succeed in task-performance ones [15].

Table 1. Related State-of-the-Art works. n.s.: Not Specified

Attack	Physical target	Targeted models	Used technique
[4]	MLaaS	CNN	TA & Regression
[21,22]	FPGA	BNN	SEMA
[11]	FPGA	CNN & ResNet	SEMA
[3]	GPU	CNN	SEMA & TA
[1]	µC	MLP & CNN	CEMA
[20]	µC	CNN	SPA & ML
ours	µC	MLP & CNN	SEMA only

3 Related Works and Contributions

A growing number of works investigates architecture extraction with physical attacks. Proposed techniques are essentially based on Side-Channel Analysis (SCA) such as Timing Analysis (TA), Simple Power/EM Analysis (SPA/SEMA) or Cache Attacks (CA) targeting various physical targets (FPGA, microcontrollers (µC), GPU and cloud services hosting DNN models referenced as MLaaS for Machine Learning as a Service). These works are summed up in Table 1. Usually, architecture and parameters are extracted distinctively, however Batina *et al.* propose in [1] to identify layer boundaries while performing parameters extraction with Correlation EM Analysis (CEMA). Correlation scores are used to deduce if currently targeted neuron belongs to the same layer as previous targeted one or not. In other words, if CEMA *fails*, then that means we pass through the next layer. Since parameters extraction with CEMA is highly challenging, as detailed in [8], this method raises practical issues and, to the best of our knowledge, has not been fully demonstrated. Other SCA techniques have been used like TA as in [4] to recover model numbers of layers. SCA are also occasionally combined with other approaches such as learning-based clone reconstruction [18,22] or ML-based classification of traces among a limited set of architectures [20][6]. More recently, [11] presents a complete analysis of the impact of FPGA implementations of CNN on the related electromagnetic activity and proposes and evaluates an obfuscation-based countermeasure.

In this work, our goal is to highlight leakages related to the architecture of embedded DNN models. We are considering models implemented on a 32-bit microcontroller thanks to NNoM library [12] that relies on the widely used CMSIS-NN module [9]. Both of these tools are open-source and as such, we logically consider the attacker has a total access to their code. Acquisitions presented all along this work are obtained with 16 averaged traces all acquired while performing the inference algorithm from a single input belonging to the test set (not used during model training). We deliberately set in such a minimalist setting to assess the level of information that could be extracted by an adversary that has almost no prior knowledge about the victim model and limited

[6] In [20], authors consider 4 variants of AlexNet, Inceptionv3, ResNet-50 and 101, for a total of 16 architectures.

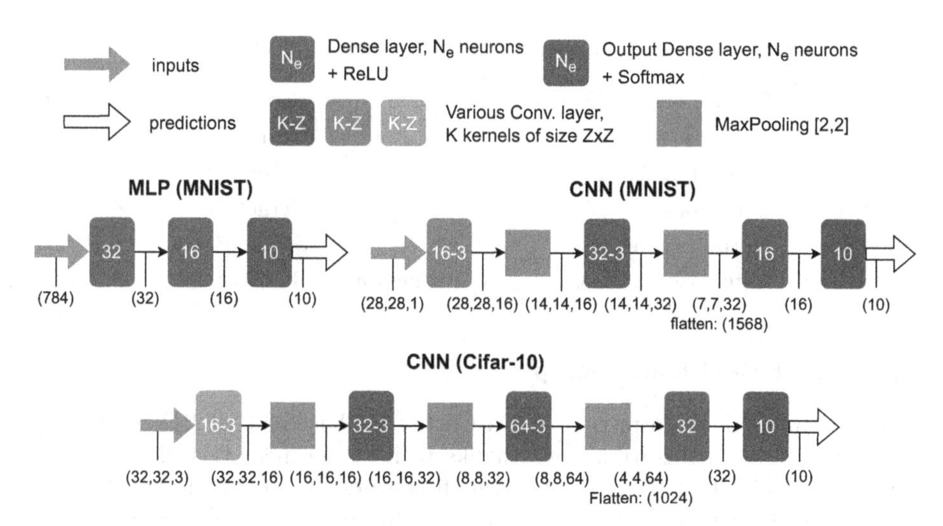

Fig. 1. Detailed architecture of considered models. Various conv. implementations are detailed in Sect. 6.1 (Color figure online)

experimental data available. This work is not intended to be exhaustive with regards to the extraction of every hyper-parameter of every possible layer types. As most of reference papers in model extraction field, we focus on most common layers and related parameters, i.e. by targeting MLP and CNN models. Furthermore, all our traces, implementations (and complementary results) are publicly available in order to foster new experiments on this topic that we claim to be highly critical in the actual large-scale model deployment context.

4 Experimental Setup

4.1 Models and Datasets

We use two traditional benchmarks MNIST[7] (28×28 digit grayscale images, 10 labels) and Cifar-10[8] (32×32 color images, 10 labels). Therefore, we have $H_{in} = 28$ and $C_{in} = 1$ for MNIST and $H_{in} = 32$ and $C_{in} = 3$ for Cifar-10. For MNIST, we trained a MLP and a CNN with the TensorFlow platform that achieved respectively 96% and 98% on the test set. For Cifar-10, we trained a CNN that reached 76% on the test set. Architectures of these models are illustrated in Fig. 1. Colors are used to identify the layer and the shape of each output tensor is shown under each layer. Dense layers (green) have N_e neurons and convolutional layers (blue, olive and cyan) have K square kernels of size Z. Classical stride (= 1) and padding value (*same*) are used, they don't impact the output shape. We only use Max pooling layer (orange) with a kernel size of 2 so that the output tensor is reduced by half. Used activation function is ReLU, except for the last layer with SoftMax function.

[7] http://yann.lecun.com/exdb/MNIST/.

[8] https://www.cs.toronto.edu/~kriz/cifar.html.

Table 2. Adversarial objective: list of the hyper-parameters to extract

Target	Parameters	Notation	Target	Parameters	Notation
$\mathcal{A}_M$	# layers	L	Conv. Layer	Output shape	H_{out}
	Type of layers	$\varnothing$		# kernels	K
Dense Layer	# neurons	N_e		Kernel size	Z
MaxPool	Output shape	H_{out}		Stride, Padding	S, P
	Filter size	Z_{pool}	Activation Layer	ReLU or not	$\varnothing$

4.2 Target Device and Setup

Our experimental platform is an ARM Cortex-M7 based STM32H7 board that can embed state-of-the-art models thanks to its large memories (2 MBytes of Flash and 1 MByte of SRAM). For this first work, interruptions are disabled during model inferences as well as cache optimization available on the board. We measured EM emanations coming from the **unopened** chip with a Langer probe (EMV-Technik LF-U 2,5, freq. range from 100 kHz to 50 MHz) that is connected to an amplifier (Fento HVA-200M-40-F) with a 200 MHz bandwidth and 40 dB gain. Acquisitions are collected thanks to a Lecroy WavePro 604HD-MS oscilloscope. Additional specifications are in the public repository.

5 Threat Model

The attacker aims to recover the architecture of an unknown quantized victim model (M_W) with as much detail as possible. Table 2 lists all the information the adversary wants to extract thanks to EM traces. The attack context corresponds to a particular black-box setting. Indeed, the adversary has no knowledge of model architecture nor parameters but is aware of the task performed by the model and the usage of CMSIS-NN module. Adversary is also expected to have appropriate Deep Learning expertise including the most classical and logical layer sequences. The attacker is able to acquire EM side-channel information leaking from the board embedding the targeted DNN model. However, we consider the attacker does not perturb program execution and collect only traces stemming from usual inferences. As we assume a minimal practical setting, adversary is restricted to a single test input. Exploited trace results from the averaging of few inferences with this specific input[9].

Furthermore, we want to point out that in such context, the adversary has access to every needed resources for profiling attacks. Target board is known as well as CMSIS-NN back-end usage, allowing to set a dictionary linking layers with various characteristics to their corresponding EM activity. Simple pattern detection tools could also be used to speedup extraction process. However, usage of such tools or profiling techniques is not in the scope of this work.

[9] Raw traces are available in the public repository.

6 Layers Analysis

At the model scale, inference is a feedforward process with the computation of each layer performed one after the other. The output tensor of one layer becomes the input tensor of the next one. In this section, we focus at layer scale with a two-step methodology. First, we analyze the CMSIS-NN implementation of each layer of our models to reveal repetitions of regular computation blocks that should appear in our EM traces. More particularly, we aim to link these *countable* patterns to hyper-parameters of the layer. Second, we experimentally evaluate if these theoretical assumptions are confirmed in our EM traces and assess the complexity of the extraction and potential limitations.

Our approach is based on the principle that the attack is performed according to the computational flow: when the adversary targets layer l, we suppose that analysis of layer $l-1$ is complete and thus its output tensor shape is known, meaning that input tensor shape of layer l is mastered as well[10].

6.1 Convolutional Layer

CMSIS-NN convolution functions are all based on the same two-step process, as presented in Algorithm 1: Im2col and General Matrix Multiply (GeMM) algorithms. Im2col is a standard optimization trick that implements a convolution through a matrix product rather than several dot-products as defined in Eq. 1. The trick is to prepare all the local areas obtained from sliding window into column vectors and expand the kernel values into rows. Then, convolution is equivalent to a single matrix multiplication that allows an important execution speedup but at the expense of an increased memory footprint. To reduce the latter, CMSIS-NN iteratively performs Im2col with small sets of column vectors [9]. Depending on input and output tensors size, three different functions are proposed[11]:

- `arm_convolve_HWC_q7_fast()` (blue conv. layers in Fig. 1) is for C_{in} multiple of 4 (due to SIMD read and swap) and C_{out} multiple of 2 (due to matrix multiplication applied on 2×2 elements). The computation is speedup for the padding management by splitting the input tensor into 3×3 patches.
- `arm_convolve_HWC_q7_RGB()` (olive conv. layers in Fig. 1) is exclusively optimized for 3 channels inputs (hard-coded condition checks).
- `arm_convolve_HWC_q7_basic()` (cyan conv. layers in Fig. 1) is used otherwise and has a very similar structure as the RGB variant.

Output shape (H_{out})

Code Analysis. Whatever the (optimization) differences between the three implementations listed above, extraction of H_{out} relies on the same principle. As presented in Algorithm 1, outer loops iterate over the tensor size H_{out} (reminder: we

[10] Obviously, we expect that the size of the inputs feeding the model is known by the adversary (e.g., 28×28 for MNIST and $32 \times 32 \times 3$ for Cifar-10).

[11] Equivalent functions are available for non-squared input tensor.

Algorithm 1. General conv. implementation

Input: I_{in} Input tensor of size $H_{in}^2 \cdot C_{in}$, ker (Kernel tensor), S, P, I_{out} Output tensor of size $H_{out}^2 \cdot C_{out}$
Output: Filled I_{out}
1: **for** $i_y \leftarrow 0,\ i_y < H_{out},\ i_y + +1$ **do** ▷ Iterate over H_{out} (y-axis)
2:   **for** $i_x \leftarrow 0,\ i_x < H_{out},\ i_x + +1$ **do** ▷ idem (x-axis)
3:     $\text{buff}_{in} \leftarrow im2col(I_{in}, i_y, i_x, S, P, H_{in}, C_{in})$ ▷ Apply im2col conversion
4:     **if** $len(\text{buff}_{in}) == 2 \times C_{in} \times Z^2$ **then** ▷ Check if 2 input columns are set
5:       $GeMM(\text{buff}_{in}, ker, C_{out}, C_{in} \times Z^2, I_{out})$ ▷ Perform matrix-multiplication
6:       $\text{buff}_{in} \leftarrow 0$ ▷ Buffer reset
7:     **end if**
8:   **end for**
9: **end for**

consider square inputs) to run the core computations with the im2col trick then matrix multiplication (GeMM). These computations are time consuming and are likely to induce clear visible EM activities, especially GeMM step (Algorithm 2 described after). GeMM is called every two iterations (line 4, `if` statement) when the buffer buff_{in} is filled with two *input columns* thanks to im2col (line 3). As a result, GeMM function (line 5) is called $H_{out} \times H_{out}/2$ times during Algorithm 1 execution. If we set Np as the number of regular patterns resulting from GeMM function over the part of the EM trace corresponding to the targeted conv. layer, then we can link Np to H_{out} as: $N_p = H_{out} \times H_{out}/2$, i.e. $H_{out} = \sqrt{2 \times N_p}$.

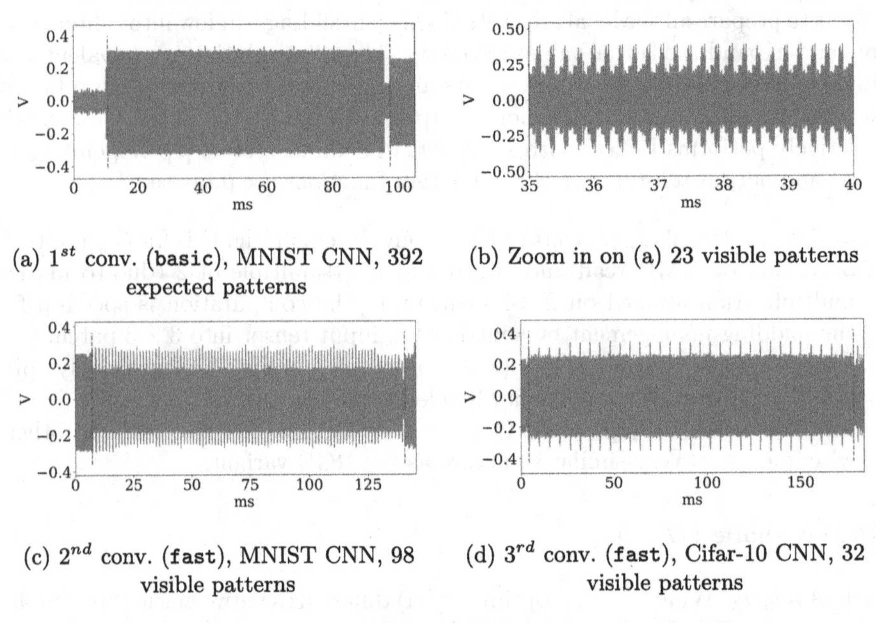

(a) 1st conv. (`basic`), MNIST CNN, 392 expected patterns

(b) Zoom in on (a) 23 visible patterns

(c) 2nd conv. (`fast`), MNIST CNN, 98 visible patterns

(d) 3rd conv. (`fast`), Cifar-10 CNN, 32 visible patterns

Fig. 2. Overviews of conv. layers EM activity related to H_{out}

Observations and Limitations. Thanks to this analysis, we assess our traces for each conv. layer of the two CNN models (MNIST and Cifar-10) and observe the correct number of patterns Np as noticed in Table 3. Figure 2a illustrates the first conv. layer for MNIST with $Np = 392$ patterns (Fig. 2b gives a zoom on 23 patterns between 35 and 40 ms). Figure 2c and 2d illustrate `fast` conv. layers of either MNIST or Cifar-10 CNN with respectively $Np = 98$ and 32 patterns. After counting Np, we easily deduce H_{out} values. According to observed layers, number of patterns can be important and hard to count *by hand.* However, since these patterns are clear and accurate (see Fig. 2b and 2d), an attacker could take advantage of basic pattern detection tools (out of the scope of this work) to make the counting easier.

Next, we zoom in one pattern to focus on the GeMM function and look for other regular patterns that may be related to other hyper-parameters, more particularly the number of kernels K and their size Z.

Number of Kernels (K)

Code Analysis. Matrix-multiplication computation is described in Algorithm 2. It mainly consists of an outer `for` loop iterating over the number of output channels (C_{out}) divided by 2, referenced as `rowCnt` and defined at line 1. We remind that for a conv. layer: $C_{out} = K$. Thus, being able to count the iterations of the `for` loop (lines 2–9) enables to recover K. If N_p is the number of regular EM patterns resulting from computations inside this loop, then we assume that $K = 2 \times N_p$.

Algorithm 2. Matrix-product for conv. (**GeMM**)

Input: buff$_{in}$, ker, C_{out}, $C_{in} \times Z^2$, I_{out}, $bias$ (bias tensor)
Output: Partly filled I_{out}
1: $rowCnt \leftarrow C_{out} >> 1$ ▷ Set $rowCnt = K/2$
2: **for** $rowCnt > 0,\ rowCnt - -1$ **do** ▷ Iterate over $K/2$
3: $sum, sum1, sum2, sum3 = init\_sum(bias, C_{in} \times Z^2)$
4: $colCnt \leftarrow (C_{in} \times Z^2) >> 2$ ▷ Set $colCnt$ as in Eq. 3
5: **for** $colCnt > 0,\ colCnt - -1$ **do**
6: $simd\_mac(sum, sum1, sum2, sum3, \text{buff}_{in}, ker)$
7: **end for**
8: $apply\_mac(sum, sum1, sum2, sum3, C_{out}, I_{out})$
9: **end for**
10: $Manage\_remainder\_if\_any(C_{out}, \text{buff}_{in}, bias, C_{in} \times Z^2)$

Table 3. H_{out} and expected # patterns

Layer	Mnist		Cifar-10	
	H_{out}	N_p	H_{out}	N_p
1	28	392	32	512
2	14	98	16	128
3	∅	∅	8	32

Observations. We evaluate the number of regular patterns appearing on our traces when zooming in the first pattern we previously extracted H_{out} from. For each conv. layer, we observe groups of new regular patterns composed of the expected number N_p. Furthermore, these groups are observed throughout the EM activity related to the entire conv. layer, corresponding to every call of the GeMM function of Algorithm 1. These regular patterns are composed of a spike (Sa) followed by a segment of lower frequency activity (La) (Fig. 3b and 3d). Figure 3a and 3c represent targeted outer `for` loop iterations for the first two conv. layers of MNIST CNN with respectively $N_p = 8$ and 16, that corresponds to $K = 16$ and 32 kernels. Such observations have been successfully made for every other conv. layers for both Cifar-10 and MNIST CNN models.

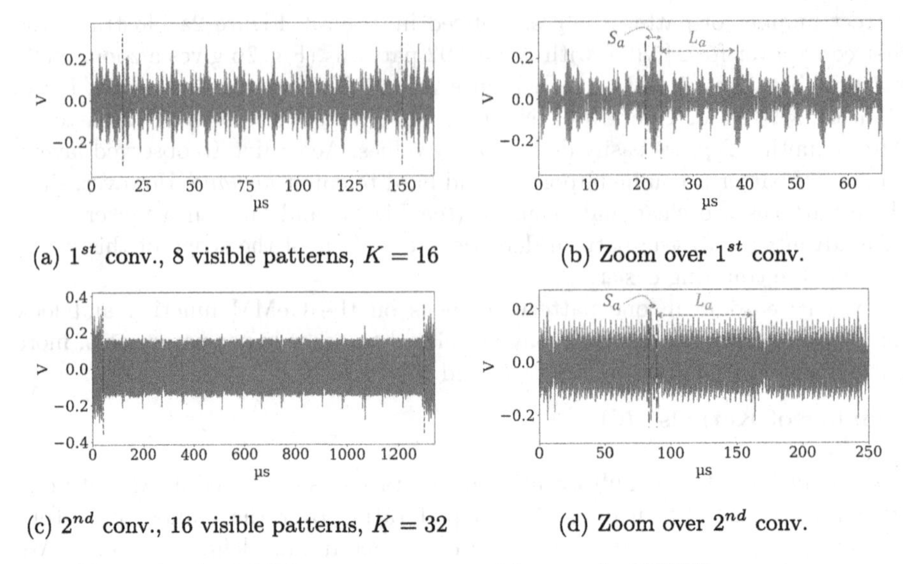

(a) 1st conv., 8 visible patterns, $K = 16$ (b) Zoom over 1st conv.

(c) 2nd conv., 16 visible patterns, $K = 32$ (d) Zoom over 2nd conv.

Fig. 3. Single GeMM execution traces with zoom for MNIST conv. layers

Size of Kernel (Z)

Code analysis. Like the number K of kernels, their size Z is also manipulated in Algorithm 2. The inner `for` loop (lines 5–7) iterates over the variable `colCnt` which directly depends on Z. This part represents the core computation of convolution (with SIMD-based accumulations, line 6). `colCnt` assignation (line 4) is as Eq. 3:

$$colCnt = \frac{1}{4}(C_{in} \times Z^2) \tag{3}$$

Thus, as before, we expect that if these very regular computations result in regular EM patterns, then we can estimate Z. However, such a statement is based on the knowledge of C_{in}. As previously mentioned, the attacker is supposed to master the dimensions of the output tensor of the previous layer, meaning that he knows C_{in}. So, if EM activity related to the inner `for` loop can be distinguished from the rest with N_p regular patterns, then Z is estimated as $\sqrt{(4 \times N_p)/C_{in}}$.

Observations. When observing EM activity between two spikes defined previously as La (Fig. 3), repetitive patterns can be spotted as in Fig. 4. EM activities related to SIMD-based accumulations of the second and third conv. layer of Cifar-10 CNN are shown respectively in traces 4a and 4b. For the second layer, $C_{in} = 16$ and $N_p = 36$ patterns can be counted, giving correctly $Z = 3$ according to previous assumption. Same correct result is obtained for the third layer with $C_{in} = 32$ and $N_p = 64$ visible patterns.

Limitations. It is important to note that dividing $C_{in} \times Z^2$ by 4 can induce remainders to be treated separately afterwards. In this case, another EM activity

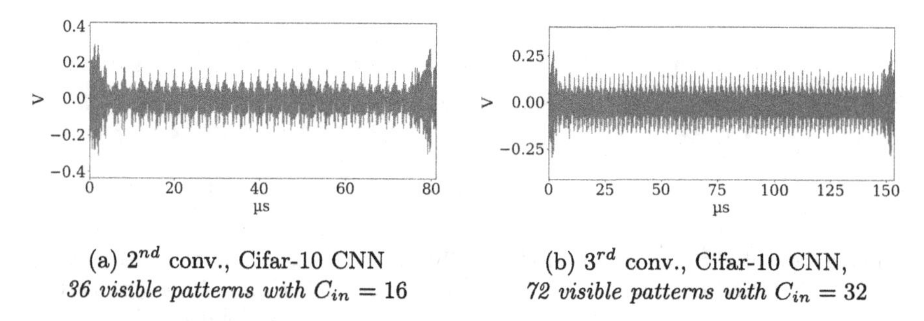

(a) 2nd conv., Cifar-10 CNN *36 visible patterns with* $C_{in} = 16$

(b) 3rd conv., Cifar-10 CNN, *72 visible patterns with* $C_{in} = 32$

Fig. 4. Zoom in Matrix-product EM activity with kernels of size $Z = 3$

will appear after the `colCnt` expected patterns. Alternatively, patterns to be seen and counted are becoming pretty small and can be hard to distinguish one another. Moreover, the value of C_{in} strongly impacts the number of patterns to differentiate. As an example, number of input channels in the first convolution layer of MNIST CNN is necessarily equal to 1 as input are grayscale images. With 3×3 kernels, only two patterns shall appear, with division remainder treated afterwards that emits different EM activity. In other words, K can be challenging to recover, especially when C_{in} is small.

Stride (S) and Padding (P)
Once H_{in}, H_{out} and Z are known, based on the fact that $P < Z$ and thanks to Eq. 2, S and P can be deduced. By computing S for possible P values, only a single integer value will result for S.

6.2 Pooling Layer: Output Dimensions (H_{out}) and Kernel Size (Z_{pool})

Code Analysis. CMSIS-NN implementation (Algorithm 3) of max-pooling relies on two computational blocks. First one (lines 1–6) handles the pooling along x-axis with two nested loops iterating over H_{in} and H_{out}. From an adversary's point of view, the second block (lines 8–11) is the most interesting since it completes the pooling over the y-axis with a single loop over H_{out} only. Pooling operation is classically performed with two steps: (1) selection and allocation of local areas, then (2) statistic computation (here, maximum). Thus, we expect to observe two distinct EM activities related to these separated loops with the last (and shorter) one directly related to H_{out}. Then we can deduce $Z_{pool} = H_{out}/H_{in}$.

Observations. EM activity of an entire MaxPool layer is represented in Fig. 5a (2nd pooling, MNIST CNN). As expected, we observe two distinct blocks of EM emanations, the second one revealing patterns that are more distinguishable. When zooming over this second part, as in Fig. 5b (1st pooling, Cifar-10), these patterns are composed of two segments, one with low amplitude activity followed

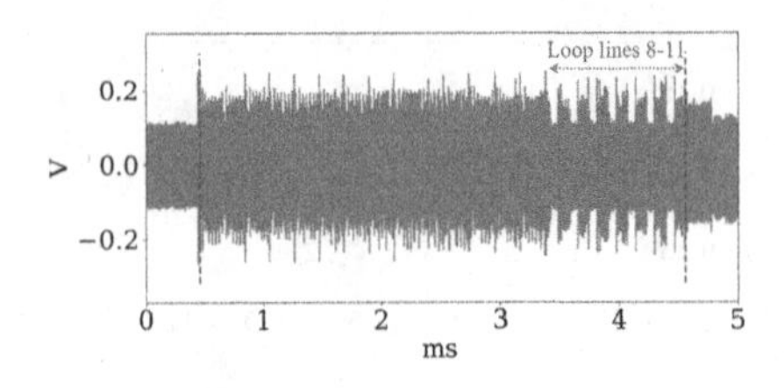

(a) Overview of 2nd maxpool of MNIST CNN, $H_{out} = 7$

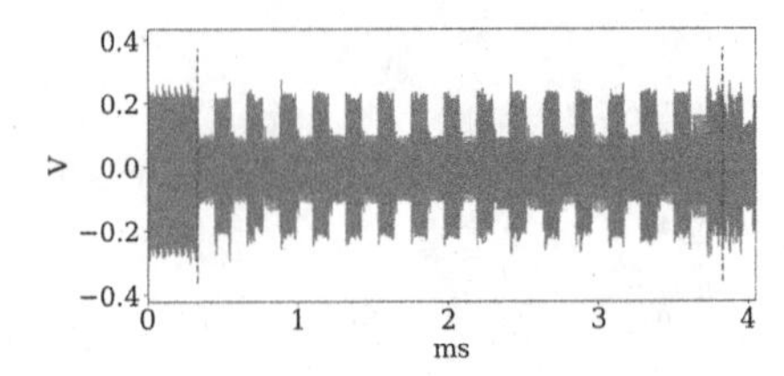

(b) Zoom over 2nd loop of 1st Maxpool of Cifar-10 CNN, $H_{out} = 16$, triggers set as in Alg. 3 at lines 7 and 12

Fig. 5. MaxPool layers of CNN models

by another of higher amplitude. We guess that it corresponds to the two steps of the pooling computation. The number N_p of these patterns can easily be hand-counted and matches H_{out} value. We have been able to do equivalent observations for all the MaxPool layers of our CNN models. Experimentally, we note that analyzing the first block to retrieve both H_{in} and H_{out} is feasible but more complex than simply counting H_{out} with the second block.

Algorithm 3. MaxPool - `arm_maxpool_q7_HWC`

Input: I_{in} input tenor of size $H_{in}^2 \cdot C_{in}$, I_{out} output tenor of size $H_{out}^2 \cdot C_{out}$, P, S, H_{ker}
Output: Filled I_{out}

1: **for** $i_y \leftarrow 0,\ i_y < H_{in},\ i_y++1$ **do** ▷ Pooling along x axis
2:     **for** $i_x \leftarrow 0,\ i_x < H_{out},\ i_x++1$ **do**
3:         $win_{start}, win_{stop} \leftarrow set\_window(i_y, i_x, I_{in}, H_{ker}, P, S)$
4:         $compare\_and\_remplace\_if\_larger(win_{start}, win_{stop}, i_y, i_x, I_{in})$
5:     **end for**
6: **end for**
7: $trigger\_up()$ ▷ Pooling along y axis
8: **for** $i_y \leftarrow 0,\ i_y < H_{out},\ i_y++1$ **do** ▷ Directly iterates over H_{out}
9:     $row_{start}, row_{stop} \leftarrow set\_rows(i_y, I_{in}, I_{out}, H_{ker}, P, S)$
10:     $compare\_replace\_then\_apply(row_{start}, row_{stop}, I_{in}, I_{out})$
11: **end for**
12: $trigger\_down()$

6.3 Dense Layer: Number of Neurons (N_e)

Code Analysis. Algorithm 4 describes function dedicated to dense layers. It is built around a `for` loop iterating over the `rowCnt` variable that is the number of neurons N_e divided by 4 (line 1), since neurons are handled by groups of 4. Indeed, `init_sum_with_bias` function (line 3) sets four sum variables. They are used for weighted sum computation (and bias value addition) through multiply-accumulate SIMD operations (`__SMLAD`), performed in `simd_mac` (line 6). We are likely to observe N_p regular patterns emanating from this loop, with $N_e = 4 \times N_p$.

Algorithm 4. Dense layer - `arm_fully_connected_q7_opt`

Input: I_{in} input vector of size H_{in}, ker weight vector of size N_e, $bias$ bias matrix of size N_e, I_{out} output vector of size H_{out}, P, S, H_{ker}
Output: Filled I_{out}
1: $rowCnt \leftarrow N_e >> 2$ ▷ Nb. neurons divided by 4
2: **for** $rowCnt > 0,\ rowCnt - -1$ **do** ▷ Iterate directly over $N_e/4$
3: $sum, sum1, sum2, sum3 = init\_sum\_with\_bias(bias, rowCnt)$
4: $colCnt \leftarrow H_{in} >> 2$
5: **for** $colCnt > 0,\ colCnt - -1$ **do**
6: $simd\_mac(sum, sum1, sum2, sum3, ker, colCnt)$
7: **end for**
8: $apply\_mac(sum, sum1, sum2, sum3, rowCnt, I_{out})$
9: **end for**
10: $rowCnt \leftarrow N_e\ \&\ 0x3$ ▷ Manage remainders if any
11: **for** $rowCnt > 0,\ rowCnt - -1$ **do**
12: $sum = init\_sum\_with\_bias(bias, rowCnt)$
13: $colCnt \leftarrow H_{in} >> 2$
14: **for** $colCnt > 0,\ colCnt - -1$ **do**
15: $mac(sum, ker, colCnt)$
16: **end for**
17: $apply\_mac(sum, rowCnt, I_{out})$
18: **end for**

Observations. We observe significant EM activity that results from neurons handling as illustrated in Fig. 6. Observed patterns are mainly composed of uniform blocks separated by clear spikes (especially visible on Fig. 6a). Logically, related EM pattern length directly depends on the number of inputs to the layer and therefore to neurons. Then, dense layers managing broader inputs are easier to analyse. By counting the N_p spike-separated blocks of each dense layers of our MLP and CNN, we checked the link between this number of occurrences N_p and the number of neurons N_e of the layer. Figure 6b and 6b illustrate the two dense layers of the MNIST MLP model with respectively 8 and 4 patterns corresponding to $N_e = 32$ and $N_e = 16$ neurons. As well, Fig. 6c and 6d show dense layers with respectively $N_e = 16$ (4 patterns) and $N_e = 32$ (8 patterns) neurons.

Limitations (Special Cases). As neurons are grouped in sets of 4, some special cases occur when remaining neurons still need to be computed. The line 10 of Algorithm 4 checks remainders and handles them throughout the following `for` loop. Neurons are then handled one by one as only a single sum is initialised then used in `mac` call at line 15. To illustrate this phenomenon, we trained an additional MLP model on MNIST (noted SP-MLP), composed of 4 dense layers with respectively 23, 18, 13 and 10 neurons. These correspond to different remainders.

Figure 7 shows EM activity of each layer for SP-MLP. For the 23-neuron layer, we clearly observe on Fig. 7a 3 blocks that stand out after the core sequence of 5 patterns. This directly matches what is expected with 5 groups of 4 neurons completed with the 3 remaining ones managed independently. However, similar analysis cannot be performed on the two other traces. This difficulty comes from input tensor shape reduction from first layer to the second and third ones. To verify that neurons are managed in the same way for both of these, triggers are raised and lowered inside the outer `for` loop over `rowCnt` (represented as

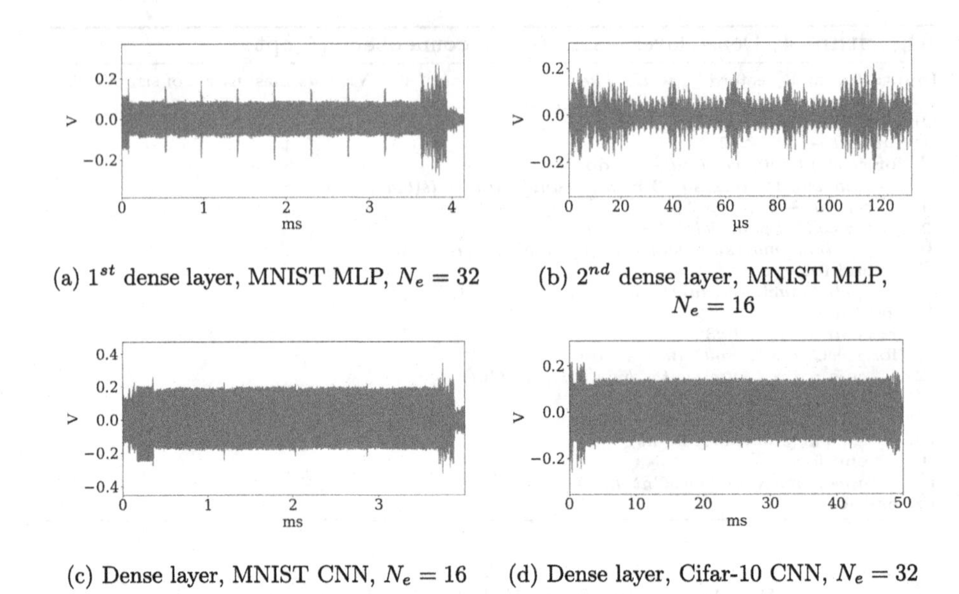

(a) 1st dense layer, MNIST MLP, $N_e = 32$

(b) 2nd dense layer, MNIST MLP, $N_e = 16$

(c) Dense layer, MNIST CNN, $N_e = 16$

(d) Dense layer, Cifar-10 CNN, $N_e = 32$

Fig. 6. Overview of dense layer EM activity corresponding to general cases

rectangles in Fig. 7d and 7e). In addition, it illustrates problematic pattern shape changes induced by usage of triggers around observed piece of code. Indeed, they become more visible in Fig. 7d and 7e compared to Fig. 7b and 7c.

6.4 Activation Layer

The nature of the Activation Functions (AF) is an important information since it strongly affects input flow from one layer to the next. It is especially useful for parameter extraction as it can modify layers input distribution and potentially their signs (e.g., $\forall x \in \mathbb{R}, ReLU(x) \geq 0$). Many different AF exist with Sigmoid, Tanh, Softmax (mainly for the model output normalization) and ReLU as the most popular. The first three imply an exponential or division computation that are time consuming. ReLU is predominantly the most popular AF because of training efficiency and its low computation requirements. Therefore, we mainly focus on the distinction between ReLU from Sigmoid and Tanh functions (i.e., the extraction goal is to answer the question: *is the AF ReLU or not?*). Measuring computation time of entire AF layer can give strong evidence (as also investigated in [1]) because ReLU function is processed faster than the other two as shown in Table (a) from Fig. 8. Such approach takes on its full meaning when considering potential template capacity of the attacker.

However, we observe that looking at the EM patterns gives additional hints that help distinguish ReLU. For that purpose, we trained three MNIST CNN models with the same architecture but different AF (one per model: ReLU, Sigmoid or Tanh). Traces of Fig. 8 are zoomed in on the beginning of each second

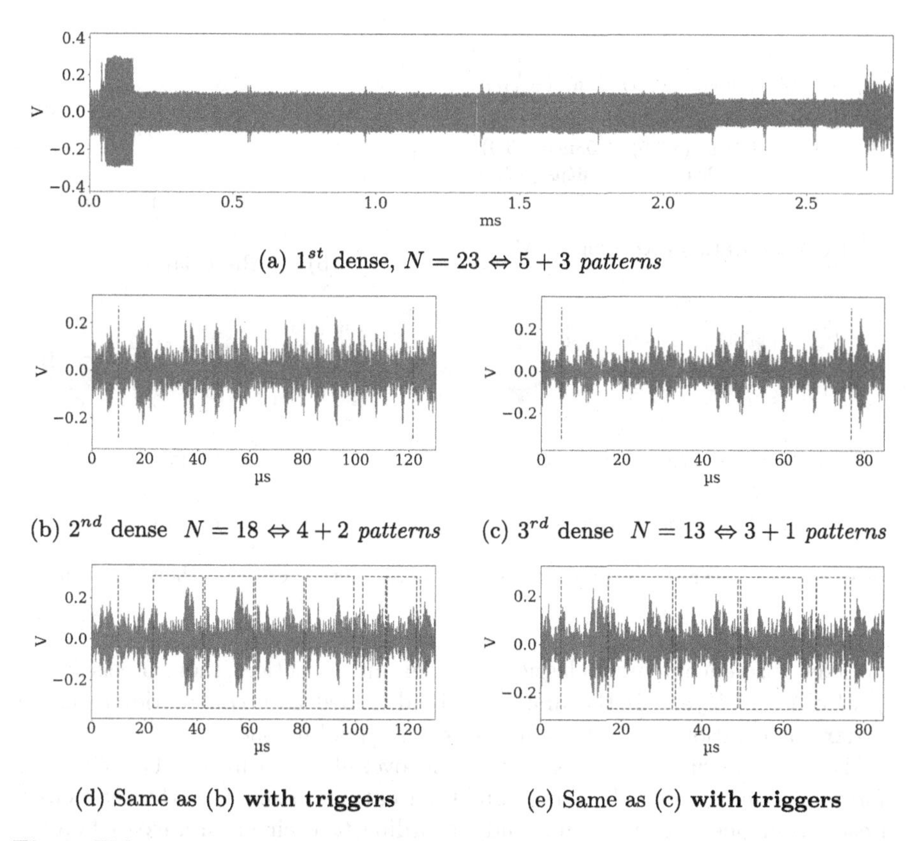

(a) 1st dense, $N = 23 \Leftrightarrow 5 + 3$ *patterns*

(b) 2nd dense $N = 18 \Leftrightarrow 4 + 2$ *patterns*

(c) 3rd dense $N = 13 \Leftrightarrow 3 + 1$ *patterns*

(d) Same as (b) **with triggers**

(e) Same as (c) **with triggers**

Fig. 7. EM activity overview of dense layers for the custom SP-MLP model. Expected patterns clearly appear for the first layer, not for the 2nd and 3rd

AF layer. Trace 8b, corresponding to ReLU, exhibits regular and distinguishable patterns (duration of few μs). Sigmoid and Tanh traces also present groups of peaks repeated throughout traces. However, these are more complex and less explicit than ones related to ReLU and the order of magnitude of their duration is greater, especially for Sigmoid traces. From these observations, distinction between ReLU and the two other AF is quite straightforward. Noteworthy, it is not a clear-cut between Sigmoid and Tanh.

7 Architecture Extraction Methodology

At first, an adversary may exploit the nature of the task performed by the model. In some cases, it can give hints about the type of the victim model. A model performing a computer vision task is likely to be built around CNN principle. However, there are no strict rules that match a task to a model type. We claim that task knowledge may provide a set of possible layer types that a first analysis of the trace overall structure must confirm. Moreover, preliminary

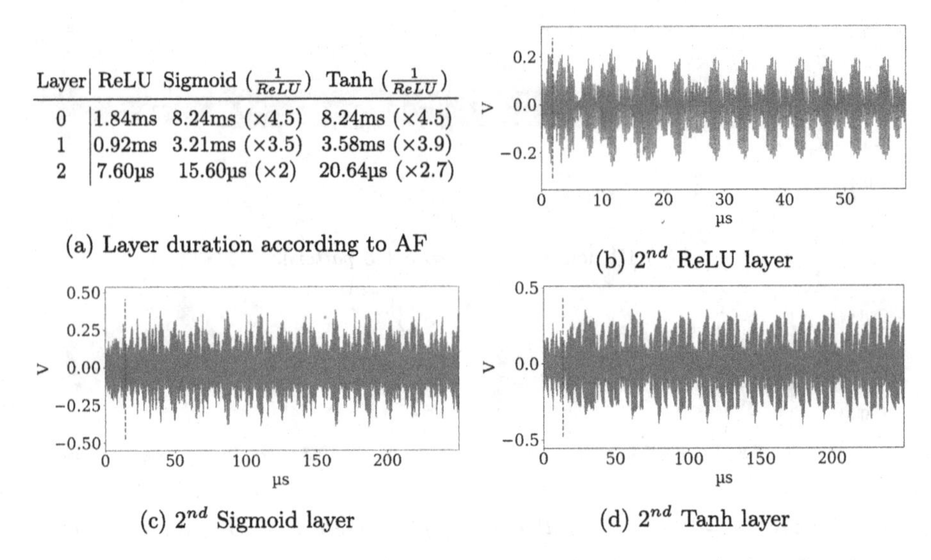

Layer	ReLU	Sigmoid ($\frac{1}{ReLU}$)	Tanh ($\frac{1}{ReLU}$)
0	1.84ms	8.24ms (×4.5)	8.24ms (×4.5)
1	0.92ms	3.21ms (×3.5)	3.58ms (×3.9)
2	7.60µs	15.60µs (×2)	20.64µs (×2.7)

(a) Layer duration according to AF

(b) 2nd ReLU layer

(c) 2nd Sigmoid layer

(d) 2nd Tanh layer

Fig. 8. Zoom over the beginning of second activation layers with their duration

knowledge also encompasses classical deep learning practices, more precisely the *logical* order of layers. If the first layer is identified as a convolutional one, a standard association is another conv. layer or a pooling one.

The first information to extract is the overall structure of $\mathcal{A}_M$ with two information: the number of layers, L and their natures. Then, the attacker focuses on each layer one after the other and, according to their nature, extracts a set of hyper-parameters easing to design a substitute model M'_Θ.

Step 1: Finding the Number of Layers. When analyzing the general shape of the averaged EM trace, the very first observation is that it can be straightforwardly split into several blocks by identifying the boundaries that separate them. A frequency spectrum can highlight the separation between the blocks. The Fig. 9d illustrates how this processing can made easier the layer splitting for the Cifar-10 CNN. Without difficulty, for the three models studied in this work, this simple analysis gives the exact number L of layers for each model architecture $\mathcal{A}_M$. Figure 9 illustrates the identified blocks for our models with the same colors as in Fig. 1. To validate that layers truly correspond to blocks and for illustration purpose only, triggers have been added to clearly mark out layer boundaries.

Step 2: Identifying Layers' Nature. Having the position of the L layers on the EM trace, the attacker knows the intrinsic layer complexity that directly influences the execution times. Typically, pooling and AF layers are far less complex than the main computational ones (in our case, dense and conv.) that they usually follow. Thus, the real challenge is to correctly initiate the extraction process by identifying the first layer as a dense or a conv. layer. A correct extraction of the first layer is fundamental, since the *course* of the attack, as well as assumptions

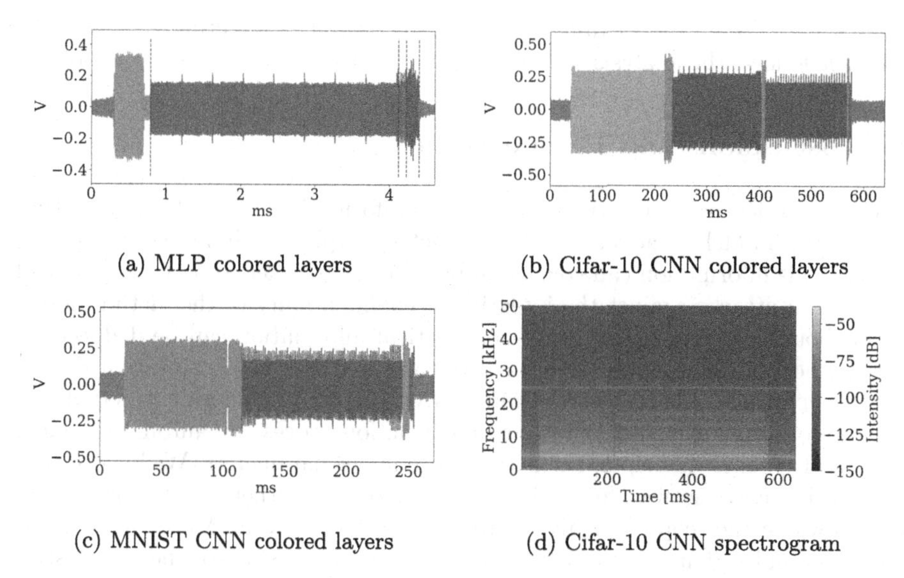

(a) MLP colored layers

(b) Cifar-10 CNN colored layers

(c) MNIST CNN colored layers

(d) Cifar-10 CNN spectrogram

Fig. 9. Average EM trace for the 3 models. Spectrogram for Cifar-10 CNN.

made on $\mathcal{A}_M$, come out of this *good start.* To successfully achieve this distinction between dense and conv. layer, the attacker relies on a twofold analysis:

- *Complexity/execution times.* First insight is provided by simple timing analysis. Although insufficient, this analysis may dispel doubts between some layer hypotheses. Complexity is estimated in different ways in the literature but a standard way is to consider the number of Multiplications and Accumulations (MACs). Such metric directly stems from computations and input/output dimensions. MAC complexity of conv. and dense layers are as follow: $MAC_{Conv2D} = (Z^2 \times C_{in}) \times (H_{out}^2 \times C_{out})$, $MAC_{dense} = H_{in} \times H_{out}$. Complexity of the first layer of both MNIST models are 112896 MACs for the CNN and 25088 MACs (x4.5 less) for the MLP. Thus, usually, conv. layer takes longer to execute than dense one (with similar inputs)[12].
- *Patterns:* the attacker mostly leverages on specific EM activities. As detailed in Sect. 6, the regularities and repetitions of patterns are very different between dense and conv. layers and confusion is very unlikely. Thus, layer's nature is identified before accurately extracted the related hyper-parameters.

Once the first layer is identified, ML expertise and usual layers sequence knowledge can provide additional hints about next layers nature. This encourages the attacker to extract $\mathcal{A}_M$ by analysing model layers consecutively.

[12] Obviously, if the dense layer had x4.5 more neurons, MAC complexity between dense and conv. would be equal but having a too large number of neurons (i.e. trainable parameters) for dense layers is usually unsuitable with classical overfitting issues.

Step 3: Extracting hyper-parameters. Once the overall architecture is known, the attacker follows the analysis presented in Sect. 6 to recover hyper-parameters.

8 Discussions and Perspectives

Our main objective is to estimate how much information about the architecture of a victim model an adversary can extract by exploiting limited side-channel traces. This information can considerably help the attacker to perform powerful adversarial attacks against the integrity or confidentiality of the victim model. Even though we demonstrate that very critical information can be deduced by the methodical analysis of both an EM trace and the implementation details of the deployment library (here, CMSIS-NN), the use of pattern extraction and recognition tools may significantly ease extraction process by automating most steps and potentially help in extracting more hyper-parameters. We believe that this work paves the way for such further analysis as other efforts that would aim to widen the scope of analyzed architectures and layers. More particularly, batch normalization or tensor arithmetic layers as in state-of-the-art ResNet models or Attention blocks as in Transformer models are relevant candidates for future works. We also highlight that variants of the standard convolution have been proposed for (computational) efficiency purpose (e.g., Depthwise Separable Convolution) [14]. To the best of our knowledge, all these variants rely on highly repetitive and regular computations as the convolution with im2col and GeMM, therefore it should not fundamentally differ from what we exposed in this work.

Moreover, the impact of experimental setup simplifications (i.e., disable of interruptions and cache optimizations) must be studied in complementary studies. Such changes could harden hyper-parameters recovery and directed research focused on protection dedicated to model architecture. To the best of our knowledge, they are very few of them, especially for microcontroller platform. Authors of [11] propose an obfuscation strategy for FPGA accelerators. They leverage on optimization parameters of convolution computation to mitigate EM emanations coming from conv. layers. Porting similar protection to microcontroller could be challenging due to limited resources and model performance to be preserved.

9 Conclusion

When dealing with DNN model extraction attack, architecture is crucial. Whatever the attacker's objective, a complete or even partial recovery provides an essential advantage. With this work, we highlight that the attack surface for such a threat is significantly extended by side-channel analysis. More importantly, regarding our application scope, we demonstrate that there is no need for complex exploitation methods (e.g., with heavy supervised profiling step) because of the strong repetitiveness and regularity of most of performed computations that make SEMA a surprisingly powerful tool. Typically, the Russian dolls effect that we exploit for conv. layers (enabling the recovery, one after the other, of several important hyper-parameters) is highly representative of this confidentiality flaw that we claim to be a very worrying concern.

Although we highlight some limitations or more complex special cases that need to be handled in future works, our concern is not based only on the relative simplicity of the attack, but also on the hard challenges related to the development of efficient *and* practical protections, compliant with the constrains of 32-bit microcontrollers. With ongoing regulatory frameworks for AI systems and upcoming security certification actions, model architecture obfuscation appears as the key defense challenge, that we urgently need to solve in order to bring robustness to the large-scale deployment of ML systems.

Acknowledgements. This work is supported by (CEA-Leti) the EU project InSecTT (ECSEL JU 876038) and by ANR (Fr) in the framework *Investissements d'avenir* program (ANR-10-AIRT-05, irtnanoelec); and (Mines Saint-Etienne) by ANR PICTURE program (AAPG2020). This work benefited from the French Jean Zay supercomputer with the AI dynamic access program.

References

1. Batina, L., Jap, D., Bhasin, S., Picek, S.: CSI NN: reverse engineering of neural network architectures through electromagnetic side channel. In: 28th USENIX Security Symposium. USENIX Association (2019)
2. Carlini, N., Jagielski, M., Mironov, I.: Cryptanalytic extraction of neural network models. In: Micciancio, D., Ristenpart, T. (eds.) CRYPTO 2020. LNCS, vol. 12172, pp. 189–218. Springer, Cham (2020). https://doi.org/10.1007/978-3-030-56877-1_7
3. Chmielewski, Ł, Weissbart, L.: On reverse engineering neural network implementation on GPU. In: Zhou, J., et al. (eds.) ACNS 2021. LNCS, vol. 12809, pp. 96–113. Springer, Cham (2021). https://doi.org/10.1007/978-3-030-81645-2_7
4. Duddu, V., Samanta, D., Rao, D.V., Balas, V.E.: Stealing neural networks via timing side channels. arXiv preprint arXiv:1812.11720 (2018)
5. Gongye, C., Fei, Y., Wahl, T.: Reverse-engineering deep neural networks using floating-point timing side-channels. In: 57th ACM/IEEE Design Automation Conference (DAC). IEEE (2020)
6. Hector, K., Moellic, P.-A., Dumont, M., Dutertre, J.-M.: Fault injection and safe-error attack for extraction of embedded neural network models. arXiv preprint arXiv:2308.16703 (2023)
7. Jagielski, M., Carlini, N., Berthelot, D., Kurakin, A., Papernot, N.: High accuracy and high fidelity extraction of neural networks. In: Proceedings of the 29th USENIX Conference on Security Symposium (2020)
8. Joud, R., Moëllic, P.A., Pontié, S., Rigaud, J.B.: A practical introduction to side-channel extraction of deep neural network parameters. In: Buhan, I., Schneider, T. (eds.) CARDIS 2022. LNCS, vol. 13820, pp. 45–65. Springer, Cham (2023). https://doi.org/10.1007/978-3-031-25319-5_3
9. Lai, L., Suda, N., Chandra, V.: CMSIS-NN: efficient neural network kernels for arm cortex-M CPUs. arXiv preprint arXiv:1801.06601 (2018)
10. Lin, J., Chen, W.M., Lin, Y., Gan, C., Han, S., et al.: MCUNet: tiny deep learning on IoT devices. In: Advances in Neural Information Processing Systems, vol. 33 (2020)

11. Luo, Y., Duan, S., Gongye, C., Fei, Y., Xu, X.: NNReArch: a tensor program scheduling framework against neural network architecture reverse engineering. In: IEEE 30th Annual International Symposium on Field-Programmable Custom Computing Machines (FCCM). IEEE (2022)
12. Ma, J.: A higher-level Neural Network library on Microcontrollers (NNoM) (2020)
13. Maji, S., Banerjee, U., Chandrakasan, A.P.: Leaky nets: recovering embedded neural network models and inputs through simple power and timing side-channels-attacks and defenses. IEEE Internet Things J. **8**(15) (2021)
14. Nguyen, B., Moëllic, P.A., Blayac, S.: Evaluation of convolution primitives for embedded neural networks on 32-bit microcontrollers. In: Abraham, A., Pllana, S., Casalino, G., Ma, K., Bajaj, A. (eds.) ISDA 2022. LNNS, vol. 646, pp. 427–437. Springer, Cham (2022). https://doi.org/10.1007/978-3-031-27440-4_41
15. Orekondy, T., Schiele, B., Fritz, M.: Knockoff nets: stealing functionality of black-box models. In: Proceedings of the IEEE/CVF Conference on Computer Vision and Pattern Recognition (2019)
16. Papernot, N., McDaniel, P., Goodfellow, I., Jha, S., Celik, Z.B., Swami, A.: Practical black-box attacks against machine learning. In: Proceedings of the ACM on Asia conference on Computer and Communications Security (2017)
17. Papernot, N., McDaniel, P., Sinha, A., Wellman, M.P.: SoK: security and privacy in machine learning. In: 2018 IEEE European Symposium on Security and Privacy (EuroS&P), pp. 399–414. IEEE (2018)
18. Rakin, A.S., Chowdhuryy, M.H.I., Yao, F., Fan, D.: DeepSteal: advanced model extractions leveraging efficient weight stealing in memories. In: IEEE Symposium on Security and Privacy (SP). IEEE (2022)
19. Tramèr, F., Zhang, F., Juels, A., Reiter, M.K., Ristenpart, T.: Stealing machine learning models via prediction APIs. In: USENIX Security Symposium, vol. 16 (2016)
20. Xiang, Y., Chen, Z., Chen, Z., et al.: Open DNN box by power side-channel attack. IEEE Trans. Circ. Syst. II: Express Briefs **67**(11) (2020)
21. Yli-Mäyry, V., Ito, A., Homma, N., Bhasin, S., Jap, D.: Extraction of binarized neural network architecture and secret parameters using side-channel information. In: IEEE International Symposium on Circuits and Systems (ISCAS). IEEE (2021)
22. Yu, H., Ma, H., Yang, K., Zhao, Y., Jin, Y.: DeepEM: deep neural networks model recovery through EM side-channel information leakage. In: IEEE International Symposium on Hardware Oriented Security and Trust (HOST). IEEE (2020)

Correction to: PQ.V.ALU.E: Post-quantum RISC-V Custom ALU Extensions on Dilithium and Kyber

Konstantina Miteloudi, Joppe W. Bos, Olivier Bronchain, Björn Fay, and Joost Renes

Correction to:
Chapter 10 in: S. Bhasin and T. Roche (Eds.): *Smart Card Research and Advanced Applications*, LNCS 14530, https://doi.org/10.1007/978-3-031-54409-5_10

In the original version of this chapter the affiliation of the second and third author, Joppe W. Bos and Olivier Bronchain, were published incorrectly. This has been corrected. Correctly it should read: "NXP Semiconductors, Leuven, Belgium".

The updated version of this chapter can be found at
https://doi.org/10.1007/978-3-031-54409-5_10

S. Bhasin and T. Roche (Eds.): CARDIS 2023, LNCS 14530, p. C1, 2024.
https://doi.org/10.1007/978-3-031-54409-5_14

Author Index

A
Albillos, Ninon Calleja 213
Alshaer, Ihab 3

B
Bauer, Sven 43
Beroulle, Vincent 3
Berzati, Alexandre 62
Bos, Joppe W. 190
Bossuet, Lilian 235
Boussam, Sana 213
Bronchain, Olivier 190

C
Cagli, Eleonora 235
Calle Viera, Andersson 62
Cler, Gauthier 107
Colombier, Brice 3

D
De Santis, Fabrizio 43
Deleuze, Christophe 3
Dufka, Antonin 169

E
Eyraud, Rémi 235

F
Fay, Björn 190

G
Gierlichs, Benedikt 23
Grosso, Vincent 235
Guillaume, Jeremy 87

H
Heydemann, Karine 62

J
Joud, Raphaël 256

L
Le Bouder, Hélène 127
Llavata, Dorian 235

M
Maalоouf, Eïd 127
Maistri, Paolo 3
Maurine, Philippe 107
Miteloudi, Konstantina 190
Moëllic, Pierre-Alain 256

N
Nafkha, Amor 87

O
Ordas, Sebastien 107

P
Pelcat, Maxime 87
Pontié, Simon 256

R
Renes, Joost 190
Rigaud, Jean-Baptiste 256

S
Salvador, Rubén 87
Sarry, Modou 127
Sluys, pcy 23
Standaert, François-Xavier 148
Svenda, Petr 169

T
Thomas, Gaël 127

S. Bhasin and T. Roche (Eds.): CARDIS 2023, LNCS 14530, pp. 277–278, 2024.
https://doi.org/10.1007/978-3-031-54409-5

U
Udvarhelyi, Balazs 148

V
Verbauwhede, Ingrid 23

W
Wouters, Lennert 23

Z
Zaoral, Lukas 169

Printed in the United States
by Baker & Taylor Publisher Services